Techniques in Ecology and Conservation Series

Series Editor: William J. Sutherland

Bird Ecology and Conservation: A Handbook of Techniques
William J. Sutherland, Ian Newton, and Rhys E. Green

Conservation Education and Outreach Techniques
Susan K. Jacobson, Mallory D. McDuff, and Martha C. Monroe

Forest Ecology and Conservation: A Handbook of Techniques
Adrian C. Newton

Habitat Management for Conservation: A Handbook of Techniques
Malcolm Ausden

Conservation and Sustainable Use: A Handbook of Techniques
E.J. Milner-Gulland and J. Marcus Rowcliffe

Invasive Species Management: A Handbook of Principles and Techniques
Michael N. Clout and Peter A. Williams

Amphibian Ecology and Conservation: A Handbook of Techniques
C. Kenneth Dodd, Jr.

Insect Conservation: Approaches and Methods
Michael J. Samways, Melodie A. McGeoch and Tim R. New

Acknowledgements

We are enormously thankful to Ian Sherman and Helen Eaton of Oxford University Press for their encouragement and support. We also thank James Pryke and René Gaigher for their assistance with preparation of the figures. We are grateful too for the wonderful artwork of some of the equipment used in insect conservation drawn by Anni Coetzer. Adrian Armstrong, Mattias Larsson, Sharon Louw, Jürgen Ott, Nils Ryrholm, Glenn Svensson, and John Terblanche, individually or jointly very kindly supplied authored boxes on their specialist fields. Mark Robertson provided helpful comments on parts of the text. Many other researchers also supplied photographs and are acknowledged alongside their contribution.

We wish you interest and enjoyment in your research, as we have had in ours over the years.

MICHAEL J. SAMWAYS
MELODIE A. MCGEOCH
TIM R. NEW

Contents

Introduction

Insects, as the most diverse group of animals with which we share our world, add substantial variety and diversity to the extraordinarily rare veneer of life on this biotic jewel we call Earth. Yet this insect diversity, the manifestation of millions of years of evolution, is under threat of attrition, as populations are lost, genetic diversity reduced, species go extinct, and the biomes and resources on which they depend succumb to human cupidity. It has been estimated that perhaps a quarter of all insect species are heading for extinction over the next few decades. In a mere blink of a geological eyelid, the ecological impact of humans is causing the demise of a vast amount of Earth's biodiversity. This variety not only has intrinsic value but also may have much practical, utilitarian value which is only starting to be explored.

Among the challenging issues in insect conservation is to establish the extent of diversity of the insect world, at genetic, population, and species levels. We then need to know which components of that diversity are being lost from which habitats and from where in the world. Insect diversity is so great and the challenge of its conservation so urgent, often we need to put in place actions that, through some value judgement, will conserve as much variety as possible rather than to treat each component species individually. To do this, we may need to use surrogates, which stand in for much of the diversity which we cannot adequately measure. Then, once we have put in place management actions that aim to conserve as many populations as possible, we need to establish how successful the conservation management actions have been.

Insects, and many other terrestrial invertebrates, present some special challenges, simply because they are so varied, so diverse, and so unknown. They are also very much part of the fabric of terrestrial and freshwater ecosystems as we know them, and indeed are often intimately associated with plants. Insect ecology and conservation therefore goes hand in hand with plant ecology and conservation.

This book aims to introduce approaches and techniques for insect ecology and conservation. While there are already some excellent introductory texts on sampling insects (although generally not with a specific conservation focus), the field of insect conservation has grown so much in recent years that there is a need for a contemporary and complementary text. Also, as there are some emerging techniques which are not in the traditional ecological techniques texts, we felt the need to produce a book that introduces students and researchers to approaches and techniques that can be used for both documenting and managing the conservation of insect variety. In general, this book is intended for the early-career researcher, with, say, a bachelor's degree in biology, entomology or zoology who intends pursuing practical insect conservation work and writing a technical report, thesis, or research paper. We do not claim to be exhaustive, as new techniques, often

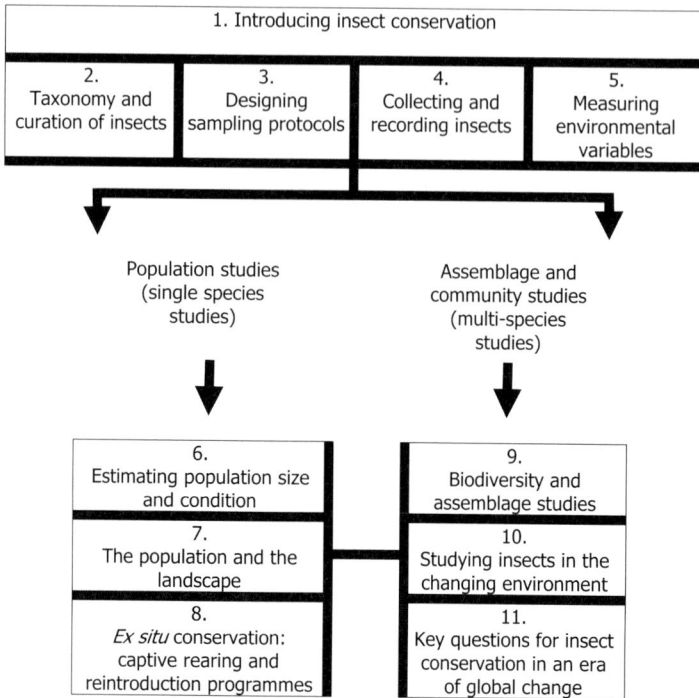

Fig. 1 Layout of this book. The introductory chapters (2–5, 11) are closely interrelated and are the foundation for good planning of your project. The project may then be on a single species or on several species. Studies on the changing environment afftect both single species and species assemblages. Bold lines and arrows represent the most important linkages.

based on new ecological concepts, are continually emerging. The approach we take is summarized in Figure 1. Besides the 'Introducing insect ecology and conservation', there are four interrelated foundation chapters. The first of these, 'Taxonomy and curation of insects', emphasizes the importance of becoming familiar with your focal organisms, especially relative to the ecological and conservation questions being addressed, bearing in mind that ecology is a stepping stone to management for conservation. The next chapter, 'Designing sampling protocols for insect ecology and conservation', would seem, intuitively, to better sit after 'Collecting and recording insects'. However, we wish to emphasize the crucial importance of planning the sampling before you go out to collect and record the subject insects. Then, as no insect lives in isolation, but rather is intimately interacting with its environment, it is essential to record appropriate environmental variables while sampling and recording insects (Chapter 5).

Your studies are then likely to be either on population aspects of single species, or, on the other hand, on the variety of insect species and their ecological relationships. Single-species studies are covered in Chapters 6–8. Assemblages (all species in one taxon, e.g. ants) and communities (interactions between species, e.g. food webs) are then dealt with in Chapters 9, 10, and 11. Of course, embedded in and sometimes even part of an assemblage/community study are aspects of population biology, hence the horizontal link in Figure 1. There is some overlap in the chapters to ensure consolidation of certain points and to strengthen linkages between chapters where it is crucial to do so. While we have made every effort to be fair and accurate, we cannot be held responsible for the effectiveness of the techniques, nor extensiveness of coverage, nor for individual actions relative to local, national, or international legislation.

1

Introducing insect conservation

1.1 The challenge: the relevance of sampling insects

Insects are the most speciose class of animals, with nearly a million described species, and probably several times this number still to be discovered, diagnosed, and named, especially in the tropics (Grimaldi and Engel 2005). Their functional roles are enormous. They play major roles in virtually every terrestrial and freshwater ecosystem. The conservation of insects has major functional significance in maintaining the diversity of ecological processes on which all life ultimately depends.

These roles are often difficult to quantify in economic terms. Nevertheless, in the USA alone, wild insects are estimated to be worth $57 billion per year by providing services such as controlling pests, pollinating flowers, burying dung, and providing nutrition for other wildlife (Losey and Vaughan 2006).

The vast number of insect species, both described and undescribed, ensures their participation in an even greater number of food web interactions (Memmott 2000). Many of these interactions are highly specific, be they between herbivores and plants, predators and prey, or hosts and parasites. Our lack of detailed knowledge of the biology of most species renders their contribution to such integrated scenarios very difficult to evaluate. To assure continuing ecological integrity, we need to minimize losses. This approach is based on the *precautionary principle*, which means conserving as much as possible just in case there is any yet unforeseen value in genes, populations, species, or their interactions. To illustrate our lack of knowledge of the functional roles that insects play, in field samples we generally cannot even associate immature stages of insects with their corresponding adults, and these life stages are often functionally very different. The challenges of studying insects begin to multiply when we consider that many species can be sampled only during short periods of the year, and many disperse widely across the landscape and may not reside in the area in which they are sampled. Furthermore, many are small, cryptic, and intrinsically difficult to find and study. These challenges differ greatly from those facing investigators of vertebrates and plants.

Past generations of entomologists have honed many of the traditional methods used to sample insects, largely through the transition of 'collecting techniques' to more quantitative or standardized methods, with many new approaches and refinements continuing to be developed (New 1998; Southwood and Henderson 2000; Wheater and Cook 2003; Leather 2005). Today, the purpose of sampling insects is not necessarily only for scientific advance. We now also sample insects for the purpose of ensuring we conserve as much insect diversity as possible (Samways 2005). We do this for insects in their own right, as well as to conserve the functionally important contribution that insects make to ecology and society.

Let us now overview some facts, philosophy, and perspectives that enable us to pursue sampling insects with some broad and general insight.

1.2 An historical perspective

The number of insect families has increased steadily over the last 400 million years (Mayhew 2007), with massive radiation during the early Carboniferous, just over 325 million years ago. The Permian extinctions took their toll on the terrestrial fauna, with loss of 8 out of 27 insect orders living at that time (Labandeira and Sepkoski 1993). Another major radiation event through the Triassic and Jurassic established the foundation for the huge insect diversity of today: 84% of the Tertiary insect families of 100 million years ago are still present today. Nevertheless, the meteorite impact that is thought to have caused the demise of the dinosaurs at the end of the Cretaceous, 65 million years ago, also affected the insects, with the implication that specialist feeders succumbed much more than generalists (Labandeira et al. 2002).

Members of only about a third of the living insect orders eat plants. A few orders, such as Lepidoptera and Hemiptera, are predominantly herbivorous. To feed on plants in the first place was a huge evolutionary hurdle, simply because plant bulk is not nutritious and the surface of the plant is a hostile environment (Strong et al. 1984). Yet those insects that did adopt herbivory have scored handsomely, with today some 37% of all animal species being insects that feed on plants. There are strong suggestions that Cretaceous insects, both as herbivores and as pollinators, fostered an enormous feedback loop (co-evolution) where more plant species provided more opportunities for insect speciation. Other interactions, such as parasitism and predation, also became involved. The resulting co-evolutionary process emphasizes that insect and plant ecology and conservation go hand in hand.

Interestingly, there seems to have been little insect extinction during the Ice Ages, which were harsh and telling times for many large mammals and birds (Coope 1995). In response to the advance and retreat of the glaciers, insect distributions shifted southwards and northwards respectively (in the northern hemisphere) (Ponel et al. 2003). However, one result, which is critical for a conservation perspective, is that there was loss of genetic diversity along the northern margins of the northern hemisphere (Hewitt 2003).

Mammals were important players on the landscape before the appearance of humans. Today we see the full effect of mammals only in a few places, such as the natural African savanna. However, shortly before tool-using humans began modifying and simplifying the landscape, there appear to have been large numbers of mammals creating a dynamic range of habitats. A resonance from these times is seen today where domestic livestock modify insect habitats, sometimes beneficially, but often not. Large mammal impacts on plants and on the physical environment, along with other factors such as fire, form an important ecological background for current insect conservation management. Indeed, for current conservation management a *sense of the past* is an important conceptual foundation (Figure 1.1).

Hunting by tool-using humans led to dramatically changed interactions on the landscape. About 6000 years ago megafauna populations in many parts of the world plummeted and forests began to disappear. Although there is little definitive evidence for all but a few insect extinctions, some estimates can nevertheless be made. Mawdsley and Stork (1995) suggest that 11 200 insect species have gone extinct since the year 1600. The future also looks bleak, with McKinney (1999) pointing out that a quarter of all insect species are under threat of imminent extinction. Indeed, the losses may be greater for insects than for many other taxa, with insects in the UK, for example, declining more than vascular plants or birds (Thomas et al. 2004).

1.3 Ethics of insect conservation

Studies of insects for conservation must begin with a firm set of values and clear objectives, not least because insects are often deliberately or unintentionally killed in the process. Furthermore, the basis for policy formulation of some conservation agencies is about the practical, utilitarian, or instrumental value of insects, with little consideration being given to the less tangible and higher values which constitute the ethical foundations of conservation. Yet, as outlined

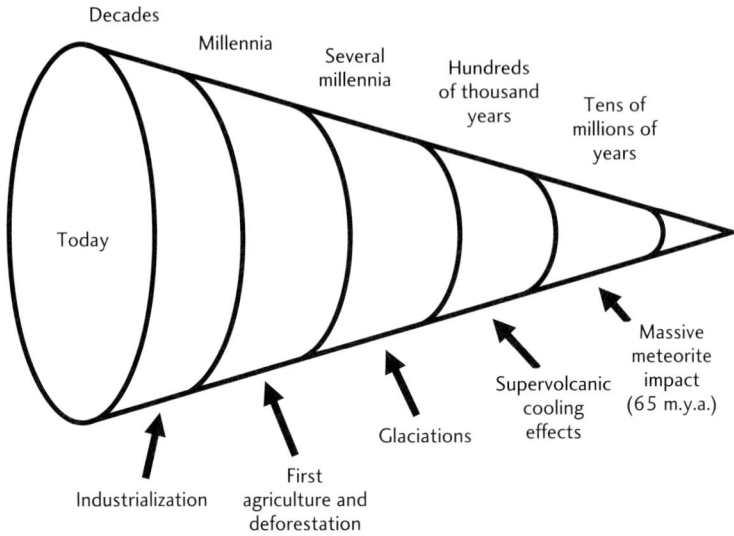

Fig. 1.1 It is conceptually valuable when undertaking insect ecology and conservation studies to have a sense of the past. In recent years, industrialization (along with intensive agriculture and urbanization) has made a major impact. By 6000 years ago, at least in Europe, deforestation and settled agriculture were well under way. Before that there were various dramatic natural events which were filters through which successful insect populations passed. These events give current species their particular biological characteristics and susceptibility to modern global environmental change.

by Haider and Jax (2007), incorporating a sound ethical base into conservation action can enhance its effectiveness.

At the outset, we may ask why we should even bother with conservation of insect species, given that by far the majority of all species that have ever existed are today extinct from natural causes. Arguably, extinction is therefore the norm. There is one major difference between the situation today and yesterday: insects are currently becoming extinct faster than at any time in the past, largely because of the huge and sudden impact of humans, whose activities essentially diminish the capacity of even rapidly breeding organisms to adapt to more gradual changes in their environments. Thus it ethically behoves us to do something about slowing, and hopefully stopping, this extinction spasm. At the basal, *utilitarian* level of values, this means maintaining all the warp and weft of life which maintains ecosystems in a relatively stable, and more importantly, resilient state (Costanza and Mageau 2000) (Figure 1.2). In short, with no insects there can be no world as we know it.

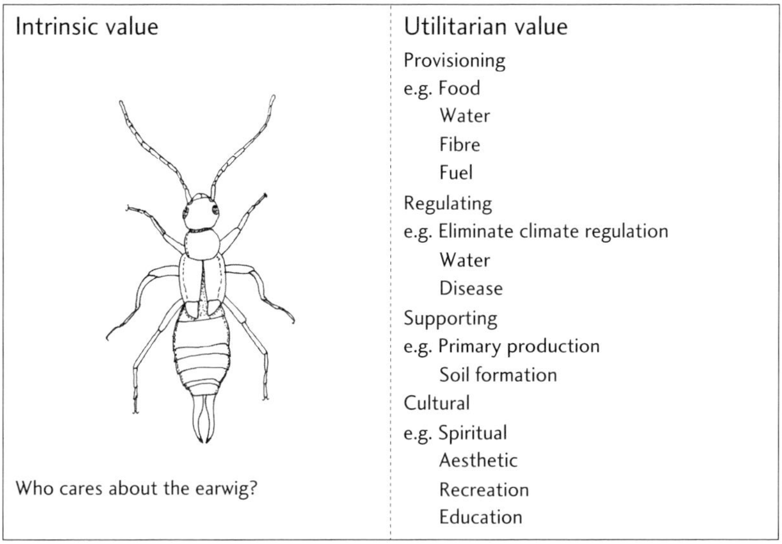

Intrinsic value	Utilitarian value

Who cares about the earwig?	Provisioning
	e.g. Food
	    Water
	    Fibre
	    Fuel
	Regulating
	e.g. Eliminate climate regulation
	    Water
	    Disease
	Supporting
	e.g. Primary production
	    Soil formation
	Cultural
	e.g. Spiritual
	    Aesthetic
	    Recreation
	    Education

Fig. 1.2 The value of insects. Intrinsic value recognizes the importance of conservation for all organisms for their own sakes, irrespective of value to humans. Utilitarian value, on the other hand, recognizes the value of biodiversity to humans. Many of the categories of utilitarian value (based here on those of the Millennium Ecosystem Assessment 2005) are consumptive, i.e. of direct use, while some are non-consumptive, i.e. used but not consumed (such as many cultural values). The Precautionary Principle should be considered when dealing with insect species with consumptive value. In other words, care should be taken when using these taxa because any organism may play a role in maintaining a functioning ecosystem.

The *utilitarian values* of insects are varied. Some insects are important commodities, but insect values are not necessarily consumptive, as in the case of crop pollination and pest management. Protected areas were originally created largely for our aesthetic benefit. While they have ecotourism value, today they have crucial conservation value as well. Mostly, the subjects of nature reserves and parks are rare and charismatic species that readily foster interest and are tourist drawcards. This non-consumptive yet utilitarian value is also playing an increasingly important role in insect conservation (Hill and Twist 1998). These values, however, remain essentially human-orientated.

An alternative value system is to consider the *intrinsic value* of organisms (Figure 1.2). This view sees insects and all their interactions as having the right to be, irrespective of their usefulness to us. We and they have entwined destinies, with rare and inconspicuous species also having a place on the conservation agenda. In turn, this approach raises the profile of insects as subjects worthy of conservation, as it also does with other cryptic organisms. However, insects are also particularly important components of many terrestrial ecosystems, and these ecosystems have systemic value. Conservation must therefore be directed towards protecting the natural biotic components and interactions in these ecosystems (Rolston 1994).

Many conservation organizations are largely driven by utilitarian values. However, the Red List, which is the most authoritative list of threatened species, and produced by the World Conservation Union (IUCN), in particular has a very strong flavour of intrinsic value. A wasp on the Red List can, at least theoretically and based on information available, receive the same amount of attention as a whale. In other words, the Red List system encourages 'equal value'. Of course, in practice, this is not always the case, simply because most people are more concerned for a blue whale than for a lesser-spotted fig wasp. With an increase in environmental awareness worldwide and a rise in collective human insight, the profile of insects is increasingly being raised in general conservation planning. Biodiversity conservation is now being more honestly brokered than in the past, as a wider range of organisms are being included, rather than mainly large and charismatic vertebrates.

1.4 Ethics of collecting insects

Collecting is an all-embracing term for taking specimens from the wild. More formal *sampling* involves taking a subset of the whole, so as to have a meaningful representation of the whole. In practice, and ethically, this can be a difficult decision for the insect conservationist, as sampling commonly equates to killing, to enable identification, naming, and counting of specimens. It is therefore important to collect only enough individuals to answer the key question being posed by your investigation, and to minimize unnecessary deaths or harm to individuals. Furthermore, when very rare or threatened species are studied, the capture or removal of any individuals may be prohibited or a serious threat.

General sampling at a site, for example to estimate species diversity, must be guided by possible harm to any threatened or susceptible taxa known to occur there. An additional consideration is that the sample must be manageable, there being no point in collecting so many individuals that they cannot be sorted and analysed. Intuitively obvious as this may seem, it is nevertheless surprising just how many studies in insect ecology and conservation flounder on this point. So, to restate our central theme here: it is critical to ask precisely *what it is that you are researching*? The answer to this question then determines your sampling protocol. How many insects you then sample also depends on the rarity of the focal species and whether it is threatened or not. The insect conservationist thus engages in *collection parsimony*, where individuals are taken or disturbed only to a minimum to answer the probing question. A corollary is that you need to bear in mind that any physical sampling method is only one window on the world. Methods need to be chosen which address your research question and give a true reflection of the world you are investigating, while not harming your subject populations.

Several general codes of conduct for insect collectors combine the ethical and practical considerations needed to ensure a responsible approach (Box 1.1 is an example). Although these codes were initially designed mainly for hobbyists, the major points included apply, sometimes even more particularly, to those engaged in scientific sampling.

Box 1.1 Code for insect collecting

With the increasing loss of habitats a point has been reached where a code for collecting should be considered in the interests of the conservation of insects. Whereas in the past collecting had only a trivial effect on insect numbers, today in some parts of the world, habitat loss has become so acute that collecting may have a critical extra impact on species.

By subscribing to a Code for insect collecting, entomologists show themselves to be a concerned and responsible body of naturalists who have a positive contribution to make to the cause of conservation. All entomologists should in principle accept the following code, and try to observe it in practice.

1. **Collecting—general**
1.1 No more specimens than are strictly required for any purpose should be killed.
1.2 Readily identified insects should not be killed if the aim is to examine them for aberrations or any other purpose: insects should be very carefully handled alive and then released where they were captured.
1.3 The same species should not be removed in numbers year after year from the same locality.
1.4 Supposed or actual predators and parasitoids of insects should not be unnecessarily destroyed.
1.5 When collecting leaf-mines, galls, and seed heads do not collect all individuals that can be found; leave as many of these habitat units as possible for the population to recover.
1.6 Consideration should always be given to digital photography as an alternative to collecting (in some cases, such as with butterflies, this alternative may even be convenient).

1.7 Collect specimens for exchange, or for sending to other collectors, sparingly or not at all.

1.8 Insects being used for commercial purposes should be either bred or obtained from old collections, and not collected from the wild.

2. Collecting—rare and threatened species

2.1 Rare species should be collected with the greatest of restraint.

2.2 Threatened species should generally not be collected at all, unless with very good reason and with all the appropriate collecting permits.

2.3 Collectors should look for new localities rather than collect local or rare species from well-known and perhaps overworked localities.

2.4 Previously unknown localities for rare and threatened species should be brought to the attention of local conservation authorities, so that species protection measures can be initiated.

3. Collecting—lights and light traps

3.1 The catch at or in a light trap should not be automatically killed.

3.2 Live trapping should be practised wherever possible, so that the trapped insects can be released back into the wild at the same habitat after examination.

3.3 Particular care should be given to catching very rare or threatened species. If this happens accidentally, traps should in future be relocated.

3.4 Light traps should be positioned so as not to annoy other people.

4. Collecting—permission and conditions

4.1 Always seek permission from the landowner or property occupant when collecting on private land.

4.2 Always comply with any conditions specified by the permit-granting agency.

4.3 When collecting (accompanied by an appropriate permit) in nature reserves, always supply the final species list to the relevant authority, as this information is often very valuable for management of the reserve.

5. Collecting—avoidance of damage to the habitat

5.1 Minimize disturbance to other wildlife as far as possible, especially to nesting birds and rare plants.

5.2 When beating bushes and trees for insects, beat the branch rather than thrash and damage the foliage.

5.3 When searching under bark and in leaf litter, always replace the material back as you found it. Always leave some bark or litter in the area, so some populations remain undisturbed.

5.4 Overturned logs and stones should always be returned to their original positions.

5.5 Aquatic plants, mosses, and liverworts should always be replaced in their original positions.

5.6 When sampling twigs, leaves, or inflorescences always use secateurs rather than simply breaking or tearing the plant part.

5.7 Be very selective in choosing sites for treacling, so as not to leave unsightly marks on trees.

5.8 Be particularly sensitive to and aware of any rare or threatened species in the area which could be disturbed by your activities.

6. Breeding

6.1 Breeding from a fertilized female or pairing in captivity is preferable to taking a series of specimens in the field.

6.2 Never collect more larvae or other livestock for rearing than can be supported by the available supply of food plant.

6.3 Unwanted insects that have been reared should be released in the original locality, not just anywhere. Alien insects should NEVER be released.

6.4 Do not attempt to establish a new colony or population in a new area without first consulting local conservation authorities.

Based on the Invertebrate Link code (www.benhs.org.uk/code.html)

Collecting also depends on assessing the time needed to process the collected specimens, and facilities available for housing them (see Chapter 2 for more details). Most multi-species studies must be supported by a *voucher collection*. This is a range of specimens for immediate and regular use as a reference for yourself, and as reference for researchers in the future. The collected species in such studies may at first not be identified, and simply be recognizable morphological forms, called *morphospecies* (see Chapter 2 for more details). It is much easier to remember nicknames rather than numbers, so it is often useful to give your morphospecies in the voucher collection vivid, personalized names in lieu of the scientific names which can be substituted later. For most studies, these morphospecies will be named by taxonomic specialists. For some taxonomic groups, and where there are good keys, it may be possible for you to do some of the identifications yourself. Nevertheless, where precise scientific identifications are required, your conclusions should be verified by a specialist. Designated morphospecies may, in fact, turn out to be two or more true species, one or more morphological or ecological forms, different life stages, or different sexes. If this happens, it does of course call into question the validity of analyses conducted on morphospecies. Finally, the voucher collection should be kept safe and curated as a formal museum collection in perpetuity (see Chapter 2 for more details).

Sometimes it is also essential to undertake *contingency collecting*. This means collecting for future potential projects. The prime example is for *barcoding* and other molecular studies. When this is the case, at least five individuals (or in the case of rare and threatened species, a single leg each from five different individuals from the same local population) should be collected and kept in absolute alcohol or acetone. Other potential uses include image analysis (for fluctuating asymmetry studies, for example) or for a pictorial database. For traditional morphological studies, ideally you should retain about 12 specimens (to illustrate variation about the mean) although for pictorial purposes only 2–3 specimens in good condition would be required.

Before any sampling takes place, it is essential to investigate the legal requirements for collecting permits for the places (e.g. national park) or taxa (e.g. protected species) you intend to sample and, if necessary, to obtain these. There are now many cases where insect collectors, including bona fide researchers, have inadvertently—or even deliberately in some instances—transgressed the local legislation regarding procurement of specimens. Always ask the local conservation authority about what is required *before* you start collecting or sampling, and allow plenty of time for any necessary permit applications to be prepared and processed, as this can take up to several months.

One further consideration is that you may wish to send specimens across international borders in which case you may need a CITES (Convention on International Trade in Endangered Species) permit. Again, start by asking your local conservation officer for advice on any import or export permits needed.

1.5 Ecological and evolutionary timescales

Sampling is a snapshot, in both time and space. Insects often have short generation times and their evolution can be very rapid (Mavárez et al. 2006). A conceptually good starting point is that short term (seconds to months) can be considered as ecological time and longer term (usually many tens of years) as evolutionary time. Over time ecological processes gradually translate into evolutionary processes. This distinction is conceptually important. Whereas ecological studies are essentially short-term, conservation must consider not only the current ecological situation, but also long-term evolutionary potential. Conservation is not just about today, but about the future as well. However, one must be very careful not to put absolute time frames on ecological vs evolutionary timescales, as not only do they blend into each other, but more and more findings are showing that evolution, especially among insects, can happen very quickly indeed. Thus we are dealing with a conceptual spectrum, and not a finite timescale.

Because sampling is such a brief episode in the complex ever-changing milieu of life, it is critical to be crystal clear on your goals. First, is it a spatial study, a temporal study, or both that you are planning (Figure 1.3)? Spatial sampling

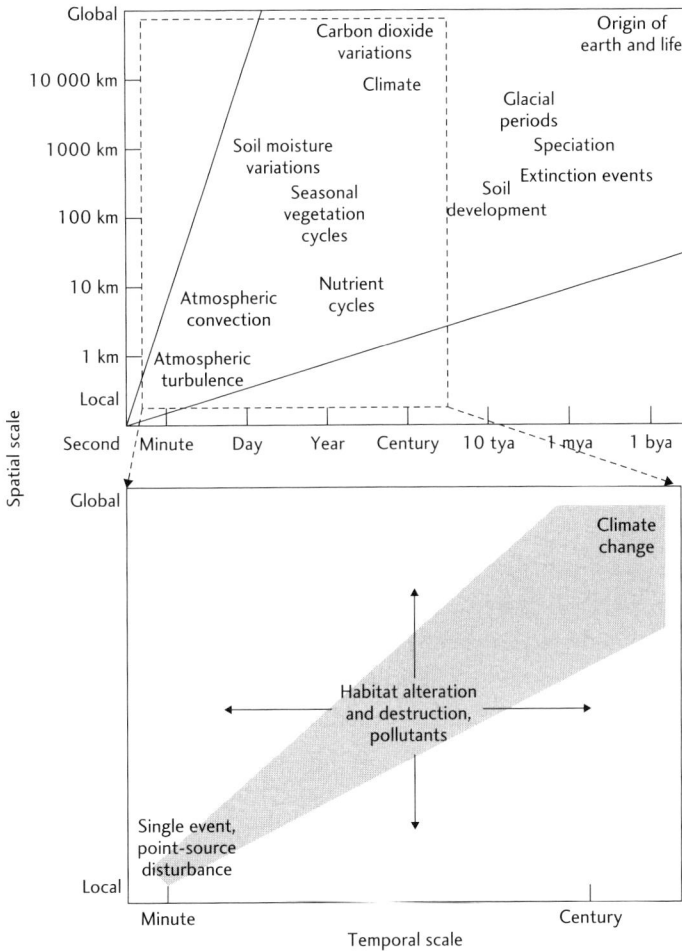

Fig. 1.3 The space–time scales of Earth system processes (above) and the scales at which studies of human-induced environmental change impact on insect biodiversity are conducted. The shaded area is the space–time scaling zone within which investigations are likely to have maximal predictive power. When starting an insect conservation research project, you need to be very clear on the aims or goal of the study. As part of the planning, you need to have a clear vision of the spatial scale of the study (e.g. small scale: comparison of insects on individual trees; large scale: comparison of aquatic insects across ponds in a region, even perhaps a global study). The spatial scale also depends on resources available, physical and financial, relative to time available. Your study may only be spatial, but ideally samples should be taken over time, at least to get some indication of seasonal changes. True temporal studies are conducted over many years. Seasonal studies and those conducted over a few years take place in ecological time. Studies examining, for example, thousands of years of subfossils of beetle elytra in soil strata, are considering changes over longer, evolutionary time (redrawn and modified from McGeoch 1998).

takes place at various scales from local to global, making it necessary to inter-rogate the scale of your investigation. Are you, for example, comparing insect assemblages on different trees in the same local area, or are you comparing insect assemblages on particular trees but at different locations? It may even be a larg-er-scale study, which then calls for careful consideration of the practicality of simply getting to the sites. Whatever the scale, the study will require adequate replication, often at more than a single scale (see Chapter 3).

Investigating temporal changes has its own set of challenges (see Chapter 3). Nature, virtually by definition, is always changing. There are nested changes, from diurnal (24 hours) to seasonal, then to roughly decadal (relative to sunspot cycles) and to longer term still. Overlying these regular changes are stochas-tic ones, where conditions may become unusually dry or wet, or hot or cold. Such out-of-the-blue impacts can be major for insects, whether directly, such as by affecting survival or behaviour, or indirectly, by affecting their mutualists, predators, parasitoids, and pathogens.

Time of day can be very critical for many species of insects, which often have strict diurnal behavioural patterns. Such behaviour may change when conditions are particularly adverse. Dragonflies, for example, may be on the wing at midday on moderately warm days, yet retreat to shade at the same time of day on very hot days. Closely related species may exhibit very different patterns of diurnal activity.

Long-term changes may be easy to track but may be difficult to interpret, because the environmental variables that affect insect populations are interre-lated (synergistic) (Figure 1.4) and change in intensity. For example, as rainfall patterns change often so does temperature, and these in turn may interact with human pressures such as intensity of agricultural practice. Such a complex of synergistic factors is believed, for example, to be driving changes in butterfly populations in Europe (Warren et al. 2001).

Despite all these causes of variation, a temporal study over a year can often provide valuable initial insight. However, such a study can be only a first approximation of seasonal changes in the focal insect population. Nevertheless, it can be very useful for focusing more detailed studies, such as measurements of population density or movement patterns. But rarely, if ever, is a single season or year of study enough to provide definitive information on which sound management plans can be based.

In summary, even studies at small spatial scales present logistical and analytical challenges. As the spatial scale gets larger, practicalities, such as simply travelling to sites on time, become a major consideration. Temporal scaling also has its set of challenges. Genuinely long-term studies need team work, although smaller, short-term studies carried out by one person, say, over three days, at different times of the year (e.g. each month), can provide real insight into seasonal

Fig. 1.4 Many impacts on insect populations, assemblages, and communities are multiple and synergistic, and change over time. These synergistic impacts need to be thought through very carefully to ensure that all environmental variables are accounted for in your study. Pictured here is the Basking malachite damselfly *Chlorolestes apricans*, a highly threatened species of the Eastern Cape, South Africa, at one of its last remaining intact habitats with an abundance of natural, low-growing riparian vegetation. Elsewhere, the species is suffering from the key threat of invasive alien trees, which continue to grow and spread along the river bank, shading out its sunny perches. The presence of the trees, in turn, leads to other adverse impacts, including shading out of the indigenous plants which are its oviposition site. The alien trees also encourage cattle to shelter under them, so breaking the stream banks and silting the water.

changes of insect populations and assemblages, essential for their conservation. Long-term studies are inherently difficult to undertake, not least because most granting agencies operate on a funding cycle of only a small number of years.

A final consideration when planning a study is that future generations may find your baseline results very valuable for comparison. This is why it is essential when writing up to record and save information on all aspects of your study in detail (exact location, sampling methods, species recorded, etc.) so that such direct future comparisons can be made.

1.6 The taxonomic challenge

The world of insects is so vast and complex that many assemblage- or community-level studies, particularly in warmer countries, are dealing with species which we currently cannot name. Yet we still aim to understand and conserve this nameless mass. This is the *taxonomic challenge*. Dealing with it involves focusing the study, both spatially and temporally. It also means being clear on the exact objectives and outcomes of the study. In practice, this usually involves doing some preliminary sampling before embarking on the full project. Although armchair planning may be the starting point, it is only when you get into the field and start some of the actual sampling that it becomes possible to assess whether the practical methodology is feasible and suitable for achieving your objectives (see Chapter 3.4). Once you have done such a *pilot study*, only then can you revise your objectives and re-plan. This may mean changing what you set out to do originally, which is far better than continuing uncritically and ending up with a pile of data of little value.

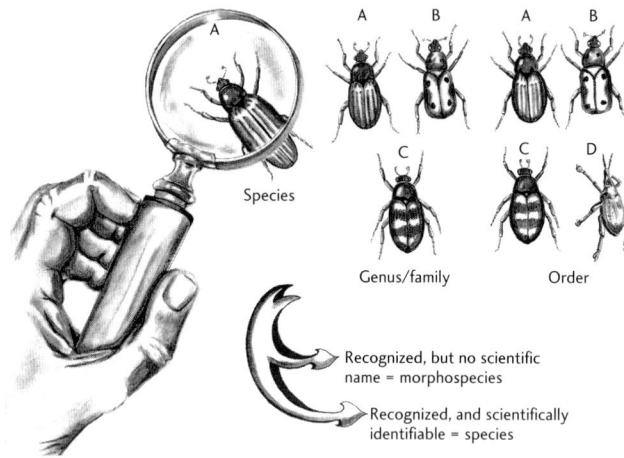

Fig. 1.5 The morphospecies concept. When sorting insect samples, it is best to start by sorting specimens into higher taxa, such as order or family/genus. Simply using these higher taxa rarely has any significant ecological value because such higher taxon groups do not provide sufficiently high resolution information for conservation. There is much more information in samples if you can identify specimens to lower taxonomic levels, particularly to species level. However, many species are undescribed and many others cannot be identified and named without consulting a taxonomic expert. Non-experts can sort (represented by the two-pointed arrow) the specimens into putative species (termed morphospecies) and give them an interim or working name, e.g. Coleoptera, Carabidae sp. 1. This is the morphospecies approach. Here, recognizable species A is sorted, with other specimens (B and C) into morphospecies (upper arrowhead). Where there is a dearth of taxonomic knowledge or expertise, they will remain as morphospecies. However, it may be possible to name the three species and assign them to known scientific species (lower arrowhead). Both these outcomes enable analysis of the ecological data, although species names clearly have much more taxonomic and ecological value than the casual names of morphospecies.

Critical to this feasibility study is the dimension of your taxonomic challenge. It may only be possible to sort samples to a higher taxon (e.g. family), labelling the component morphospecies (specimens recognizable as morphological entities and thought as possibly true species; see Chapter 2.5.1 for more details) of that higher taxon as sp. 1, sp. 2, . . ., sp. *n*. Such a study has two weaknesses. First, among insects, higher taxa are generally poor surrogates for species patterns and responses to change. Second, when doing assessments for whatever reason, the use of specimens formally identified to species level makes results repeatable. This is not possible with morphospecies (but see Chapter 2 for some solutions). Using higher taxa also has reduced value for conservation, especially of endemic species. Again, this is because without species names, it is not possible to compare samples from one study to another and so see how the status of species is changing relative to environmental change. Nevertheless, morphospecies can be excellent when undertaking a comparative, one-off study of, let's say, contrasting landscape patches and how the profile of the insect assemblages is changing. Indeed, in geographical regions where the insect fauna is poorly known, use of morphospecies may be the only reasonable option. Another possibility is to use a mixture of actual named species and morphospecies. This at least provides some taxonomic resolution yet also enables the project to continue without being held up by taxonomic challenges.

1.7 Summary

Insect diversity has increased over geological time, with insects today being more speciose than any time in the past. Their success has been associated with that of plants, meaning that insect conservation goes hand in hand with that of plants. Insect extinctions are becoming more widespread, making it important that we do something now to mitigate this trend. At the outset of a study, it is important that we are aware of the ethical foundation for why we are doing it. On the one hand, motivation for insect conservation can be utilitarian (we conserve because it is useful for us to do so), or, on the other hand, we conserve because of higher values which respect insect survival in its own right. This translates into whether we are conserving, say, an attractive butterfly species per se, or whether the reserve area for that butterfly is also significant for a whole range of other insect species which may not be glamorous or even particularly visible.

We sample insects to get a meaningful picture of the whole without actually measuring that whole. We choose a subset to provide ecological information and to enable us to make conservation decisions. As insects and their world are so complex, it is essential when starting a project to be absolutely clear on the exact research and conservation goal. We also need to be very sure about exactly how much time is available to achieve that goal. Sampling thus has to be done bearing the end in mind, so that we sample effectively and leave enough time for analysis and writing up. Collection of data and specimens is therefore done parsimoniously, but may also involve collecting for unforeseen eventualities, such as for future DNA studies.

Although insect ecological studies are largely about the present, insect conservation also considers the future and evolutionary possibilities for insects. As sampling needs to be tuned to the research question, it means being clear on the spatial and temporal scales of our study from the outset. We also need to have realistic expectations and ensure that our time and funding is sufficient to undertake the planned work. This means developing an initial plan but then going out into the field and doing some preliminary sampling to see what is actually feasible. Insect ecology and conservation studies inevitably have major taxonomic challenges (see Chapter 2). Higher taxon surrogates, such as orders or families, rarely provide good information for conservation planning. This means that it is essential to identify sampled specimens to species level. Where this is not possible, morphospecies can be used, bearing in mind that they have far less currency than actual, identified species. In biodiverse localities where only some species are identifiable, it remains useful to use a mixture of clearly recognizable morphospecies alongside identified species for purposes of conservation planning.

2

Taxonomy and curation of insects

2.1 Introduction: essential planning

Many insect species are unknown to science. Even species that have been described are often difficult to recognize and identify. This is the basis of the taxonomic challenge raised in the previous chapter. This dearth of knowledge is a major consideration for much of what we do in insect ecology and conservation. Whereas surveys of birds or mammals can often be undertaken without collecting and keeping specimens, this is usually not the case for insects. Apart from some well-known groups, such as certain butterflies and dragonflies, generally it is necessary to capture and preserve *voucher specimens* for identification by a specialist. Ecological insect studies often involve generalized collecting techniques that automatically capture a whole suite of species, including many *non-target taxa*, which may not be an intended part of the study. Together, these non-target taxa are known as *bycatch*. This generalized approach has associated ethical and practical problems, as discussed in the previous chapter, and presents risks to rare and threatened species. Indeed, the focal taxa of your study may well represent only a small proportion of the total insect sample, with the remainder never being used to advance knowledge. Even targeting only a single species may involve accumulation of bycatch. This remainder, rather than being discarded, should be kept for later study, perhaps by someone else in the future. This has particular value when, for example, a remote location has been surveyed or unusual specimens have been captured.

Collecting and sorting samples is very time-consuming and laborious, and it is essential that collation of samples is planned with much thought and care, both for ethical reasons and for efficient use of time. In other words, you need to develop a strict *sampling protocol*. When sampling is done strategically and efficiently, important ecological information on spatial distribution and temporal

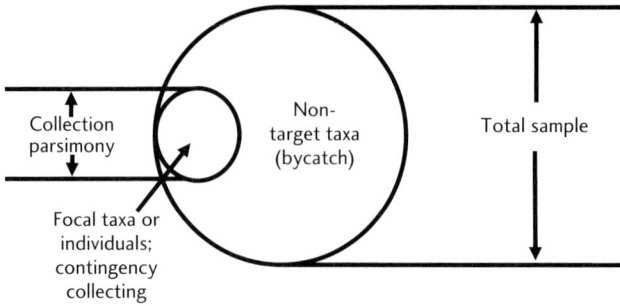

Fig. 2.1 Most collection techniques capture a range of taxa (the total sample) beyond those being targeted (the focal taxa, or other individuals e.g. functional types, size classes). These non-target taxa constitute bycatch that should be kept for possible future study. Sometimes it is appropriate to engage in contingency collecting (see Chapter 1), where taxa and individuals (which may or may not be the targeted, focal taxa) are collected for future studies, e.g. morphometrics or DNA. For both ethical and ergonomic reasons, insect ecologists and conservationists should always practise collection parsimony (see Chapter 1), where only those individuals essential to fulfil the project are collected.

activity patterns will be obtained. At the outset, it is important to establish (1) which insect group(s) is/are to be studied, and (2) the level of taxonomic interpretation that is needed to answer your research question. The position and numbers of samples and the *sampling extent* (see Chapter 3.3.2) to be examined must be clearly defined. As discussed in Chapter 3, the aims of the sampling programme (i.e. what you want to show) are integral to this. In short, your sampling should as targeted as possible. This theme is a major component of Chapter 3, but is introduced here to emphasize the importance of essential planning for interpreting insect samples, that, in turn, must be properly referenced in a collection (the subject of this chapter).

In contrast to generalized trapping of many species at a time, conservation surveys may be focused on particular species. These are often conspicuous or distinctive taxa, which need not necessarily be killed for identification. Indeed, for single-species conservation studies, it is often essential *not* to kill or even damage them—for example, by handling during mark–release–recapture studies in population dynamics (Chapter 7). Murphy (1989) raises the sobering question 'Are we studying our rarest butterflies to death?', which summarizes the dilemma that we need data on threatened species but we must gather them without harming the individuals or the population.

In this chapter, we outline some ways in which the insects in samples can be assessed, and explore the importance of *knowing* the insects you are studying. As well as being able to recognize each insect species (or equivalent) reliably, knowledge of its ecology and seasonal pattern of development helps you to decide how and when to sample. We include discussion of the selection of which groups to concentrate on to gain good information (section 2.2); the extent of taxonomic interpretation needed (section 2.3); possible substitutes for taxonomic species (section 2.4), necessitated by lack of taxonomic knowledge in diverse faunas; approaches to rapid assessment using less specialized labour (section 2.5); and the importance and care of collections of voucher specimens as reference points for your study section 2.6).

It may seem strange that this chapter has come before Chapters 3 and 4, which are on sampling procedures and methods of catching insects. We do this to emphasize that the aims, scope, and outcomes of your project must be carefully thought through *before* you start collecting.

2.2 Selection of insect groups for study

Inventories that are aimed at sampling all taxa in a particular area (all-taxon biological inventories, ATBIs), although noble in intent, are usually impractical. A more practical approach is to survey particular taxa (focal groups). Simply selecting groups because they reflect the individual interests of participating scientists requires caution, because 'favourite' taxa may not provide the best possible information for the particular ecological or conservation research question you are asking. Focal groups ideally should be easily sampled, reasonably diverse, have acknowledged or suspected indicator or other surrogacy values, and be sufficiently well understood taxonomically that many or most of the species can be recognized and identified. Thus, water quality is commonly appraised on an *EPT index*. This involves a focus on the larvae of Ephemeroptera (mayflies), Plecoptera (stoneflies), and Trichoptera (caddisflies) (Resh and Jackson 1993), many of which can be recognized to species level through availability of user-friendly keys and illustrated handbooks. All of these taxa may change in abundance as a reflection of changing water quality. In geographical areas where these EPT taxa are less well known, adult Odonata can be excellent surrogates (Smith et al. 2007).

It is entirely natural to gravitate toward the relatively few 'well-known' groups of insects for assessments and ecological analyses, both for ease of taxonomic interpretation of material collected and for the amount of information they consequently furnish. Indeed, many terrestrial insect surveys specifically select certain focal groups such as ants or carabid beetles. Groups such as these are chosen because (on the basis of substantial literature evidence) their species richness and assemblage composition may be linked in some predictable way with habitat quality, as well as to changes in that quality in response to anthropogenic disturbance.

Thus, the richness and ecological diversity of ants in Australia has led to delineation of a series of *functional groups*, whereby ants can be allocated to a particular ecological category, and the balance between the groups changes as conditions change in the wider environment (Andersen 1995b, Majer et al. 2004) (Boxes 2.1 and 2.2). The groups are founded on genera or other higher taxa (such as major species complexes), so obviating the need for species-level identifications (and allowing for inclusion of undescribed species), and the scheme has led to ants becoming the most frequently used insect 'indicator taxon' for terrestrial ecosystems in Australia. In many parts of the northern hemisphere, carabid beetles have been shown to be a useful focal taxon (Niemelä et al. 2000), but in many ecosystems in southern Africa and Australia (but see Horne 2007) their diversity is comparatively low and their taxonomy poorly understood. There is thus geographic variation in the suitability of particular taxa, and it is important to ensure that the chosen focal group is appropriate for the geographical area you are working in (see Chapter 10).

Box 2.1 Groups of ants used in environmental change assessments

---

Functional groups of ants (Hymenoptera: Formicidae) are listed below, with comments based on Majer et al. (2004). Their representation in Australia, North America, and South Africa is shown in Table 2.1.

1 **Dominant Dolichoderinae (DD).** These behaviourally dominant ants are predominant in productive ant communities. The role is dominated by members of the subfamily Dolichoderinae.

2 **Generalized Myrmicinae (GM).** These ants are often abundant and assessed as subdominant to DD. They tend to have lower activity, smaller colonies, and smaller foraging territories, and are broadly distributed in relation to stress and disturbance. They tend to predominate in moderately (rather than high) productivity environments for ants.

3 **Opportunists (OPP)** are unspecialized ants, often with wide habitat distribution, and are poor competitors. They thus dominate on sites with low diversity, such as those where stress or disturbance limits diversity of other ants.

4 **Subordinate Camponotini (SC)** are based on the vast genus *Camponotus*, which tend to be behaviourally subordinate to DD.

5 **Climate specialists (CS)** comprise taxa with distributions centred on the arid zone (Hot Climate Specialists, HCS), cool temperate regions (Cold Climate Specialists, CCS), or the mesic tropics (Tropical Climate Specialists, TCS).

6 **Cryptic species (CRS)** are those tiny ants (mostly Myrmicinae or Ponerinae) that nest mainly in soil, litter, or decaying wood, and are essentially removed from ant competitive hierarchies. They tend to be most abundant in forests, and may be a major component of litter ant communities.

7 **Specialist predators (SP)** are medium-sized to large and strongly predatory ants. They tend to forage singly (the group was at one time called 'solitary predators') and have little competitive interaction with other ants.

The importance of *knowing your insects* extends well beyond simply being able to identify them. It also includes understanding their responses to habitat change and other environmental changes (such as climate change). In this broader context, insects become 'tools' in conservation assessment where they are used to indicate a change in the environment (see Chapter 10). Whether insects are the focal taxa of the conservation exercise, or whether they are just 'tools', accurate identifications and consistent recognition of the particular taxa involved is still necessary. In studies of single species, a 'name' helps to retrieve existing information on the taxon and, conversely a 'wrong name' yields misinformation (Thompson 1997). Studies focusing more on assemblages or communities, or even on the whole fauna of a particular site, require consistent recognition of many species to document and track changes in species richness and composition over time.

2.3 Taxonomic knowledge

Even closely related species of insects, such as those in the same genus, may differ greatly in their abundance, basic biology, and responses to environmental change. Consistent recognition of species (or some equivalent working unit, as noted later) is the foundation for interpreting how a community works. However, many species cannot be named or recognized even by specialists, particularly in the tropics. Even for the best-known regional insect faunas, such as that of the UK, substantial proportions of very diverse groups such as parasitic Hymenoptera, Diptera, and many small beetles can be named only with much difficulty and with advice from taxonomic specialists (Barnard 2000). A field ecologist cannot assume that all insect groups collected in a survey can be identified reliably to species level, even in the better-documented places. Some form of triage (the selection of particular insect groups for study, disregarding others), however ethically

Table 2.1 *A key to groups of ant functional groups between Australia, North America and southern Africa. Major genera are listed in general order of ecological importance. Tropical rainforests are excluded because this habitat is present only in Australia. From Majer et al. (2004). See Box 2.1 for explanations of 'groups'*

Functional group	Australia	North America	Southern Africa
DD	*Iridomyrmex*	*Dorymyrmex* (bicolor gp)	Absent (except invasive aliens)
	Anonychomyrma	*Forellus*	
	Papyrius	*Liometopum*	
GM	*Monomorium* (several)	*Pheidole*	*Monomorium* (most)
	Pheidole	*Crematogaster*	*Pheidole*
	Crematogaster	*Monomorium*	*Crematogaster*
OPP	*Rhytidoponera*	*Aphaenogaster*	*Tetramorium*
	Tetramorium	*Myrmica*	*Lepisiota*
	Paratrechina	*Dorymyrmex* (insana gp)	*Paratrechina*
	Aphaenogaster	*Aphaenogaster*	
		Paratrechina	
SC	*Camponotus*	*Camponotus*	*Camponotus*
	Polyrhachis		
	Calomyrmex		
	Opisthopsis		
HCS	*Melophorus*	*Pogonomyrmex*	*Ocymyrmex*
	Monomorium (several)	*Myrmecocystus*	*Tetramorium* (solidus gp)
	Meranoplus	*Messor*	*Monomorium (setulosum gp)*
		Solenopsis	*Messor*
			Anoplolepis (some)
CCS	*Notoncus*	*Formica* (several)	*Anoplolepis* (some)
	Prolasius	*Lasius*	
	Stigmacros	*Leptothorax*	
	Monomorium (leae, kiliani gps)	*Prenolepis*	
	Podomyrma	*Stenamma*	
TCS	*Oecophylla*	*Neivamyrmex*	*Dorylus*
	Podomyrma (some)	*Attini*	*Myrmicaria*
	Tetraponera	*Pseudomyrmex*	*Cataulacus*
	Monomorium (some)		*Meranoplus*
CRS	*Solenopsis* (Diplorhoptrum)	*Solenopsis* (Diplorhoptrum)	*Solenopsis* (Diplorhoptrum)
	Hypoponera	*Acanthomyops*	*Plagiolepis*
		Hypoponera	
SP	*Myrmecia*	*Polyergus*	*Pachycondyla*
	Cerapachys		*Leptogenys*
	Pachycondyla		*Plectroctena*
	Leptogenys		

challenging the concept may be, is almost inevitable in collating the material accumulated from insect surveys (New 1998). Taxonomic tractability is a very important practical aspect when selecting particular focal groups.

Insects can be divided into three broad groups relative to taxonomic knowledge on them (New 1999):

- **Well-known groups.** In these groups many of the species are named or recognizable, with illustrated keys or handbooks available to help identification. Unfortunately, few insect groups come into this category, although they include butterflies, dragonflies, and some groups of beetles (such as carabids and tiger beetles). A bonus of working with well-known groups is that improved taxonomic knowledge is usually correlated with improved biological knowledge.

- **Catch-up groups.** These have a strong established taxonomic framework, perhaps to generic level, and many of the species have scientific names. Some identification guides may be available, but with significant gaps in coverage, and so they are not suitable for use by most non-specialists, because errors can easily be made. Such groups could usefully be made the focus for taxonomic studies because they could be converted to 'well-known' with relatively little work, and could then augment the portfolio of insect groups that could be interpreted meaningfully in surveys.

- **Black hole groups.** These are the groups of diverse, small, and inconspicuous insects, for which taxonomic knowledge is relatively poor, and in which many species have not been recognized, diagnosed, or named. The taxonomic literature for these groups is mostly specialized rather than widely available as identification handbooks to non-specialists. The incompleteness of knowledge renders almost all identifications by non-specialists unreliable. Nevertheless, many such groups have massive ecological importance. Disney (1986), for example, considered the values of flies and parasitic Hymenoptera in ecological interpretation to be high, but the groups are simply intractable at present, and can be employed only in very approximate terms. Likewise, poorly understood parasites, even if not numerically abundant, can have disproportionately important roles in ecosystem functioning (Hudson et al. 2006).

The goals of your study must be very clear, so that you can select your focal group bearing in mind the extent of taxonomic expertise or support available. Indeed, pragmatic criteria will be fundamental to selecting and narrowing down the groups to be studied. For a freshwater insect survey in Australia, Cranston and Hillman (1992) selected taxa on (1) availability of specialist taxonomic expertise to validate identifications, (2) availability of a good local identified reference collection, including earlier voucher material from the study site, and (3) acknowledgment by others that the groups have special merits as biological indicators.

It is useful, wherever possible, to seek advice from more experienced workers or taxonomists, particularly where there might be doubt about the level of tractability of any group proposed for survey, and you should do this well before a study starts. A relevant specialist may, for example, be able to give advice on selection and adequate sampling of the insects and their subsequent treatment. Such a specialist is also likely to have substantial background biological knowledge and ability to recognize any especially interesting species in the material that you should look out for at your study site.

There is also an economic consideration. If assistance is required for sorting, mounting, and identifying the insects collected, this should be costed realistically in planning the project budget. Also, there should be clear agreement with such collaborators on the intellectual property issues involved (such as co-authorship of papers), where material will be stored, and where any type material will be permanently deposited. It is not good practice to take specialist aid for granted, or to expect an expert to identify large arrays of inadequately processed or poorly preserved material without prior arrangement or subsequent acknowledgement. In many cases, taxonomic institutions are obliged to charge fees for identifications to cover the considerable time required. Some taxonomists however, may be very interested in your collection, and may request that they keep your material in exchange for identifications. This is usually a fair deal, but before embarking on it, you must clear it with your own institution to ensure that this is an acceptable arrangement.

We discuss below (section 2.5) some of the ways in which samples can be interpreted without having to undertake formal species identifications. Nevertheless, it is always preferable to have material properly identified, as it then has much greater value for communication and for comparison with already published data (Thompson 1997). The history of applied entomology, in particular, includes many cases in which misidentification of key pests, or confusion between related or similar species, has hampered management programmes. Similar situations are also beginning to arise in conservation in species-rich areas of the world.

For many insect groups, only the adult stage can be recognized and diagnosed with confidence. Most surveys of terrestrial insects are limited to adult insects. Yet immature insects often predominate in samples, but most of these have not been associated with corresponding adults by direct rearing, so cannot be identified reliably. Freshwater insects have a rather different bias, as the larvae of orders such as mayflies, stoneflies, and caddisflies in some geographical areas are well known and distinctive at the generic or species level, and occasionally may be more easily diagnosed than the adults. Nevertheless, very few surveys have compared adult and larval samples of insects in a given site or habitat, although Hawking and New (1999) compared the distribution of larval and adult dragonflies along an Australian river. Andersen et al. (2002) make the interesting observation that sampling just large-sized ants (those more than 4 mm long) reduced sampling times normally devoted to all ants down to only 10% and yet produced sound results. It may be possible to use subsets of taxa where appropriate and still get good results (that might not necessarily be simply large-sized individuals of other taxa). Indeed, this idea of subsets as surrogates is an exciting field of research in its own right and can enable substantial time saving when working at the critical and sensitive level of species, rather than at insensitive, higher-taxonomic levels, such as families (see next section). However, the conclusions drawn from using species subsets has been little tested (McGeoch 2007). Caution is needed so as not to make unjustified assumptions.

2.4 Taxonomic resolution

One aspect that can help overcome lack of species level information is to identify material to a lower resolution, i.e. to only higher taxonomic levels, such as by order, family, or genus, as well by functional groups, as mentioned for ants above. It is almost inevitable that such steps reduce the detail and value of the biological information obtained. Studies of insects categorized to only ordinal level are often difficult to use because such umbrella-groupings are virtually ubiquitous, and little information is conveyed. Classifying insects to orders is similar to classifying vertebrates simply as 'birds', 'mammals', 'reptiles', and so on, with each category containing a multitude of ecologically disparate forms. Such approaches are deceptively attractive, because they are simple and cheap to pursue, require little specialist discrimination, and little training to provide consistent results. A broad picture may indeed be gained, but the finer brush-strokes needed for refined management and critical assessment of diversity are almost inevitably missing. Danks (1996) notes that even family-level separations in insect samples are generally inappropriate because of extensive speciation at lower taxonomic category levels, which is often where there is the most subtle ecological sensitivity and discrimination.

Any such approach using higher taxonomic groups, or focal groups, as surrogates for wider biodiversity is part of the general trend to seek *rapid biodiversity assessment* (RBA) methods to gain information at a lower cost than when undertaking more complete assessments. Ward and Larivière (2004) discuss four main categories of RBA involving insects, as follows:

- **Sampling surrogacy** (reduced sampling intensity), such as trapping for shorter periods or reducing the number of individual trap units. This can be very attractive in reducing later costs, but care is needed to ensure that the regime remains scientifically rigorous (Chapter 3).
- **Species surrogacy:** as noted above, this is where higher groups are substituted for species-level analyses.
- **Taxon surrogacy:** the use of morphospecies (below) rather than formally identifying all components of the samples.
- **Taxon focusing:** the selection of particular taxonomic groups as surrogates of wider biodiversity.

The practical savings in time and costs gained from such broad approaches may still allow worthwhile assessment of complex insect samples, and set a foundation for more detailed, perhaps later, taxonomic analysis. Tests of the value of higher-level surrogacy have provided mixed results, so that it is usually not clear whether, for example, generic diversity strictly parallels species diversity within an insect group across samples or sites (see biodiversity indicators, Chapter 10). One instructive example involves ants in Australia. Most of the hundred or so genera (Shattuck 1999) can be recognized reasonably easily from user-friendly published keys, but many species identifications are precluded by lack of recent revisions and the presence of many undescribed taxa. Generic and species richness are sometimes closely correlated (Andersen 1995a), although this may sometimes result from the presence of many species-poor genera in samples (Neville and New 1999). Because ant functional groups (Box 2.1 above) are specified at levels above species, the ecological information obtained from these functional groups is largely valid. Also for the objectives involved high taxonomic resolution (identification to species level) is not necessary. Any such surrogacy measurement should be employed only on the basis of clear understanding and initial testing of the exact relationship between the higher and lower taxa (New 1999). Without this, serious errors may occur. For example, when there is an environmental disturbance rare and threatened species are often lost, and common and widespread species are gained or retained. In this case, if assessments are done at higher taxonomic levels this effect would not be detected and the approach would be counterproductive for conservation assessment. In summary, we strongly recommend that you do your study at the species level, rather than at higher taxonomic levels.

2.5 Morphospecies and parataxonomists

Most of the costs associated with insect sampling are incurred by the laborious sorting and identifying. Yet these must be done carefully to ensure consistency, which is important when quantifying samples, or comparing sites, or comparing the same sites over time. With a shortage of taxonomists relative to the diversity of insect life, it is imperative to use alternative methods so as not to slow the urgent task of insect diversity conservation. To this end, two tactics are widely employed in insect studies: use of non-named but consistently recognizable entities (morphospecies; section 2.5.1, Fig. 1.5), and employment of biodiversity technicians (*parataxonomists*) to replace or precede more specimen identification by specialists (section 2.5.2). These two methods commonly go hand in hand.

2.5.1 The morphospecies approach

The term *morphospecies* (see Figure 1.5) is used to denote a consistently recognizable entity thought to be equivalent to a species, but not named formally as such. The underlying principle of their use is that morphospecies are (1) deemed to be true surrogates for 'real' species, and so (2) may be used to classify and represent insect diversity without compromising accuracy, reflecting that the most commonly used measure of diversity is species richness (the number of species).

Morphospecies can be separated on any character or suite of characters that are sufficiently constant to distinguish amongst the entities involved. A variety of equivalent terms, such as *recognizable taxonomic unit* (RTU) or *operational taxonomic unit* (OTU), have been used, and are essentially synonyms (however, as a precaution, you should read the background definition to any such term you come across in a particular research report). Use of any such entity depends on consistency of recognition, and that each morphospecies contains all the representatives of a given type, and only those individuals. Sorting is therefore subject to the same constraints as for named species with neither *splitting* (the excessive: there should be division of entities on minor or non-significant variation to inflate the number of species or morphospecies present) nor *lumping* (the amalgamation of different forms through failure to recognize consistent differences, so lowering the number of real taxa). However, it is always better to err on the side of splitting, because, should you find at the end of your sampling that what you thought was two species (or more) now turns out to be one, it is easy to combine figures in your data matrix. But if you have one species and it turns out to be two or more, it means that you will have to sort through all your samples again to separate the individuals into their respective species.

It is important to recognize the advantages and limitations of morphospecies. Different groups of insects pose different problems for consistent recognition. In one comparative trial using ground-dwelling forest invertebrates, Oliver and Beattie (1996) found that ants and some beetles showed very high correspondence between morphospecies and species. In contrast, other families of beetles were much more difficult to identify, and morphospecies:species equivalence was much lower. The morphospecies approach allows inclusion of poorly known taxonomic groups into assessments of diversity. The ideal, of course, is that a morphospecies is fully equivalent to a real taxonomic species in such estimations. You may well need guidance and training to ensure that this ideal is attained. This may involve some preliminary sorting to gain practice before embarking on a full sorting exercise. Constant referral to *voucher specimens* (i.e. reference specimens as representatives of every different entity that you distinguish) is crucial at this stage. When working in biodiversity-rich and poorly studied areas, it is highly unlikely that all the specimens collected will be able to be named, even by specialists, and so it makes very good sense to build a well-curated collection of voucher specimens, which may contain both identified and unidentified (morphospecies) specimens. Clearly however, one must also recognize the long-term limitations of using morphospecies: for example, it precludes comments on endemism, threat levels to specific species (e.g. Red Listing), and detailed autecological studies.

Inevitably, separation of species (or morphospecies) based on morphological characteristics reflects a typological approach to species distinctions, and may mask unknown levels of variation. As Heywood (1994) emphasized, a 'species' may consist of many populations with differing genetic constitutions and adaptive potentials, and the lumping vs splitting dilemma becomes important in assuring reasonable consistency in sorting. There may indeed be a temptation to produce long species or morphospecies lists as inventories, because of the political message of high diversity they convey, but in other situations the emphasis may be on 'how few species' rather than 'how many species' are present.

In most cases, quality control over the consistency of recognition can be improved in two main ways:

- Continued cross-referencing to a complete set of the voucher series, sorted and accumulated progressively as the work proceeds.
- Progressively accumulated notes, sketches, and digital images for each putative morphospecies and its diagnostic features (which would not be a formal taxonomic description), providing a working tool for consistent recognition.

Such accounts would also be supplemented by the voucher collection. These records are important, because there are differences in perceptions among different researchers. Good quality multidimensional digital images can help overcome these observer differences.

The use of morphospecies in your study should be considered very carefully. Some of the problems and possible limitations of their use are summarized by Krell (2004), who emphasizes the need for effective quality control. However, that summary is naive in terms of practical conservation sampling, especially in the tropics. It overlooks the real value of morphospecies as stepping stones to later taxonomic verification. It also ignores the fact that in many parts of the world insect species simply cannot be named because there are so many species, many of which may well be undescribed, or there is no scientific specialist available to do the job.

A final point is that where the study covers a range of well-known, catch-up, and black hole groups, you may, in your data matrix, be able to insert taxonomic names for the well-knowns and morphospecies names for the catch-ups and black holes (e.g. Buprestidae sp. 1, Buprestidae sp. 2, …). This is a particularly useful approach where you are working in lesser-known geographical areas such as the tropics.

2.5.2 Parataxonomists

The term *parataxonomist* was used initially for technical staff employed in Costa Rica under that country's developing national biodiversity inventory (InBIO) to sort samples of insects and other organisms to morphospecies. Parataxonomists are usually employed on very large projects involving teams of researchers. The approach helps overcome lack of specialist capability or availability (Janzen 1997) by employing and training local people to collect, process, and undertake preliminary identifications of insects based on morphological recognition.

This approach does not solve all the problems of sampling, because, as pointed out by Basset et al. (2000), deployment of a team of parataxonomists can produce a huge amount of data but it still needs to be adequately analysed and written up by appropriately skilled researchers. Also, where parataxonomists are used, there must be standardization between individual workers, otherwise the value of parataxonomy breaks down. Problems in sorting may arise from factors such as sexual dimorphism (so that males and females of the same species are treated as different taxa); polymorphisms (various 'forms' of the same species); small size (obscuring small morphological differences); presence of very similar species; separation or amalgamation of different instars or growth stages; as well as individual observer traits (such as colour blindness). Several studies have indicated, for instance, that an individual's sorting acuity can lessen over a period of several hours of work as they become bored or careless with increasing tiredness.

Nevertheless, with adequate initial training and continuing supervision and monitoring, highly reliable results can be obtained. For very diverse samples of tropical insects, in particular, Basset et al. (2000) noted the benefits from their activities as including (1) increased efficiency of year-round sampling, (2) rapid processing of high quality specimens at low cost, (3) enhanced interpretation of local ecological information associated with the specimens collected, and (4) increased local interest in biodiversity. In short, subject to good quality control, their activities can markedly enhance appraisal and understanding of insect surveys. Essentially, employment of non-specialists has accelerated the collection and processing of specimens to lead toward inventories of rich insect faunas in countries with little resident entomologist expertise. Even in entomologist-rich parts of the world, non-specialists have vital roles to play (Goldstein 2004).

These approaches are often advocated as contributions to rapid biodiversity assessment (RBA), sometimes without adequate emphasis on the need for consistency. The practical balance needed is that between (1) reducing costs and time through use of non-specialists and imprecise species categories, and (2) incurring increased error risk and frequency through reduced specialist involvement. Parataxonomists are not an alternative to professional taxonomists, but enhance their capacity and capability (Lewis and Basset 2007).

Wilkie et al. (1999) proposed a basic sorting protocol, summarized in Fig. 2.2, in which feedback on quality control is provided as training at several stages: during sorting invertebrates to class, to order, and to morphospecies levels. This form of process control sampling was regarded as the better general strategy to adopt, not least because it is proactive rather than reactive, and serves to minimize errors during the entire sorting process.

2.5.3 Voucher collections

Such controls need, and demonstrate the practical values of, progressive accumulation of a *voucher collection*, i.e. a reference set of the morphospecies distinguished during sorting. A voucher collection (or reference collection) is built so that the individual designated as 'morphospecies no. 16' (for example) has a reference point against which other individuals can be compared. In many ways, a voucher series serves the same purpose for an ecologist as do type specimens for a taxonomist. Both are the ultimate reference points for accuracy, and vouchers must therefore be preserved, documented, and archived safely (e.g. see below) for future study. The uses of voucher collections include: (1) checking identifications as systematic knowledge advances, (2) detecting changes when re-sampling sites at some later date, (3) validating any reinterpretation or reanalysis of the data, and (4) comparing sites, such as those under different land use (Lee et al. 1982). Without such vouchers, errors of identification and interpretation may never be detected or verified.

The responsible deposition of voucher collections, preferably in a state or regional museum where proper curation (section 2.7) can be assured, should be an integral part of planning an insect survey, and the costs of this (and for subsequent maintenance) should be included in the project budget. At present, many investigators neglect this aspect of their surveys. The concept of *virtual vouchers* (see below) is an invaluable adjunct in sample processing, because of the easy availability of high-quality comparative images. However, it does not replace the need for actual voucher specimens that can, if necessary, be dissected or otherwise more closely examined in the future.

A similar responsibility applies to the bulk samples from surveys. Danks et al. (1987) referred to these as *ecological collections*, currently of unknown value for future investigation and for monitoring ecological changes at the sites from which they were derived. These ecological collections should be clearly and permanently labelled and stored responsibly as long-term archival collections, for genetic studies as well as for traditional taxonomy. Scudder (1996) expressed a much more widespread view, that on both ethical and more pragmatic grounds such collections should not be discarded: not only have the insects been killed, but there may have been substantial investment in obtaining them. In some cases, samples may be from remote localities that may be very difficult to revisit. From the taxonomist's point of view, such places may be of special interest in yielding unusual forms of variable taxa or members of sibling groups important for demonstrating patterns of evolution or endemism. Indeed, the deliberate accumulation of ecological collections may be a wise investment in documenting our invertebrate biodiversity. As Yen (1993) put it, 'if this kind of collection had been made more widely in the past, we would probably have many more answers to conservation management issues—but it is not too late to make ecological collections for use over the next century'.

2.6 Bioinformatics

Biological informatics technology, or *bioinformatics*, is the application of information technology to solving biological problems, and it is becoming an important theme in the description and identification of organismal biodiversity. Many aspects of technology are rapidly becoming tools or components of

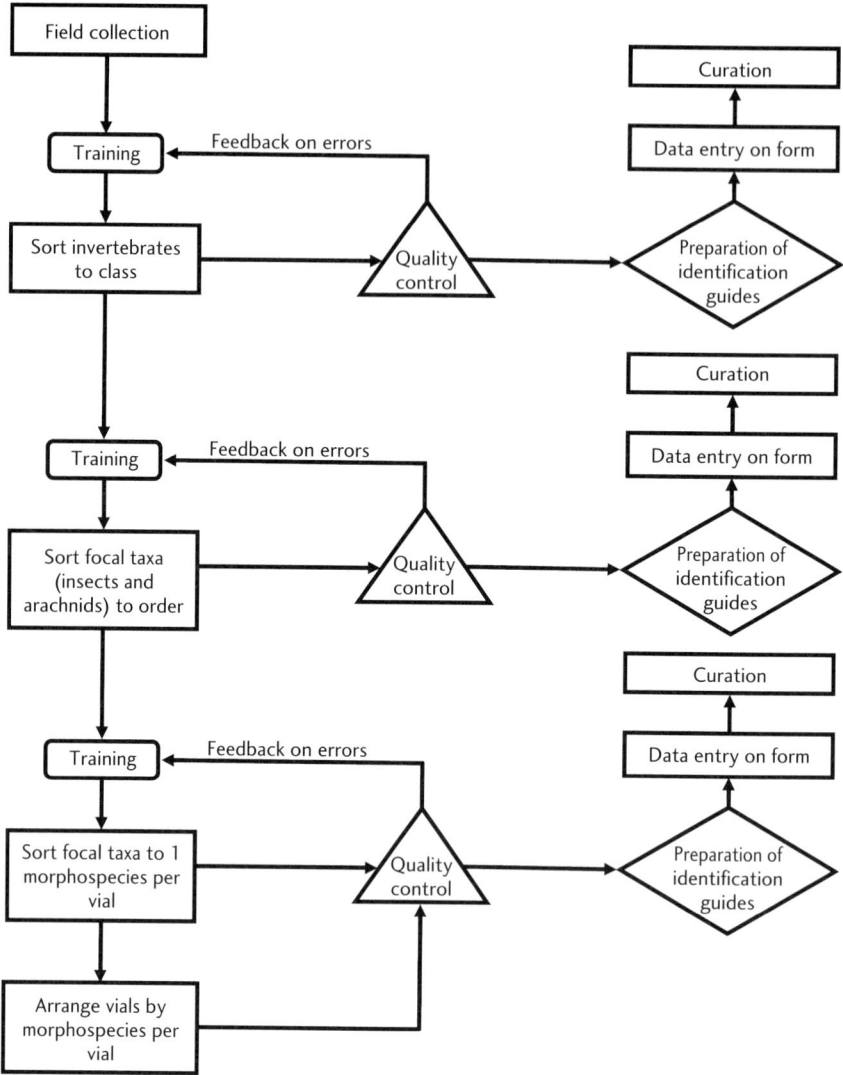

Fig. 2.2 A basic laboratory sorting protocol for insect specimens. Intermediate levels (e.g., sorting to family) may be inserted in the appropriate location using parataxonomists. NB: Each new level inserted has to have its own step of quality control and training (from Wilkie et al. 1999).

insect sample determination, in providing additional viewpoints for assessing diversity, and streamlining transmission and management of the information accumulated (Johnson 2007). The following section outlines ways in which bioinformatics is applied.

2.6.1 Virtual voucher collections

In almost every case, the voucher specimens are dead insects preserved for study by the methods outlined later in this chapter. However, alternative (or additional) approaches are gaining popularity. The development of accessible, high-quality digital photography, and accompanying software that allows photographs to be downloaded and transmitted within minutes, has opened up the potential for *photographic* or *virtual vouchers* as photographs in which recognition or diagnostic features are displayed clearly.

Oliver et al. (2000) showed how such systems may be used in assessments in ways that can be of practical value to conservation planners and land managers. Local *virtual voucher collections* (as digital image compilations) of species or morphospecies may be established, including images of the morphological characters by which each morphospecies is recognized and diagnosed. Such collections also facilitate rapid transfer of information, so that different laboratories and workers can use the same baseline for assessments, and so achieve fully standardized and comparable results.

Importantly, such systems should use accessible and generally affordable software to improve the speed and accuracy of data acquisition and manipulation. Barcode technology is used for sample and specimen tracking, and a high-resolution image-capture system is used for reference specimens and specific characters as each new entity is added. This virtual voucher collection, with associated taxonomic information, is used for onscreen recognition/identification of species, with potential to achieve wide usage on the World Wide Web as a directory of biodiversity (such as the Web of Life, http://tolweb.org/tree). Macadam (2006) notes the technical approaches to achieving high-quality voucher photographs. Importantly, however, he emphasizes that a voucher photograph can never be a full substitute for the preserved insect, but is a complement to more traditional vouchering. In his example and words 'there is no reason why a voucher photograph of a live moth should not accompany a set specimen in a collection'. The software package *Biota: The Biodiversity Database Manager* is designed specifically for biodiversity databasing and includes a facility for image storage.

2.6.2 Sample tracking

Allocation of a unique barcode label (see section 2.6.3) to each sample and subsequently sorted or mounted specimen increases the speed of data acquisition and helps raise quality by eliminating errors due to label transcription. Onscreen identification (1) is fast, (2) can be undertaken simultaneously by more than one person, and (3) obviates the need for repeated handling of real, fragile reference specimens. It is also claimed to increase accuracy and consistency of specimen allocations, and transmission of images may help in gaining advice from specialists far from the survey laboratory. A taxonomist could in theory, and increasingly in practice, work more from images accumulated from collections throughout the world, with digital images of *holotypes* and other critical museum material reducing the need for posting (and risking) such valuable specimens. The usefulness of this approach depends on the quality of the images and auxiliary information, as well as the costs of obtaining them.

2.6.3 Taxonomy using barcoding

Molecular data, particularly through the use of DNA barcoding, is a powerful tool for detecting so-called *cryptic species* (insects that resemble each other very closely so that they cannot be separated on morphological grounds alone). Molecular data thus helps to estimate species richness, to indicate intraspecific variation, or to delimit entities as biologically and taxonomically distinct (Wilson 2008). The process involves the sequencing of a single gene region, generally a 688 bp piece of mitochondrial cytochrome oxidase subunit-I (COI or coxI, for short). Cheap, rapid processing is available to do this. Specimens are not destroyed in this process, and analyses can be made from a single leg, or small part of a wing, so that the remainder of the insect can be kept in a normal collection, with suitable cross-labelling to the genetic material. However, specimens killed by chemicals such as ethyl acetate may have degraded DNA. To prevent this they should rather be killed by freezing or immersion in absolute alcohol.

The barcoding approach has already proved its value in separation of sibling species of pest insects. Perhaps the largest and most pertinent example to note here involves an inventory of the larger Lepidoptera of a major reserve in Costa Rica, a region where many of the species have not been described, and based on a survey that has continued for almost 30 years. It also illustrates the critical roles played by parataxonomists (section 2.5.2) in such exercises. Since 1978, every caterpillar found in the forest of the Guanacaste Conservation Area has been kept (background in Miller et al. 2006), given a unique voucher code, photographed, and reared to produce an adult or its *parasitoid*(s). Adult Lepidoptera are killed by freezing, and genetic analysis is based on legs of selected specimens. By the end of 2004, the project had accumulated 264 370 individual records, representing more than 1000 species, and providing unique insight into the richness and variety of a tropical insect group (Janzen et al. 2005, Box 2.2). Most similar projects will be on a much smaller scale. For example, a putative single species of skipper butterfly was shown using DNA characteristics to comprise 10 distinct species (Hebert et al. 2004).

Box 2.2 The process and background used to provide an inventory of Lepidoptera in Costa Rica based on DNA barcoding

1 Foundation: many individually vouchered and databased specimens have been accumulated as museum material over two decades from all main ecosystems of the Area de Conservacion Guanacaste.

2 Each insect voucher specimen is available automatically as a barcode voucher specimen.

3 The frozen and then oven-dried specimens have not been processed by methods known to destroy their DNA.

4 Each adult insect reared has three pairs of legs, and one member of a pair can be removed for sequencing, with 'reserves' available if needed.

5 The specimens are already identified to some level, often putative species, by more conventional taxonomic procedures, before entering the barcoding process.

6 When barcoding generates taxonomic questions, the inventory process can be modified to generate more specimens of the taxon in question, although this will of course involve a time lag maybe of up to a year through the process of discovery and rearing.

7 All species examined are either sympatric within the area or, if restricted to different ecosystems, parapatric at the meetings of those ecosystems over distances of a few hundred metres.

8 The specimens being studied are usually in excellent condition, as freshly reared rather then flight-worn adults.

9 Because they are all reared, it is often possible to know whether a pair of specimens is likely or actual siblings, and even to use barcodes of siblings and parents to examine intra-population variation and to confirm the accuracy of sequencing.

(Janzen et al. 2005)

2.7 Specimen treatment

Processing insect samples can be much more expensive than collecting them in the field, because of the large numbers of specimens accumulated. One published estimate suggested a sampling:processing cost ratio reaching 1:40. Thus, capturing and killing insects implies both a duty of care for the material, and the ability to track every specimen from a sample to its original source by adequate documentation (such as labels) and to keep it in good, examinable condition. Where temporary labels with information such as 'trap 46, transect 7' or 'grassland pitfall trap no. 12', have been used, it is essential to replace these temporary codes immediately with labels carrying information more meaningful to others. Similarly, using simple codes in notebooks can also carry risks (notebooks can get lost), and so, whether labels or notes, it is far better right from the start to use full information that is meaningful to any other researcher. This is one of the advantages of digital data loggers, where at least spatial information (such as geographical coordinates) is automatically recorded. In remote areas, where local place names are few and far between, and can change as the political climate changes, it is essential to record exact spatial coordinates. With the widespread availability of inexpensive geographic positioning systems (GPS), these data should progressively be collected as part of normal good practice. The data still need to be downloaded, and where appropriate, added to the actual specimens. In short, the aim is to produce a voucher collection that is informative and meaningful to any future generation of researchers.

At the time of retrieving each sample from the field, it should be provided with fuller information, in addition to any working code, with such labels placed routinely into each container of specimens. There are practical issues when doing this. For example, where the specimens are in alcohol, it is important to place the information inside the bottle, not outside, from where it may eventually peel off. Labels should be written in pencil and not in ink because some types of ink dissolve. Computer-generated labels on thin card can be prepared in bulk in conjunction with planning the sampling programme, so that each sample has a specially prepared label awaiting its preservation. Ensure that the ink is not soluble in the alcohol or in other preservatives (some workers recommend baking the labels in an oven at 120 °C for half an hour before use). A type size of 4–6 points for individual vials, and larger labels in 10–14 point type for bulk containers are generally suitable. Each subsample must have the same locality and ecological data on labels as the parent sample. Very careful attention should be paid to the type of bottle or vial (soda glass, because plastic is not suitable for long-term storage) used for storage of wet specimens (i.e. those in preservative). The bottles must have good seals that must not perish when in contact with the preservative, and must prevent evaporation. Bear in mind also that when the specimens are sent to a specialist via airmail the preservative must be drained, otherwise the bottles will not be handled by the air freight company. After the preservative (usually 80% alcohol) is poured out, the specimen must be plugged in the bottle with preservative-moistened cellulose wadding (as used for wound dressings in hospitals), and not with cotton wool that tends to snag the legs and antennae. The packaging for dispatch must also be labelled for customs purpose with appropriate wording such as 'Preserved scientific insect specimens, of no commercial value, and without free inflammable liquid'.

Details of label data will differ slightly for different purposes, but should normally include: (1) country; major political entity (state, territory, region); (2) name of nearest major settlement (with distance and direction from this), or name of national park or reserve (if relevant); (3) latitude/longitude or other recognized national grid reference or GPS reading; (4) major habitat; (5) collecting method (with any sample number, if relevant); (6) date of collection (day/month/year, cited as '10 March 2007' or '10.III.2007'—not '10.3.2007'); (7) name of collector (Fig. 2.3). These label cards are usually about 15 × 25 mm.

The country name is necessary because many specimens are exchanged between specialists or sent to specialists overseas for identification, with their eventual home in museums far from their point of capture. In the absence of a country name, a specimen simply labelled 'Victoria', for example, could be derived from a state of Australia, a region of western Canada, East Africa, or several other widely separated places. The sequence of date notation is recorded as 10.III.2007 (or even better, with the month in full), because this is unambiguous to all nationalities, which is not the case with 10.3.2007, which could be interpreted as either 10 March or 3 October. If many samples are to be processed, and to remove transcription errors, barcode labels can be used, as an innovation from RBA approaches (section 2.4). The ideal label should allow the reader to return to the collection site, and use the same sampling method at the same season to re-sample there.

```
SOUTH AFRICA,
 Kruger National Park, Skukuza
24° 59′ S   31° 35′ E
292 m a.s.l
12 January 2008
Beside still reaches of rocky savanna river
(Sabie), among fig trees, sweep netting
Coll. Jessica Jones
```

Fig. 2.3 An enlarged example of a data label that has been prepared to go on the insect pin just beneath the specimen (see Figure 2.16). Such labels usually measure about 25 × 15 mm. If there is not enough space on a single label, two labels can be used with one placed directly below the other on the pin. The second label usually contains the ecological or habitat information.

2.7.1 Sequence of steps for treating insect samples

The following sequence outlines the major steps needed to process insect samples, be they single species or many species. Single-species studies may impose additional demands, such as needing specimens for genetic analyses, requiring particular storage or transport modes, but the procedures are otherwise similar. Many specialist taxonomists have developed their own preferred techniques, such as slide-mounting and dissections, for close examination of the insects they study. This means that it is essential that you contact the specialists ahead of time to make sure that you supply them with the material prepared in the right way. More detailed information is readily available (Upton 1991; Uys and Urban 2006).

Although most insect sampling involves handling and identification of dead insects, it is sometimes necessary to cage and transport living insects for laboratory study, or experiments, establishment of breeding stocks, translocations (Chapter 8) or other purposes, such as rearing larvae to adults to facilitate recognition. Living insects may suffer or die from many causes, for example if: (1) they are exposed to temperature extremes, (2) deprived of suitable food, (3) overcrowded in small containers, or (4) not sufficiently confined so they can damage themselves on cage sides. More rarely, long-term travel, such as by air, may be needed, and it is then wise to arrange any documentation needed well in advance to avoid possible delays in quarantine or customs.

The essential steps in preparing insect specimens for examination are described below.

2.7.1.1 Preservation

Techniques for preserving and maintenance of insect specimens have been developed for well over a century and are described in a wealth of texts (see Oldroyd 1968, Smithers 1982, Upton 1991, Uys and Urban 2006, Cooter and Barclay 2006). Table 2.2 summarizes many of these standard and well-tried methods for each insect order, and the following notes augment this. First, the specimens must be killed humanely, then preserved in a suitable way, prepared for study as necessary, and finally labelled and stored responsibly. Studies for conservation or environmental assessment often involve taking of bulk samples by various types of traps, while surveys for particular species or guilds (such as pollinators) may involve collection of just a few individual insects.

Bulk insect samples, normally of recently captured or killed insects, are commonly collected passively and directly into liquid preservative (as in pitfall traps, Malaise traps, or suction traps; see Chapter 4). Actively caught insects (with

Table 2.2 *Recommended preservation methods for different orders of insect. Formulations of fixative chemicals as in Table 2.3*

Order	Methods
Archaeognatha	All stages: fix and preserve in 80% ethanol
Thysanura	As above
Ephemeroptera	Fix all stages in Kahle's, Carnoys, or FAA and preserve in 80% ethanol: some workers recommend adding 5–10% glycerol. Adults can be pinned through centre of thorax
Odonata	Larvae: kill and preserve in 80% alcohol; exuviae in alcohol or glued onto cards. Adults killed and colour-fixed in acetone, and once dry, placed in transparent envelopes
Blattodea	Preserve dry: pin through centre of metanotum; usually do not set, but left wings can be spread if needed
Isoptera	Collect and preserve in 80% ethanol; series of all castes should be collected wherever possible to facilitate identification
Mantodea	Pin between base of wings. Extend one foreleg so spines on femur and tibia visible; spread left wings of vouchers. May be necessary to 'gut' large specimens, by slitting along the abdominal pleural membrane and removing contents with forceps (avoiding damage to sclerotized structures); the abdominal cavity can then be wiped dry, dusted lightly with talcum powder and boracic acid powder (3: 1), and a small wedge of cellulose wadding or tissue inserted before specimen is pinned and set
Mantophasmatodea	As above, pin through centre of thorax
Phasmatodea	Pin through base of mesothorax; gut large specimens as above. May be useful to fold legs and align antennae along the body, to save storage space. Spread left wings of vouchers
Orthoptera	Adults: pin vertically through prothorax, to right of midline; spread left wings of vouchers. Soft-bodied species and nymphs: fix in Pampel's and preserve in 75–80% ethanol
Dermaptera	Adults; pin through right elytron or equivalent position to right of midline of mesothorax. All stages: Pampel's and alcohol as for Orthoptera
Embioptera	Winged adults may be pinned through centre of thorax and wings set. Otherwise: all forms in 75–80% ethanol.
Grylloblattodea	All: 80% ethanol
Zoraptera	As above
Plecoptera	Adults: pin through centre of thorax and set. Otherwise, all stages: fix in FAA and store in 80% ethanol.
Hemiptera (Heteroptera)	Large adults; pin through scutellum to right of midline; smaller specimens: stage or card point dry. Spread wings of vouchers when necessary. Nymphs: 80% ethanol
(Auchenorrhyncha)	Large adults; pin through scutellum to right of midline; smaller specimens: stage or card point dry. Spread wings of vouchers when necessary. Nymphs: 80% ethanol

Table 2.2 *(continued)*

Order	Methods
Coccoidea	Males: collect and preserve in 75–80% ethanol. Females: collect and preserve in lactic–alcohol, to facilitate treatment for later slide-mounting. Scales and mealybugs; dry and card as necessary
Aphidoidea	Collect into 80% ethanol or lactic–alcohol; may be useful to have some specimens in a cytological fixative
Aleyrodoidea	80–95% ethanol
Psylloidea	Adults: mount on card points or preserve in 80% ethanol; may be necessary to dry-preserve galls and lerps on foliage
Phthiraptera	80% ethanol, probably for later slide-mounting
Psocoptera	As above
Thysanoptera	Fix in 60–70% ethanol, probably for later slide-mounting
Megaloptera	Adults: pin through centre of thorax and spread wings, at least on left side of vouchers. Can fix all stages in FAA or 80% ethanol for storage in the latter
Neuroptera	Adults: pin representatives to right of midline through mesonotum, spread wings of left side or both sides. Others, and early stages, fix in FAA or Carnoy's and store in 80% ethanol
Raphidioptera	As above
Coleoptera	Large adults: pin through right elytron about one third from anterior and one third from midline; usually not set. Smaller adults: stage, or mount on card rectangles or points. all stages in 85 % ethanol, after KAA or Carnoy's
Strepsiptera	Males can be carded, but more usual to preserve all stages in 80% ethanol; if host available preserve this for association
Mecoptera	Adults: pin through metathorax to right of midline; wings may be spread. Others and early stages in 80%% ethanol after KAA
Diptera	Larger adults: pin to right of midline of mesothorax. Small adults, stage. Larvae: fix in KAA for storage in 80% ethanol
Siphonaptera	Collect and store in 75–80% ethanol for later slide-mounting
Trichoptera	Adults: can pin through mesonotum to right of midline, and spread wings, or collect and preserve in 80% ethanol. Larvae (with cases), KAA before storage in 80% ethanol
Lepidoptera	Adults: pin through centre of thorax and set. Larvae: KAA and store in 80–90% ethanol.
Hymenoptera	Larger adults; pin through mesothorax to right of midline; set as required, but commonly not done. Smaller adults on card rectangles or points: ants on card points with head to right. Otherwise store in 80% ethanol, possibly for later mounting on card or slides. Larvae in 80–90% ethanol

baiting, sweep netting, benthic grabs, suction samples, etc.—again see Chapter 4) may be removed from the net or trap and also placed in a liquid preservative. Liquid preservation is the most common option, with the exceptions of (1) light-trap samples of moths and some other nocturnal insects, (2) other types of insects needed for specific purposes, or (3) in the case of specimens to be collected alive.

An alternative to wet preservation is dry preservation. This is not suitable for all insect groups, or for larvae (Table 2.3), but dry preservation is essential for adult Lepidoptera, because scales are easily washed from the wings and body in liquid. Dry preservation is the conventional method for many insect groups, particularly for larger and harder insects. For some groups, a small series of dried and pinned or staged specimens (see below) can be accompanied by bulk material stored in alcohol.

For liquid preservation of adult insects, 70–80% ethanol (ethyl alcohol) is the most common option, although more specialized fixatives may be needed when the material is to be used for genetic or histological studies. Larvae are better

Table 2.3 *Major fixatives used for preservation of insect larvae before storage in alcohol*

Fixative	Formulation (volume units)	Uses
Carnoy's liquid	Glacial acetic acid: 1 95% ethanol: 6 Chloroform: 3	Rapid penetration without hardening. As larval fixative, use for at least 2 hours, not more than 24 hours
KAA	Glacial acetic acid: 2 95% ethanol: 10 Kerosene (undyed): 1	Larvae; kill in fixative, then as for Carnoy's.
FAA	Glacial acetic acid: 1 Distilled water: 20 Formalin: 5 95% ethanol: 25	Particularly useful for aquatic insects. Can store indefinitely, but insects may harden.
Pampel's fluid	Glacial acetic acid: 2–4 95% ethanol: 15 Distilled water: 30 Formalin: 6	As for FAA
Lactic–alcohol	95% ethanol: 2 75% lactic acid: 1	Aphids and scale insects. Kill and store

treated in slightly more complex fashion, and may be fixed and stored as in Table 2.3. At the time of initial retrieval from traps, insects should be transferred into fresh preservative, because the liquid used in traps may either be unsuitable for long-term preservation, or may have become overly dilute or dirty. Furthermore, loss of body fluid from insects can lead to dilution of alcohol preservative, so that it is good practice to replace the alcohol at least once more (after about a week) before permanent storage. At initial preservation, it is useful to remove (and preserve separately) any bycatch (Chapter 2.1), as well as of course removing contaminants such as vegetation and other debris. It is also very important that the catch from each single trap (or other defined sampling unit) at any one time be kept in a separate bottle, so that the unit remains intact. Also, it is essential at this time to put the data label, written in pencil or alcohol- and waterproof ink, inside the storage bottle.

Individually captured larger, adult insects should be killed humanely in a killing bottle or jar as soon as possible after capture. Odonata, however, may be placed in a paper envelope and immersed in acetone overnight to partially preserve their colours, while also leaving them in a condition suitable for DNA studies. Larger-bodied orthopteroids should be kept dry after killing. A killing bottle is a wide-mouthed glass jar with tight-fitting screw top or cork plug lid, containing a layer (1–2 cm) of plaster of Paris (gypsum) on the bottom (Fig. 2.4). This is impregnated with an anaesthetic agent, and the insects confined in the jar are killed by exposure to the fumes. Two killing agents commonly used are ethyl acetate and potassium cyanide. A few drops of ethyl acetate on the plaster layer serves as a fumigant for at least several days before replenishment. Alternatively, crystals of potassium cyanide (**care: extremely dangerous:** check with your local authority whether you are allowed to use it) may be sealed under the plaster at the time of making the killing jar, preferably under a layer of vermiculite or similar expander. Potassium cyanide has the advantage that it is long lasting without replenishment, and (particularly in remote areas) obviates the need to carry liquid ethyl acetate in a separate, sometimes fragile, container. However, use of cyanide means that any discarded jars must be disposed of according to local regulations regarding disposal of toxic substances. The jar should anyway be wrapped securely in masking tape to guard against contamination from accidental breakage before disposal. Ethyl acetate has the advantage that the specimen is left very pliable, making it easy to set on a setting board (see below). Its disadvantage is that it can leave some specimens pink in colour, especially the legs of grasshoppers. Killing jars should be carried in a knapsack or belt-pack, not in the pocket where they are easily broken and could cause injury.

(a) (b)

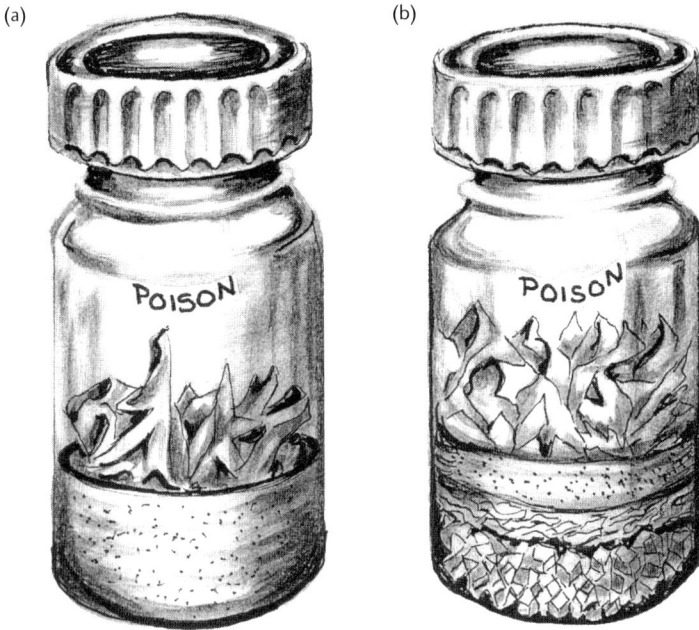

Fig. 2.4 Designs for a killing jar for selected captured insects. (a) Plaster of Paris is set into the bottom of the jar and thoroughly dried. A teaspoonful of killing agent such as ethyl acetate is added to the plaster of Paris base, and before use some crumpled strips of tissue paper are placed in the bottom. (b) This is a permanently charged jar, with a layer of cyanide crystals, sealed in with a layer of vermiculite and then a layer of plaster of Paris left to dry. Tissue paper is placed in the bottom just before use. These jars must be labelled as containing poison, and also with a skull and cross-bones (not shown here for clarity). They must never be placed in the sun or in the pocket of clothing next to the body. Wrapping the jar in adhesive tape is also recommended, to prevent injury if the jar breaks. Shown here are narrow-neck jars which are easy to handle and carry on a belt with a specially designed pocket. For insects with large wings (e.g. some butterflies) a wide-necked jar may be preferred. The cyanide jar is usually surrounded by some duct tape as an additional guard against breakage (not shown here for clarity).

While carrying out a survey (depending on the objective and target group/s), it is wise to carry several killing jars for different sites, for different samples, or for different insect groups. A jar used for killing Lepidoptera is not suitable for other insects, because it becomes contaminated with wing scales and is often difficult to clean subsequently. Insects should not be left in jars for longer than necessary, because addition of other, live, insects can lead to damage through trampling and tumbling around. Placing one or two loosely crushed paper tissues in the jar can help prevent this damage. Some workers opt to transfer stunned insects to envelopes and place these in a larger, reservoir killing jar to liberate the normal killing jars for repeated use. Tiny or very delicate insects such as microlepidoptera can be collected individually into small glass vials and these, plugged loosely with cellulose wadding, can be placed into a killing jar; or a drop of ethyl acetate added directly to the plug before sealing these in a jar. As noted earlier, killing by freezing is needed for material to be used for genetic studies: in remote areas, a car fridge or polystyrene beer-cooler is a useful tool. Cellulose wadding is much preferred to cotton wool, as it does not snag on insect body parts.

Depending on the freshness of the chemical killing jar, 1–3 hours is usually enough to ensure death, with the insects becoming comatose after much shorter periods. Specimens should then be transferred for storage or further preparation, such as pinning. Dry storage may entail papering (for butterflies, dragonflies, and other large winged insects), or *layering* for beetles and others. Papering involves placing the insects individually with wings folded back into triangular or other envelopes, preferably translucent, so that the contents are visible, and with data details written on the outside or enclosed with the specimen (Fig. 2.5). Where such envelopes are not available, or where conditions are very humid, envelopes can be made out of newspaper. Specimens can be kept like

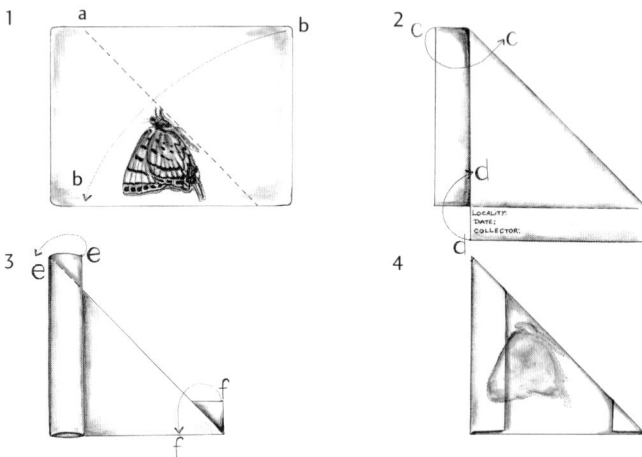

Fig. 2.5 Folding procedure for making a temporary storage envelope for dead insect specimens. Note that there is only one specimen per envelope, and the envelope is labelled with all the appropriate data.

this for long periods for future permanent mounting and storage (as long as they are protected from the ravages of pests). Envelopes can be stored in a rigid card or wooden box or cabinet drawer, with the outer container containing a small amount of naphthalene to deter damage from foraging insects such as ants, particularly at night in field camps.

Layering entails using a plastic container, such as a sandwich box, in which a thick (~5 mm) layer of paper tissues is placed. A layer of the collected, dry insects is placed on this, a label is added, and the whole covered with more tissues without crushing. The process is repeated until the box is filled (Fig. 2.6), so that the insects are kept secure, uncrushed, and firmly enclosed so that they cannot get shaken around. If field sorting time is limited, it is useful to put like with like where feasible, so that all beetles or all flies are together in one box or in one layer. This shortens later sorting and possible damage of handling dry insects before they are prepared further. Specimens from different samples or sites should always be layered separately to avoid possible mixing. Also, they should not be left in such a sealed box for more than a few days without some sort of extra drying, otherwise mould will set in. In the tropics in particular, it is essential to add

Fig. 2.6 Loose insect specimens layered in a field storage box. Papered specimens (see Figure 2.5) can also be placed in such boxes. Specimens are kept like this until they can be prepared for permanent storage.

a mould suppressant such as thymol and a desiccant such as silica gel. Both these substances are normally available at well-stocked pharmacies.

2.7.1.2 Mounting

Methods for mounting individual insects are summarized in Table 2.2. Other than slide-mounting, which is needed for critical identification of very small insects such as thrips and booklice, the main technique is *direct pinning*. This is often accompanied by some form of *setting* or spreading to display the various structures (such as wings) neatly and effectively, and *staging*.

Direct pinning is the most frequent mode for many larger insects, using specially made entomological pins of stainless steel with lengths around 38 mm. Regular dressmaking pins are no good as they are normally too short and thick, of poorer quality steel, and corrode very quickly. Various thicknesses of entomological pin are available ranging from fine (0) to extra thick (5), with the

(a)

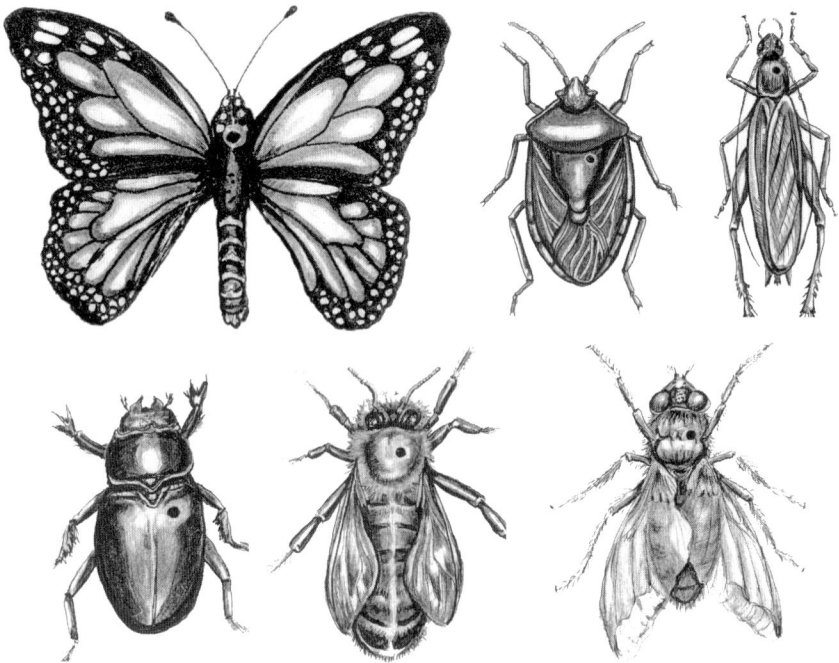

Fig. 2.7a Position of the entomological pin, viewed from above, when preparing certain insect groups for permanent storage. Clockwise from top left: butterfly (or moth) (pin central), bug (pin slightly to right of midline), cricket (and grasshopper) (pin slightly to right of midline), beetle (pin through right elytron), bee and fly (pin to right of midline). (Specimens not to scale.)

(b)

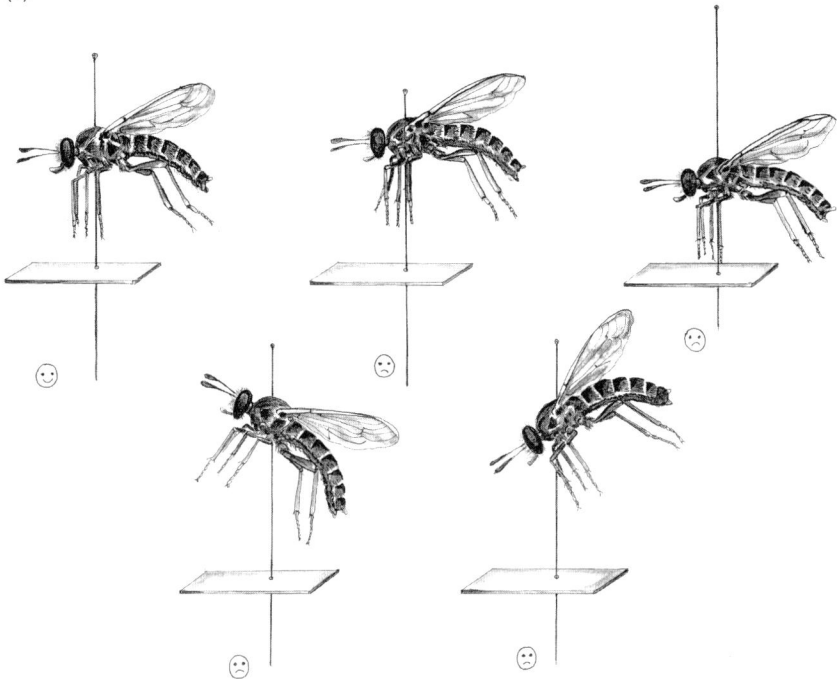

Fig. 2.7b Position of the entomological pin, viewed from the side, when preparing an insect (in this case a fly) for permanent storage. Only the specimen in the top left is correctly pinned; the other four are incorrectly pinned.

intermediate grade (3) being the most useful single category. Finer pins tend to bend, and thicker ones are needed only for the very largest insects. The position of the vertical pin differs for different insect groups. Lepidoptera, for example, are pinned through the midline of the thorax, but groups such as Orthoptera and Hemiptera are commonly pinned slightly to the side of the midline, to preserve midline features of possible taxonomic significance. Beetles are pinned through the right elytron, about one third distance from the midline and the same distance from the anterior margin. Figure 2.7 indicates some of these variations.

Points to note for standardization (Fig. 2.7b) are: (1) that the pin is vertical, (2) that the position of the insects on the pins is standardized (to facilitate easy scanning under a microscope without need to continuously refocus) and leaving enough space above the insect for handling, and enough below for labels, and (3) that a suitable-sized pin is used. To standardize height of the pinned insects on the pins, a pinning stage is used, which can easily be made at home from a wooden or aluminium block (Figure 2.8).

Fig. 2.8 A pinning stage to ensure correct and consistent pinning heights of insect specimens. This can easily be made out of a small aluminium or wooden block. The shorter shafts of the stage can also be used for maintaining consistent heights of the data labels.

Insects can be pinned only when they are soft and pliable. When dead for more than a few hours, insects rapidly become dry and stiff. Attempts to pin them in this state may damage them. Note, however, that many insects undergo a period of rigor mortis, becoming temporarily stiff for a few hours and softening for a short period after about 10–24 hours. Subsequent stiffening is permanent. However, long-dead insects, for example in paper envelopes or layered in boxes, can be relaxed by exposing them to a humid environment for a day or so (for tough beetles, more rapid treatment by immersing them in hot water is sometimes an option).

A suitable *relaxing box* can be made from a plastic sandwich box with a layer of clean sand (covered with paper to prevent contamination of insects) or tissues (without any soluble colour) that can be dampened with water to create a humidity chamber once the lid is closed (Fig. 2.9). Insects to be relaxed are placed in the chamber, on top of an additional layer of tissues to prevent direct wetting. They are then left overnight, by which time many will already be sufficiently softened to permit pinning. If not, test again after a further 6–8 hours, or after a second night. Specimens in this state should usually be handled with very soft forceps, which can be home-made from the steel straps used around wooden shipping cases (Fig. 2.10). If left unattended for longer periods, specimens may become mouldy. To prevent this, a small amount of thymol or other fungistat placed in the relaxing box will help to prevent this (NB: the use of chlorocresol, mentioned in some early texts, is not recommended as it is hazardous to health).

Fig. 2.9 A relaxing box, made from a plastic sandwich box, layered at the bottom with damp white tissue. Specimens that have hardened during temporary storage are softened (and thus made ready for setting) by placing them for a day or two in such a relaxing box. A pinch of thymol prevents any build-up of mould.

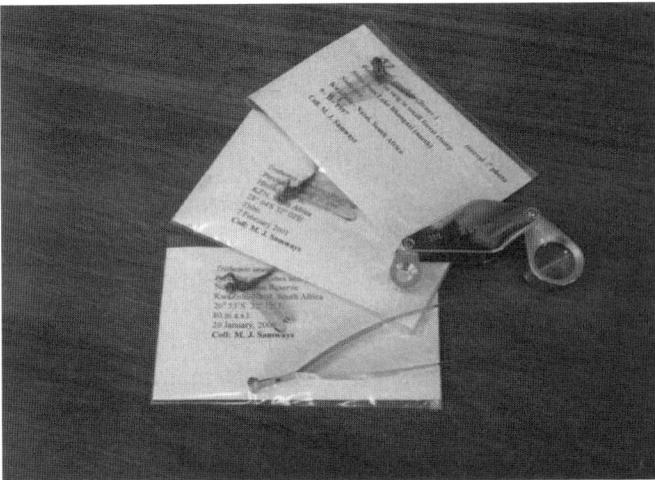

Fig. 2.10 Very sensitive forceps, home-made out of packing case steel, for handling delicate insects, e.g. putting in and taking out of dragonflies from transparent envelopes. Also shown is a simple but very versatile and useful piece of field equipment—a good-quality 10× and 20× hand lens.

Very small insects can be pinned using tiny *micropins* (usually only 15 or 20 mm long, and of very fine stainless steel) for subsequent *staging* (see below). As above, the specimens must be thoroughly soft when pinned. Micropinning is a very delicate operation, usually undertaken with the aid of a binocular microscope. The specimen is supported with fine entomological forceps on a piece of lint-free cloth such as a handkerchief, or a block of expanded polystyrene or expanded polyethylene (*plastozote*) or other commercially available plastic foam used to line insect cabinet drawers, covered with fabric to prevent movement. Micropinned insects can be set (see below) on boards, or the micropins inserted into plastic foam in shallow plastic trays with the body in contact with the foam. Wings are then gently moved to the required position and, if necessary, supported by pins, and the insect left to dry over the next 3–7 days. Labels should be pinned alongside for later staging with the specimens.

Many pinned insects, of all sizes, need some manipulation before they harden, to increase accessibility of taxonomic features and to augment their aesthetic appeal. To achieve this, they must be set, which means spreading their wings and legs (as appropriate) to expose as many of their significant morphological features as possible. Lepidoptera are usually set so that the wings of both sides of the body are extended horizontally and symmetrically, but many other groups of insects conventionally have wings extended on only one side. Where there is a long series of obviously the same species, not all specimens need be set in this way. This saves both time and storage space (see below). Besides, if needed in the future, additional specimens can always be relaxed and reset.

Setting involves use of a *setting board*. This is a wooden board with cork or plastozote facing, with a central groove to accommodate the insect's body, and a flat surface to support the wings and other appendages during drying (Figure 2.11) for 1–3 weeks, depending on size. The wings are carefully moved into position with fine mounted needles (*setting needles*), held in place by covering *setting strips*, ideally of translucent paper, and secured in position by pins, as shown Figure 2.11. Setting boards should be kept safe over the period of use. If left exposed, the insect specimens will readily be eaten by other insects or mice! Labels should accompany the insect on the setting board, and afterwards be attached to the specimen pin beneath the specimen itself. *Partial setting*, which involves simply parting the wings from the body, and moving other appendages as necessary for greater visibility, is another way of arranging insect specimens for ease of study.

Staging refers to the insect being glued on to a card mount (or stage) and further supported by a normal entomological pin placed through the mount

Fig. 2.11 Setting an insect, in this case two butterflies and a moth, to expose maximum area of the wings, while at the same time having some sort of naturalness. The butterfly at the top right still needs its wings moved up slightly, whereas the moth is its final position. The wings are spread with a setting needle, and held in position with paper setting strips and pins. Notice that the main pin through the body is exactly at 90° to the wings to ensure an orderly appearance. The data label, which may already be prepared in final form, accompanies the specimen. The set insects are left in this position, ideally for a few weeks, in a dry, protected, well-aerated place.

rather than through the insect. The insect can also be micropinned to the stage. Many small insects, too small or too hard to pin, are mounted directly on to card triangles or rectangles (Figure 2.12), a process termed *carding* or *card-mounting*. Micropinned insects are usually pinned near one end of short lengths (8–12 mm) of square cross-section foam, such as plastozote or similar compound plastic, and a regular pin inserted near the other end. Small beetles, ants, and parasitic wasps, for example, are normally card-mounted. They are stuck onto the tip of a small triangular or centre of a rectangular piece of card with water- or alcohol-soluble glue, so that specimens can be soaked off if need be. Do not use any glue that may blob or is not transparent, or any type of super-glue that damages the specimens. Ants are normally placed across the tip of card triangles (Figure 2.12), but beetles are usually placed on a firmer, rectangular mount. If the material is to be deposited in a specific institutional collection, it is useful to seek specific curatorial advice on the procedures preferred by that

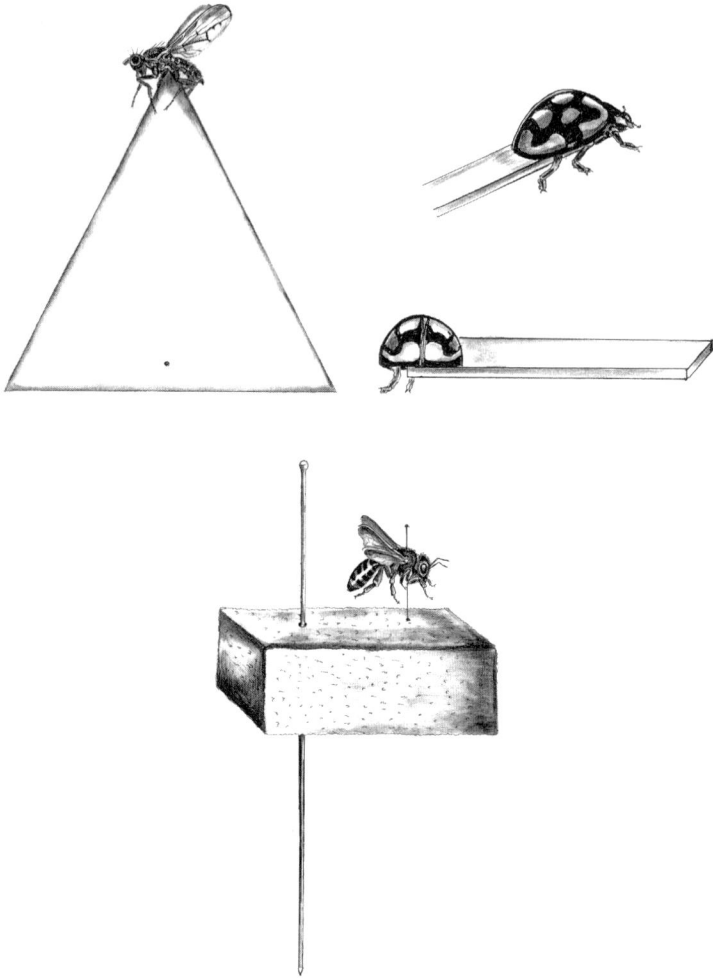

Fig. 2.12 Staged insect specimens may be mounted on the point of a card triangle, in the centre of a card or on the end of a strip for larger specimens, or micropinned onto a small block of plastozote. The data label sits below the specimen on the same pin. A wide card triangle is shown here, to leave space for any parts that accidentally break off or are deliberately removed to be water-glued to the card.

institution. All of the materials mentioned above are available from dealers in entomological supplies.

2.7.1.3 Storage

Adult insects with important, diagnostic soft parts and many larvae may be unsuitable for dry preservation and so are preserved in liquid preservative. Conversely, there are a few insects that should not be preserved in liquid as important diagnostic features, such as the white waxy covering of mealybugs, are lost.

Insects in liquid preservatives (e.g. 75% ethanol) are stored in soda glass jars with tightly fitting lids (such as fruit-bottling jars with rubber seals) to stop evaporation (Figure 2.13). Individual insects can be stored in small soda glass vials and placed upside down in such jars. Regular inspection to top up preservative as needed is a continuing responsibility for the curator of the collection. Specimen data labels should be placed inside the jars with the specimens. Fire or other safety issues may dictate that the alcohol containers must be stored in special metal or fire-proof/fire-resistant cupboards, and may also impose other conditions, such as on the total volume of alcohol that may be stored.

Pinned or staged insects are kept in wooden entomological storage boxes with tight-fitting lids, or in specially made insect cabinets, consisting of a series of glass-topped drawers (Figure 2.14). You should visit your local museum or other institutional collection to see how storage works in practice. The main purpose of storage is to protect the specimens from direct physical damage by exposure, from moisture, from excess light (that may cause fading of colours), and from damage by pests such as museum beetles (*Anthrenus*) which rapidly enter collections and eat their way through the precious specimens. The storage containers must therefore be virtually airtight. A small amount of a recommended, non-hazardous, solid insecticide, with an active ingredient such as d-allethrin (see, for example, the UK Health and Safety Executive website), can be placed inside the storage container and secured in position with a pin or a small cardboard box as necessary. This is not a permanent solution, and the storage container (box or drawer) must be inspected every few weeks to check that there are no invasive museum beetles. Should any mould appear on the specimens, it must be dealt with immediately by placing thymol crystals inside the storage box or drawer. When maintained well, a collection will last for many decades, and indeed, some museums have specimens that are well over 200 years old.

It is worth spending the time to make up or buy an *examination stage*, which is a base with a swivelling arms and a rotating stage on which is glued a small piece of plastozote (Figure 2.15). This enables the pinned insect to be turned and viewed from various angles, especially under a binocular microscope, to assist with observation and identification. A cheap, simple alternative, although it

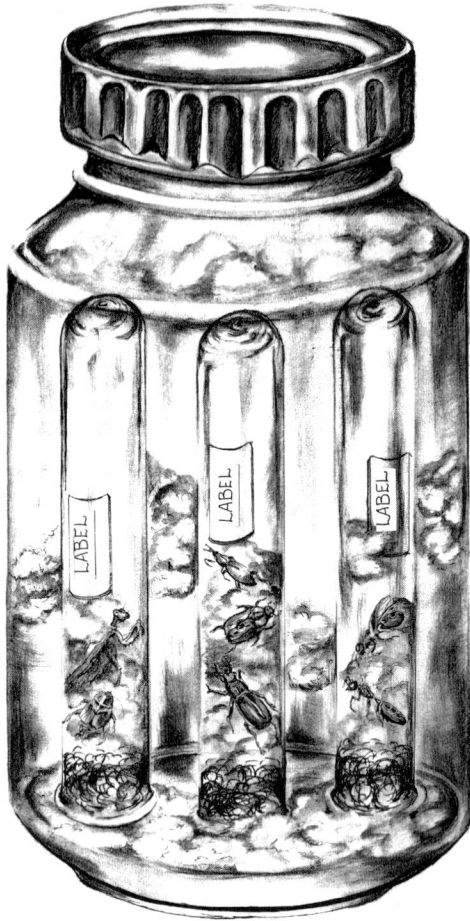

Fig. 2.13 Some insect specimens require wet storage, i.e. in a preservative such as 75% ethanol. Individual specimens, or small groups collected at the same time in exactly the same habitat, are placed in small vials, with their data labels, and plugged with cellulose wadding. The tubes are placed upside down (having expelled all the air so that they sink) in a large jar with a good seal, and the whole jar topped up with the same preservative. The vials should sit on a layer of cellulose wadding, and, if necessary, can be held in position by more wadding throughout the jar. Traditionally, wide-necked jars were used, but these tend to gradually lose the alcohol over time, so tightly sealed preservative jars with strong caps and robust seals are now favoured.

requires more handling of specimens, can be made from a blob of plasticine on a microscope slide or similar substrate that fits on the microscope stage. Again, most of the material needed for generating and curating insect collections are available from dealers in entomological supplies.

Fig. 2.14 A well-curated permanent storage collection. This is a drawer from an insect cabinet, and has a very tight-fitting lid. Some d-allethrin is placed in the corner (to keep out museum beetles), along with a small tray of thymol (to keep out mould).

Fig. 2.15 An examination stage can be made by a local metal worker. The aim is to provide a stage on to which the insect is pinned for close examination, particularly under a binocular microscope, by turning through the vertical axis (with the lower knurled knob) and through the horizontal axis (with the upper knob).

If it becomes necessary to send dried pinned insects by post or courier to a museum or specialist, they must be protected carefully. Individual insects must be placed in a tough cardboard or wooden box with a base of firmly affixed plastozote (do not use polystyrene as it does not grip the pin sufficiently and the specimens will shake free). They must be *cross-pinned* to prevent them coming loose or swivelling on the pin (Figure 2.16). The boxes containing the insects should be as small as practicable and sent within larger boxes containing expanded polyfoam chips, bubble wrap, or similar buffer to surround the insect box by 5–10 cm on all sides. Packages should be sent by the shortest direct route, and clearly marked 'FRAGILE', on the outside. For international travel, further precautions may be needed to avoid damage from opening in customs: any paperwork needed must be clearly available, and inspection of the contents without direct handling may be facilitated by covering the insect box before closing with a clear plastic wrap such as kitchen cling-film (Glad wrap, Saran wrap, or some similar product).

Finally, it is worth emphasizing again that before embarking on preparation and storage of important insect survey material, it is really worth the time and effort to discuss your project with an experienced curator. Even limited consultation may save considerable work, avoid frustration, and avoid important errors. It might also be important to have access to a major collection during the identification phase of a study, and museum staff may be able to give advice on

Fig. 2.16 When pinned insect specimens are sent by post or courier, each entomological pin with its specimen must be placed firmly into the plastozote base. The specimens are then cross-pinned to hold them firmly in position. A layer of stretch-wrap kitchen film should be wrapped across the box before it is closed. Finally, the whole box is covered with bubble-wrap or immersed in polyfoam chips inside its outer postal box.

relevant taxonomic literature during the planning stages. In turn, specialists may be able to provide most valuable aid as the work proceeds. Non-entomologists sometimes find it difficult to appreciate that, unlike workers on birds, reptiles, or mammals—all of whom may have a good working knowledge of, and be able to identify, most members of the group with which they are familiar—entomologists may specialize in only one order of insect, or even in only one family of a large order such as Coleoptera, Diptera or Hymenoptera. Usually they are still able to provide much general insight even when it is not their speciality, and where appropriate, recommend where you may be able to go to seek appropriate advice on other taxonomic groups.

2.8 Summary

Before starting any field work, it is essential to plan well. In addition to designing a sampling protocol (Chapter 3), this includes how you are going to collect insects (see Chapter 4), and how you are going to preserve and store them (this chapter). In other words, planning involves thinking through all the practical steps *before* embarking on work in the field. This plan may be refined by doing some preliminary field work (a *pilot study*, see Chapter 3) to ascertain whether the plan is workable, and to more closely define what resources, both time and money, are going to be needed.

Choosing the insect group for study depends very much on the precise aims of the project and how well the species are scientifically known in the geographical area in which you are going to work. We strongly recommend working at the species level, and not at higher taxonomic levels, such as family. This is because there is far more variation and sensitivity to environmental variables at the species level than at higher levels. It may be possible to actually identify sampled species, such as butterflies and dragonflies, even in tropical areas. But for many taxa it will only be possible for you to work on recognizable morphospecies, a collection of which is likely to be a close approximation of the real species assemblage. In geographical areas where some taxa are relatively well known and others less so, you may choose to do a study using both morphospecies and identified species. Where there are large numbers of samples to process, it may be possible to employ local parataxonomists, who can be trained to sort captured specimens into recognizable taxa, and in some cases, even identify the taxa. After sorting and thorough labelling of the specimens, it will be necessary to develop a voucher collection, which is effectively a permanent voucher collection to which you, and others after you, can refer. Such voucher collections, besides housing the actual selected specimens, may also contain additional information such as digital images, so that there is a supplementary virtual voucher collection.

It is critical for any study that the specimens are properly processed. While recommendations are given here for various insect taxa, we strongly advise that you contact the museum or other institution to whom you might send specimens for identification. They will advise on the exact method should be used for preserving, arranging and storing the collected material. For most taxa there are standardized techniques that should normally be adhered to, unless advised otherwise by a specialist. Finally, it is critical that your collection is carefully stored to enhance its long-term value. This involves, among other things, keeping out museum beetles and mould. All these considerations are part of the initial planning of the project, and should not be left until the specimens have already been collected.

3

Designing sampling protocols for insect conservation

3.1 Introduction

Insects, by virtue of their variety, small size, and large number, do not readily lend themselves to being counted or their diversity quantified. They most often first have to be conned, cajoled, or otherwise extracted from their crevices and other hiding places using a selection of the techniques that are described in the next chapter. However, once the target group or taxon (for example ground-dwelling or aquatic insects, such as ants or aquatic beetles) and most appropriate collection method(s) have been selected, several other questions arise. Ecologists (and entomologists in particular) are almost never able to obtain a complete measure of reality, and must rely on samples to 'represent' that reality. However, clearly we must do all that we can to ensure that the samples taken, and thus the outcome of the study, are as reliable (trustworthy) as possible. In other words, when sampling, we have to minimize bias and sampling error (Table 3.1). These arise from poor practical techniques (measurement or investigator error) or poor sample design (for example too few samples or poor distribution of samples as we will discuss in this chapter).

Most of the following questions should be answered well before venturing into the great outdoors. In fact, asking and correctly answering the following set of questions will make the difference between the success (meeting the object-ive, obtaining reliable, information-rich results, successful publication of the research, and producing viable recommendations for management and conser-vation) or failure of your study. The important questions you should ask are:

- When should I sample?
- Where should I sample (precisely rather than generally)?
- How often should I sample?
- How many samples should I take?

Table 3.1 *Decisions to be made when designing a sampling protocol*

Question requiring a decision	Further discussion
What are the study objectives, key questions, and null and alternative hypotheses?	This chapter
What is the target taxon (taxa) or group (e.g. functional group)?	Chapter 2
Is the taxonomic expertise available and accessible so that the species sampled can be identified?	Chapter 2
What are the variables and relational (environmental) variables to be measured?	This chapter and Chapter 5
Where will the study be conducted and what will the extent of the study be?	This chapter
What is/are the most appropriate sampling techniques?	Chapter 4
What is the grain of the study?	This chapter
How many samples and replicates are needed (power analysis)?	This chapter
Where should the samples and replicates be placed across the study area (extent)?	This chapter
What is the duration, timing and frequency of sampling?	This chapter
Is a pilot study necessary?	This chapter

The answers to these questions will determine the sampling design of the study. These answers are dictated by four things: (1) the biology of the target group or taxon; (2) the sampling technique chosen (e.g. pitfall vs sweep-net) (see Chapter 4); (3) the structure and physiognomy of the habitat and landscape; and, importantly, (4) the objective of the study. Let's start by considering the study objective.

3.2 Defining the conservation question

How you go about sampling depends very much on why you intend to sample. In fact, the most important ingredients to the eventual success of any study are clear and precise statement of the study questions and objectives, and, thereafter, good planning (Figure 3.1). If you set out on your bicycle with no specific destination in mind, don't be surprised where you land up! The same applies to research studies without clear objectives. Therefore, the more clearly and more detailed the objective is at the start of the study, the greater the likelihood that the outcome will achieve that objective. In conservation studies, management-related questions and information requirements form the priority objectives.

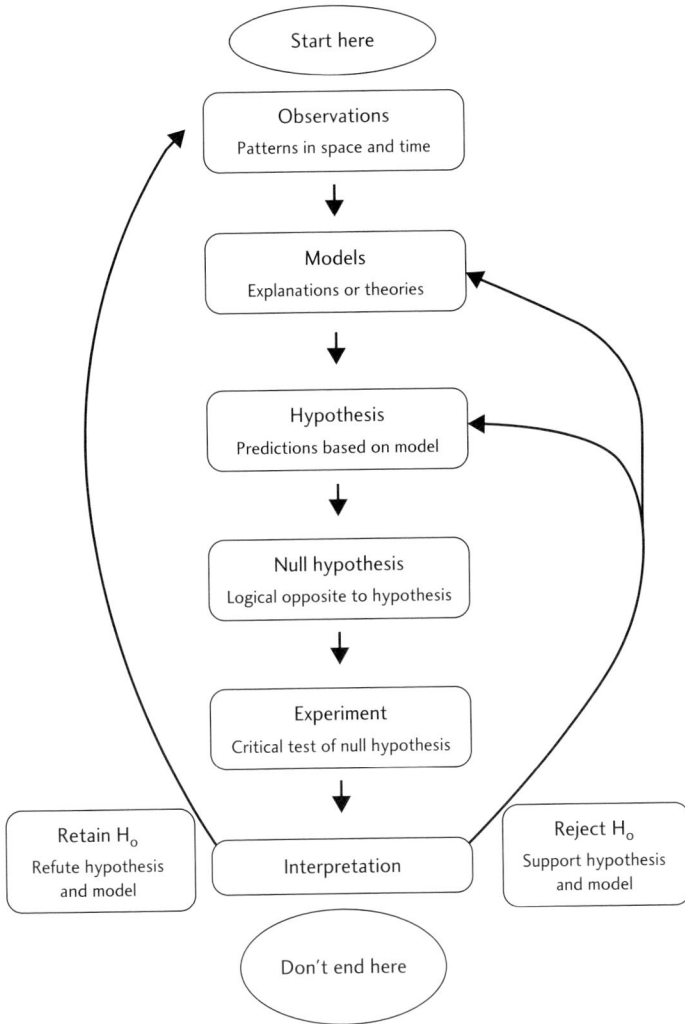

Fig. 3.1 Generalized scheme of logical components of a research programme. H_0 represents the null hypothesis (from Underwood 1997).

3.2.1 Broad objectives in insect conservation

There are generally only four broad objectives when sampling insects to answer ecological or conservation-related questions. These objectives are:

1) **To assess (inventory) the species richness and species composition of an area.** This will require a survey, and the outcome will be a species list for the area. For example, a nature reserve may want a butterfly list to supply to visitors of the reserve. In a survey, one will often use as many

different sampling techniques as is practical to obtain as complete a list of species as possible. A survey may also be more targeted and merely involve confirming the presence or, with more difficulty, the absence of a species across a series of sites, or to sample specimens for genetic analysis.

2) **To understand the habitat or environmental associations (or habitat requirements) of one or more species.** This objective requires more than a survey, and a range of habitat types (e.g. forest, grassland) or environmental states (e.g. early and late succession habitats, polluted and unpolluted) will have to be sampled (and as we will see later, adequately *replicated*). The outcome of this objective will be a species list (and usually associated abundance or density values) for each habitat type or environmental state. Here, the *same* sampling techniques and sampling effort must be applied to each habitat type or environmental state to ensure *comparability* of results between them.

3) **To understand the impact of environmental change on one or more species.** One or both of the following approaches (*sensu* Hurlbert 1984) can be adopted here. First, the study may be *mensurate*, i.e. quantifying the environment in its natural or current state (also called a *mensurate experiment*). Alternatively, the approach may be manipulative, where the environment is deliberately manipulated, or altered, in a desired way. For example, a study in which you sieve the tree bark for *saproxylic* beetles in existing stands of logged and unlogged boreal forest would be mensurate. However, if you were to negotiate in advance with the forest managers to log pre-selected forest stands at different intensities (say cleared stands, stands with half the trees removed, and other stands with no logging) and then sample these, the study would be manipulative or experimental. Insect ecophysiology studies are often experimental, where you would, for example, measure the developmental rate of a lepidopteran species at a range of temperatures in the laboratory. Insect conservation studies are commonly mensurate, although where possible manipulative studies are often more convincing (indeed arguments based on a combination of the two approaches are particularly powerful, see Dunne et al. 2004). For many rare insect species, however, when only one or a few populations are known, manipulation may be too risky.

4) **When dealing with rare, threatened, or ecologically important species,** the objective may be to monitor changes in the population size (or structure of a community when more than a single species is involved) over time (e.g. forming part of an adaptive management strategy for a rare species). One may also be interested in assessing how the population size

of the species responds to changing climate, wet vs dry seasons, different levels of harvesting or other disturbances, perhaps over long periods. These sorts of studies involve collecting *time series information* (measurements collected, usually at regular intervals, over time). They are also sometimes called *longitudinal* studies.

3.2.2 Refining the objective

Establishing which of the above broad objectives is the case (and a research programme may well encompass all four) is only the first step. Objectives can and should be described in as much detail as possible. In most studies, the broad objective becomes a lot more complex and detailed once you start thinking about it. Take, for example, the following objective: to quantify the effect of over-browsing by elephants on insect diversity. First, quantifying 'insect diversity' in an African savanna (or most other habitats) is no small task, and you would probably want to select one or more groups of insects to work with, say dung beetles or herbivorous insects. This decision may be based on previous studies showing that, for example, dung beetles are particularly sensitive to browsing or changes in habitat structure. Alternatively, it may be based on logistic grounds, e.g. dung beetles are comparatively easy to sample in a quantifiable fashion, and the dung beetle fauna for the area of interest is well known. When making this decision, it is important to make sure that you have a good understanding of the biology and behaviour of the taxon (or taxa) concerned, and that you have, or will readily be able to attain, the taxonomic expertise to identify the samples that are collected (see Chapter 2). The taxon (taxa) chosen will influence the decision on when to sample, how long to sample for and at what intervals, and which sampling technique to use (Chapter 4). The 'area of interest' also needs to be defined and may be, for example, a country, region, biome, habitat, or series of nature reserves. Next, are we interested in impacts that are short-term (a few days to one month after browsing) or longer-term (increasing loss of forest canopy or decline in tree densities over time), or impacts within or across seasons? The answer to all of these questions will determine the way in which the study is designed (Table 3.1). As a result, time and careful thinking must be dedicated to clearly defining the study objective and considering the best ways to approach it. Such planning and refinement is a sound investment for the future of your work.

In some cases, the precise objective of the study may be imposed by contractual demands, or the recommendations of action plans. You should carefully assess such objectives, and any constraints on the sampling design set by such objectives, before committing to a study. Sometimes contracts and action plans are not compiled by ecologists, and will need further development or modification to ensure that the desired outcome is achieved.

3.2.3 Hypothesis formulation

It is most useful if one has a formal hypothesis in mind (mindless data-gathering seldom results in any concrete achievement) (Figure 3.1). For example, previous research may have shown that over-grazing in savanna causes a decline in the species richness of understorey herbivores. The hypothesis would then be that over-browsing by elephants results in a decline in canopy and middle-storey insect herbivores. The null *hypothesis* (or hypothesis of 'no effect') in this case, would be that elephant over-browsing results in no change to the insect herbivores (all good hypotheses have a null and alternative form; see Shrader-Frechette and McCoy 1992; Figure 3.1). In another example, an increase in elephant activity will result in increased dung concentration in the area, which may result in an increase in dung beetle species richness in the short term (note that to test this hypothesis one would want to test whether dung concentration is indeed higher in over-browsed areas, as well as whether dung beetle species richness is higher). Browsing has also been shown to increase insect gall densities in some systems by increasing the vigour of new shoot growth (to test this hypothesis one would need to measure, for example, both new shoot density as well as gall abundance). This example relates to a well-known formal hypothesis in insect ecology, namely the *plant-vigour hypothesis* (that insect herbivory is higher on high-quality hosts, Price 1991a), and an excellent entomological example of the importance of hypotheses in the development of ecological theory (Price 1991b).

These examples also illustrate that in most instances data on both direct variables (also called *dependent variables*) as well as relational variables (also called *independent variables*) must be collected. Relational variables usually include weather, climate or soil parameters, measures of habitat quality (plant diversity of cover), or as in the example above, shoot density or elephant browsing intensity (see Chapter 5 for further discussion of relational variables). One form of relational data that will be discussed in this chapter is spatial referencing (see section 3.3.5). The process of defining the objective and formulating hypotheses helps to identify the variables that must be measured, as well as the timing, duration, spatial position, and spatial extent of the study (sampling or study design). Therefore, the more clearly the objectives are described (most formally as hypotheses), the greater the success of the study will be.

Table 3.2 *Possible outcomes of statistical hypothesis tests (from Scheiner and Gurevitch 2001)*

Reality ('truth')	Decision: based on study results	
	Do not reject null hypothesis	Reject null hypothesis
Null hypothesis is true	Correct $(1 - \alpha)$	Type 1 error (α = the probability of making a type I error)
Null hypothesis is false	Type II error (β = the probability of making a type II error)	Correct ($1 - \beta$; also considered to be the statistical power of the test)

Schrader-Frechette and McCoy (1992) discuss the importance of hypothesis formulation in ecology. They also point out that in applied sciences, such as conservation, it is often preferable to incorrectly reject a true null hypothesis (type I error, see Table 3.2) than not to reject a false null hypothesis (type II error, see Table 3.2). For example, take the following hypothesis: an upstream industrial development will change the habitat quality for a rare mayfly species (the null hypothesis would be that the development would not change the habitat quality). It would be better (an ethically based decision) to make a type I error than a type II error during the analysis stage of the study. In this case making a type I error (rejecting the null hypothesis) is equivalent to erring on the side of caution. This is preferable in this case to making a type II error (concluding that the development does not affect habitat quality) and being responsible for the extinction of a mayfly species! If the hypotheses in a study are not clearly formulated, it is difficult to evaluate these sorts of implications and consequences of the outcome of a study. Note that type I errors are not always preferable, and in fact scientists generally place more emphasis on avoiding type I errors, or prefer to err on the side of ignorance (Schrader-Frechette and McCoy 1992).

3.3 Designing a sampling protocol

Once the objective of the study is known and questions and hypotheses have been clearly formulated, a *sampling protocol* must be designed that will allow the questions to be answered and the hypotheses to be tested. The principles below are broadly applicable, and relevant, for example, to studies quantifying the spatial genetics and population structure of a species, to those quantifying species diversity, or manipulating the environment to examine disturbance effects on

Box 3.1 Elements of sample design

The basic elements of sample design (*sample grain, sample number,* and *sample extent*) are illustrated below. When these elements are combined they determine the sampling coverage and sampling intensity of a study. In A, sampling coverage is 100% using either the smaller or larger of the two grain examples shown. *Sampling intensity* is, however, greater with the smaller of the two grains. The extent divided by the grain (both measured as area) is called the *scope* of the study. Progressing from B to C to D is equivalent to a decline in coverage and intensity (as extent increases, while sample number and grain remain constant). Different combinations of grain, sample number, and extent therefore alter the coverage and intensity, or scope, of a study.

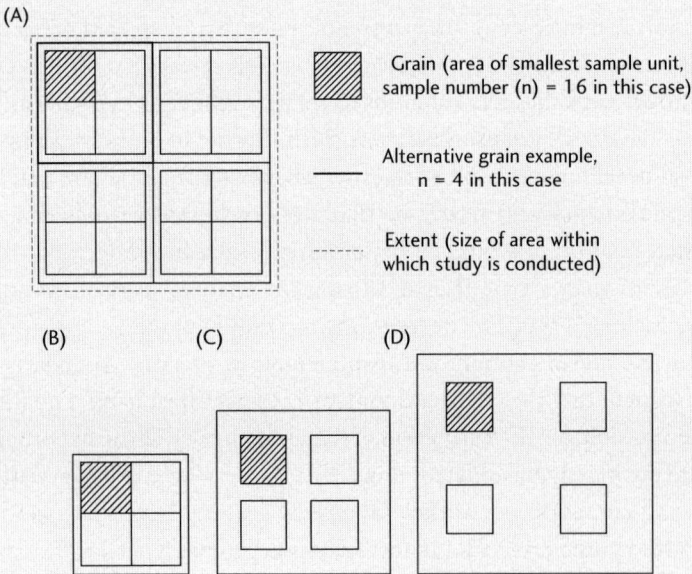

(A)

Grain (area of smallest sample unit, sample number (n) = 16 in this case)

Alternative grain example, n = 4 in this case

Extent (size of area within which study is conducted)

(B)   (C)   (D)

diversity. The three basic elements of sample design are sample grain, sample extent and sample number (Box 3.1).

3.3.1 Sample grain

Sample grain is the physical size of the minimum resolvable sample unit (Box 3.1). The grain of the study is usually determined by both the study objective and the appropriate sampling technique or method. There are two ways in which the grain of a study may be determined. Grain could be identical to the size of the physical sample unit used, such as a 1 m² botanical sampling grid, length of transect, or area covered by one or more pitfall traps. Alternatively, grain may be the area of the unit to which occupancy is extrapolated, e.g. the presence or absence of species in geographic grid squares (representation-based sample units) (Gaston 1994). The latter represents a coarser sampling resolution (lower sampling intensity) than the former. Often, however, more than one trap or sampling event is used to make up a single sample (i.e. the grain of the study is made up of more than a single physical sample unit). For example, a grid of 10 pitfall traps (say in a 2 × 5 array) may be used as a single sample (usually to obtain enough individual insects to represent the insect diversity at that point). The data for these 10 pitfalls are pooled during analysis, and the effective grain is then the area covered by the grid of 10 pitfall traps, and not the area covered by a single pitfall. In other cases, the area covered by a single pitfall could form the grain of the study. The sample grain is equivalent to the *sampling unit* when the data are entered into a *data matrix*. Ludwig and Reynolds (1988) give an excellent overview, which is well worth consulting.

Most importantly, although you may pool the data during analysis, **never** pool the actual individual samples themselves. If you do, it will not be possible to calculate, for example, between-pitfall variation in catch. You can always amalgamate data on a spreadsheet, but you can never separate the individuals from carelessly pooled samples.

3.3.2 Sample extent

Sample extent is the physical size of the area within which the study is conducted, or the area of inference. It is important to decide what the extent is, because this is the area to which the result of the study may be extrapolated. Your sampling extent may, for example, be a game reserve, or a woodland, or a series of mountain valleys. Often, the study extent is predetermined, either by contract or by biology. Many insects of conservation concern are known only from a single small area, sometimes little more than a hectare or a few such areas (e.g. the threatened Brenton blue and Eltham copper butterflies). For example, if the question has to do with a particular wetland or farming region, then the area of the wetland or of the farming region forms the extent of the study. Once this context is known and specified, then a decision can be made about where the samples will be placed within and across the area, and how many samples are needed to provide a representative estimate for the area. If this process is completed correctly, then the results of the study may be inferred to hold for the full extent of the study area (although extrapolation beyond the focal taxon or habitat would of course not be valid).

3.3.3 Sample number

The *sample number* is the number of individual units with the same grain used in the study (e.g. if the area of one pitfall is the grain of the study, then the number of pitfalls provides the sample number; if an array of 10 pitfalls is the study grain, then the number of such arrays is the sample number for the study). The sample number in a study is usually a compromise between what is logistically possible (this sets a maximum limit—more sampling means more sorting time and greater cost) and the minimum number of samples needed to provide a precise estimate of the variable of interest (where the variable may be, for example, the number of butterfly species in the area, or the number of individual host plants) (Table 3.3). As a rule of thumb, three is the minimum number of samples needed for statistical estimation. However, generally speaking, the more samples taken, the more precise the estimate and thus many more than three samples are usually a good idea (10 replicates is comparably safe). Also, the more heterogeneous the environment in which you are sampling, the more samples will be needed for a precise estimate, in other words more samples will be needed to capture the variability in that environment (Table 3.3).

In insect ecology and conservation, we are often interested in knowing how many insect species there are in an area or habitat. A small sample size is likely to fail to record the presence of some rare species within the sample area (McArdle 1990). A small sample size will also underestimate the real level of occupancy of some species and overestimate that of others. Overall, an increase in sample number tends to bring about an increase in the number of rare species found (McGeoch and Gaston 2002), because the generally large number of rare species in assemblages will only be captured with greater sample effort (see taxon sampling curves in Chapter 9).

Sample size thus has an effect on the precision of estimates (e.g. means, slopes of relationships) obtained (Table 3.3). However, it also affects the likelihood of detecting statistically significant differences between groups (or treatments), or significant relationships between variables, for a specified significance level (α = 0.05 or α = 0.01; note that α here is the significance level and is not the same as that in Table 3.2). For example, the greater the variability associated with a particular variable, the more samples are required to detect statistically significant differences between groups, or relationships between variables (if such differences or relationships indeed exist). Therefore, if the sample size in a study is not large enough, the wrong decision may be made about the null hypothesis (Table 3.2). Power analysis is a formal process by which the implications of using different sample sizes in a study may be explored (Box 3.2).

Table 3.3 *Definitions of bias, precision and accuracy (from Hellmann and Fowler 1999; Walther and Moore 2005)*

	Definition	Implication for research result
Bias	The difference between the expected value, sample mean, or test result, and the true (usually unknown) value. Non-random or directed error	Bias leads to underestimation or overestimation of the true value
Precision	The statistical variance associated with the sample data, as a consequence of measurement error and inherent variation in the variable of interest. Also referred to as 'repeatability'	Lack of precision leads to instability in the value of repeated measurements of the same object or variable
Accuracy	The less biased and more precise an estimate (such as the mean of a sample or a species richness estimate), the more accurate the estimate	Low accuracy leads to potentially misleading research results

Box 3.2 Statistical power and power analysis (a rudimentary knowledge of statistics is assumed here)

Deciding on the number of *replicates* (number of independent samples, or sample size) in a study is an important component of designing a sampling protocol. Statistical power analysis is a tool that is used to make decisions about sample size (amongst others) during this process. It is used to explore how components of the prospective study will affect the likelihood of the study achieving its objectives. The power of a test is the chance that the test will detect a departure from the null hypothesis $(1 - \beta$, Table 3.2).

The *statistical power* of a test is the probability of correctly rejecting the *null hypothesis* given that the alternative hypothesis is true (Scheiner and Gurevitch 2001) (see Table 3.2).

The statistical power of a test increases with an:

- Increase in *sample size* (n)
- Increase in *significance level* (α)
- Decrease in *variance* (σ^2)
- Increase in *effect size* (ES)

$$\text{Power} \sim (ES\alpha\sqrt{n})/\sigma$$

Power is proportionately related to the relevant parameters, because the way in which power is calculated varies depending on the statistical test being used (Quinn and Keogh 2002).

Effect size (ES)

In power analysis, efffect size is the difference between the null hypothesis and a specific alternative hypothesis (Scheiner and Gurevitch 2001). For example, if the null hypothesis is that there is no difference in insect species richness between organic (S_o) and traditional (S_t) tomato fields ($S_o - S_t = 0$) and the alternative hypothesis is that insect species richness is 25% higher in organic than traditional tomato fields (where S, the total species richness = 100), then effect size is $25 - 0$, i.e. $ES = 25$ species. Deciding on effect size can be difficult, but should be driven by what is considered to be a biologically meaningful difference in the situation of interest. Effect size is specified by the researcher and is a 'hypothetical value that is determined by the null and alternative hypotheses' (Scheiner and Gurevitch 2001). Nonetheless, statistical tests are more likely to detect large effects.

Power analysis

When used during the sampling design phase of the study, power analysis is used to determine:

- The number of replicates or samples needed to ensure a specific level of power for tests of the null hypothesis, given a set effect size, α, and variance.
- The power that a particular test will have when the maximum number of replicates logistically possible is used, with set effect size, α, and variance.
- The minimum effect size that can be detected, given a target level of power, α, variance, and sample size (Scheiner and Gurevitch 2001).

When conducting power analysis, by convention α is set at $P = 0.05$ (although the P-level may be selected as, for example 0.1 or 0.01; see discussion in Underwood 1997). A power of 80% is also often used by convention ($1 - \beta = 0.8$) (Quinn and Keough 2002). Estimates of the variance may be obtained by similar or related studies published in the scientific literature, or from a pilot study.

A variety of statistical software that conducts power analysis is available, e.g. GPOWER and PASS (see review by Thomas and Krebs 1997)

(summarized from Scheiner and Gurevitch 2001; see also Underwood 1997, Quinn and Keough 2002)

3.3.4 Sample independence

It now becomes important to make the distinction between *sample* and *replicate*. When two samples are *independent* of each other then they may be considered to be replicates. If two samples are not independent of each other, they are not true replicates. The concept of *sample independence* is an extremely important one in sampling design, and relates to the problem of *pseudoreplication*, i.e. inadequate or incorrect replication (Figure 3.2). When a study is pseudoreplicated,

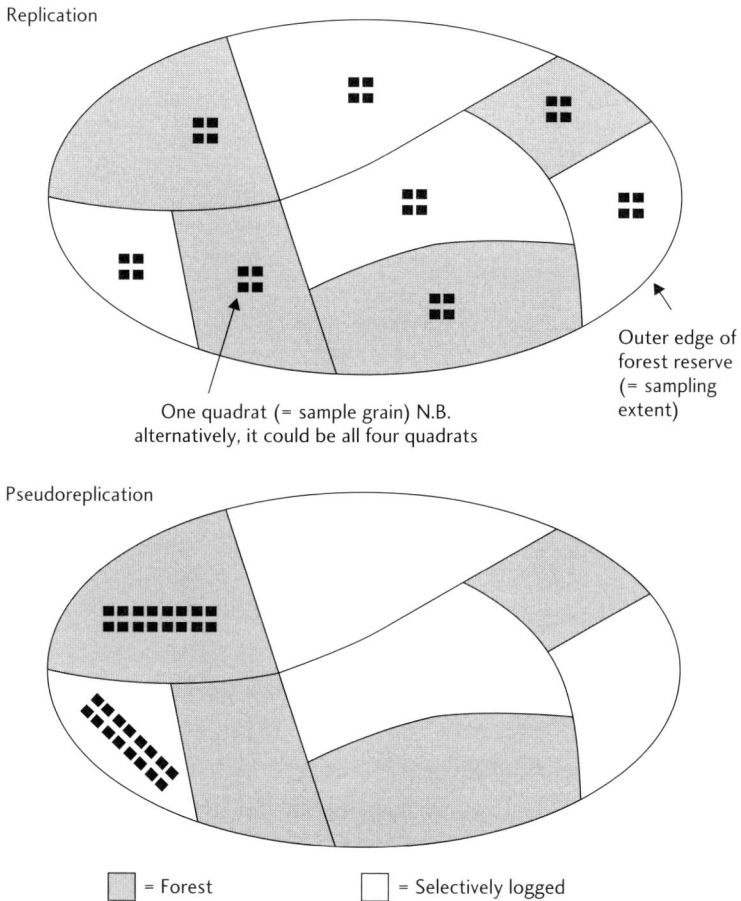

Fig. 3.2 Example of a well-replicated (above) and pseudoreplicated (below) study. The objective of the study is to examine the effect of logging on forest insects. There are two treatments, i.e. unlogged (the control, shaded) and selectively logged (unshaded). The area of interest or inference, i.e. the sample extent, is the whole forest reserve. In the well-replicated example, traps are placed across several logged and unlogged blocks of forest. In the pseudoreplicated example, trapping effort is concentrated in only a single block of logged and unlogged forest. Although many more traps are placed in these single blocks, the study is not replicated at the correct level, i.e. at the level of forest treatment (logged and unlogged). The above example provides much better inference for the effects of selective logging across the whole reserve (sampling extent). The smallest unit, say a single quadrat or trap, is the sample grain or sampling unit. Note that in ecological and conservation studies it is not always possible to have a strictly randomized block (as is the case here), as it is with planned agricultural trials.

the significance of statistical tests will be incorrectly interpreted, because the true value of α (see Box 3.3) is unknown (Hurlbert 1984). This means that the study objective will not be met and the investment in it wasted. It is clear that pseudoreplication should be avoided at all costs!

The total number of replicates needed in a study is also dependent on the number of *treatments* (and may also be assessed using *power analysis*, Box 3.2). Although the word *treatment* is more formally used in manipulative experiments, it is equally useful in many mensurate (non-manipulative) field studies. Treatment may be, for example, different forest successional stages, elevational bands, disturbed vs undisturbed habitat (all mensurate), or different temperature manipulations, genetically modified vs unmodified crop varieties, or different larval diets (all manipulative). A certain number of replicates will be needed for every treatment, and when treatments are not replicated this also constitutes the 'crime' of pseudoreplication (Box 3.3).

Box 3.3 Pseudoreplication

Pseudoreplication is the use of statistics to test for treatment effects in experiments (*manipulative* or *mensurate*; see text) where:

- replicates are not independent of each other (see also section 3.4.5), or
- treatments are not replicated (Hurlbert 1984).

Replicates are not independent of each other

when those replicates influence each another in some way, or are influenced in the same way by an external factor that is not of primary interest to the research question. To avoid non-independence, replicates in mensurate studies should be *isolated* from each other in space or time (see also section 3.3.5, Figure 3.3).

Example

Consider each of the sample units in Figure 3.3 to represent a pitfall trap. Each catch of ground-dwelling organisms is a sample. Samples from individual pitfalls that are isolated from each other by a physical or geographic barrier are *independent* of each other, and are thus true replicates. This is all very well when useful or meaningful physical barriers exist in the study area. Far more often, individual pitfall catches (or other sample units) are taken to be independent of each other when the pitfalls are separated by a certain minimum distance (see section 3.3.5). Deciding what this 'minimum distance' is, is a tricky business. The decision is usually based on some understanding of the dispersal distances or capabilities of the target taxon. The idea is that for each pitfall to be independent, it should draw on a pool of fauna that is separate from, unique or independent of the faunal pools being drawn on by other pitfall traps in the same study. A general rule of thumb is that the farther apart two samples are from each other the more likely they are to be independent. Independence can be quantified by measuring the spatial autocorrelation of variables across the study area (3.3.5).

Treatments are not replicated

when there is only a single replicate per treatment (per successional stage, elevational band, etc.). From the example below, it is clear that replication is possible at several levels. The appropriate treatment level at which replication must be used to avoid pseudoreplication is dependent on the question the research aims to address. Replication is usually also used at other levels to capture natural variability in the system and improve the precision of estimates (such as the mean) (see Table 3.3).

Example

Question	Pseudoreplication	Replication
Is there a difference in the insect diversity on trees growing …		
… at the edge vs the centre of Forest X?		
(treatment = tree position with two levels, i.e. edge and centre)	One tree is sampled at the edge of Forest X and one in the centre.	Several (independent) trees are sampled at both the edge and at the centre of Forest X
… at the edge vs the centre of forest patches in Nature Reserve Y?		
(treatments = forest patches and tree position)	One forest patch is sampled in Nature Reserve Y (with several trees at the edge and in the centre of the single patch)	Several (independent) forest patches are sampled in Nature Reserve Y (including several trees at the edge and in the centre of each patch)
… at the edge vs the centre of forest patches in nature reserves in Country Z?		
(treatments = nature reserves, forest patches and tree position)	One nature reserve is sampled in Country Z (including several forest patches in the reserve and several trees at the edge and in the centre of each patch)	Several (independent) nature reserves are sampled in Country Z (including several forest patches in each reserve and several trees at the edge and in the centre of each patch)

3.3.5 Spatial autocorrelation

Two samples are *spatially autocorrelated* when they are more similar to each other than expected by chance (Legendre and Legendre 1998). This means that the samples are spatially non-independent (samples may also be temporally autocorrelated when collecting is repeated over time). Measuring the spatial autocorrelation amongst a spatial series of samples in a study is thus a way of quantifying the non-independence discussed in Box 3.3 and section 3.3.4, i.e.

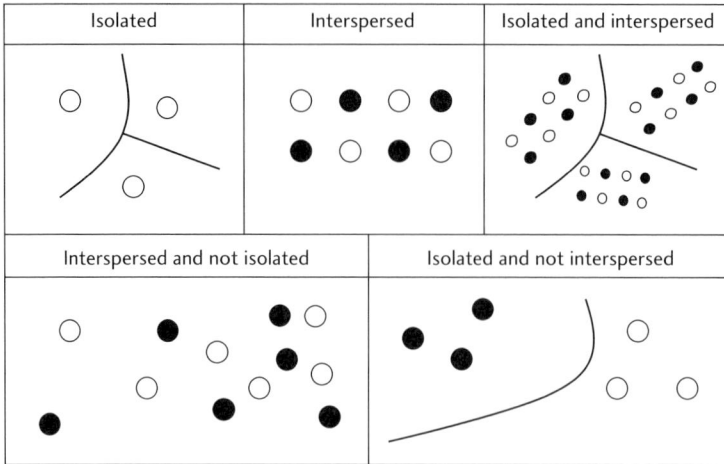

Fig. 3.3 Illustration of the concepts of isolation and interspersion of replicates for one or two treatments (black and white). Here the 'isolation' is used to achieve independence of samples. This isolation may be in the form of a geographic barrier to dispersal of individuals, or merely distance (the farther away two samples are from each other, the more 'independent' they are likely to be). Top row from left to right: Isolated (1 treatment, 3 replicates); interspersed (2 treatments, 4 replicates); isolated and interspersed (2 treatments, 3 replicates); interspersed and not isolated (2 treatments, no replicates); isolated and not interspersed (2 treatments, 1 replicate).

if two samples are spatially autocorrelated, then they are not independent and cannot be considered replicates. If spatially autocorrelated samples are treated as replicates then their inclusion as such in statistical analysis unfortunately qualifies as pseudoreplication.

One of the most common measures of spatial autocorrelation is Moran's *I*, the values of which are usually presented in the form of a *correlogram* (Figure 3.4). Moran's *I* lies approximately between −1.0 and +1.0, and values significantly greater than zero (or not significantly different from zero) demonstrate the presence of significant spatial autocorrelation. Values that are significantly less than (or no different from) zero identify distances over which values of the variable are significantly less similar (or no more similar) than expected by chance (and thus independent). Spatial or geographic coordinates (x and y or latitude and longitude) must be taken for each sample, forming a relational variable, in order to quantify the spatial autocorrelation across samples. As a rule of thumb, a minimum of 32 samples with unique spatial positions is needed to quantify spatial autocorrelation.

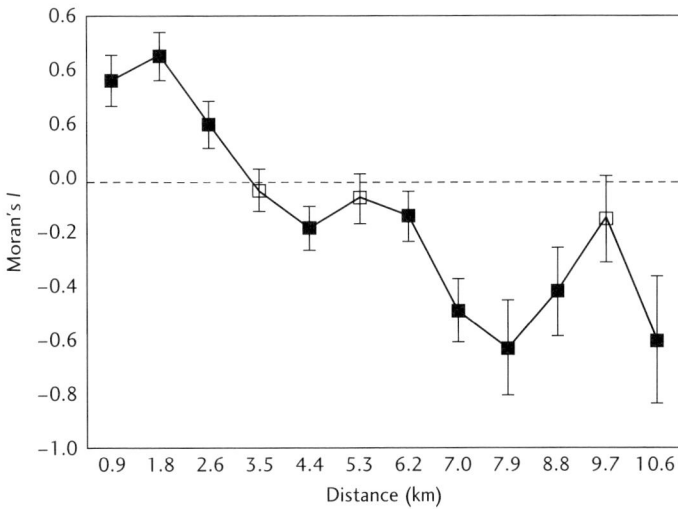

Fig. 3.4 Correlogram of the density (number of galls per plant) of a leaf-galling sawfly species (overall correlogram significance, P < 0.0001). Closed symbols (Moran's *I* values ± s.d.) in the correlogram are significant at P < 0.05 (after progressive Bonferroni correction (α = 0.05)). This correlogram shows that gall density on plants within approximately 3 km of each other is more similar than you would expect by chance. In words, gall density measured from plants within 3 km of each other cannot be considered independent. (From McGeoch and Price 2004).

The measurement of spatial autocorrelation is used in another very interesting way in studies of insect populations or assemblages and relational variables that encompass sampling at several points in space. Here, spatial autocorrelation is used to estimate the distance over which values of a particular variable (e.g. species abundance, species richness, temperature, soil nutrient content) are significantly more similar to each other than expected by chance (or the distance at which they become significantly more different from each other than expected by chance). This provides a measure of *patch size* (or gap size), and can be used to generate hypotheses about the processes structuring the population or assemblage. For example, leaf gall abundance (expressed as gall density) in Figure 3.4 is positively spatially autocorrelated over distances of 2.6–3.5 km (an estimate of leaf gall patch size). Because sampling took place in a linear fashion along a river drainage system, this result shows that, on average, stretches of river drainage approximately 3 km long encompass similar densities of leaf galling sawflies (McGeoch and Price 2004). Measuring the spatial autocorrelation of variables is thus useful both for quantifying areas across which samples are non-independent, as well as for quantifying the spatial structure of populations

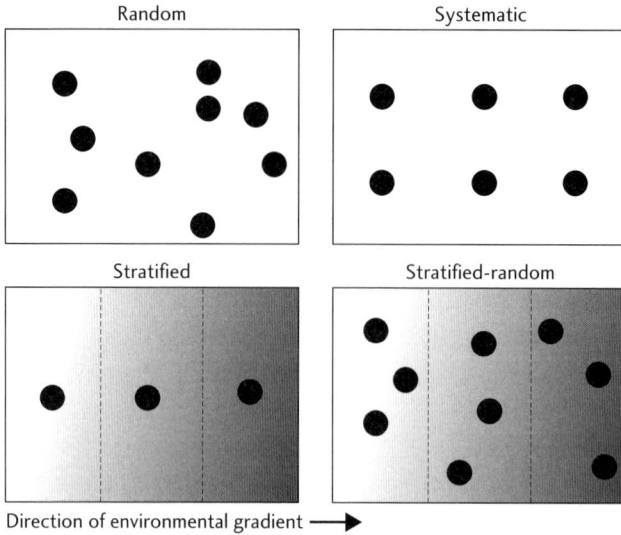

Fig. 3.5 Alternative arrangements for sample placement across the extent of a study area. When there is a dominant environmental gradient in the study area, the extent is stratified perpendicular to the direction of the environmental gradient.

and assemblages. Spatial analysis in ecology has become a field in its own right (see Fortin and Dale 2005), and spatial position of sites and samples must now always be considered in studies conducted over space (Legendre 1993).

Therefore, always record the spatial position of each sample in the study and remember that you may have to quantify the spatial autocorrelation of variables across the study area to:

- identify the distance at which two samples (or values of a variable) become independent of each other
- justify or demonstrate that samples are spatially independent of each other, or
- describe the spatial structure of a variable across the extent of the study.

3.3.6 Sample placement

Where you put your traps, or where you collect insect individuals (i.e. the position of each sample in the study area) is *sample placement*, and several terms are used to describe the way in which this is done. These terms (*random, stratified, systematic*, and also *interspersed*) all have to do with the way in which samples are arranged within the *extent* of the study (Figure 3.5). Randomization is the random assignment of samples to particular treatments, or the random arrangement of replicates within the extent of the study. Randomization prevents

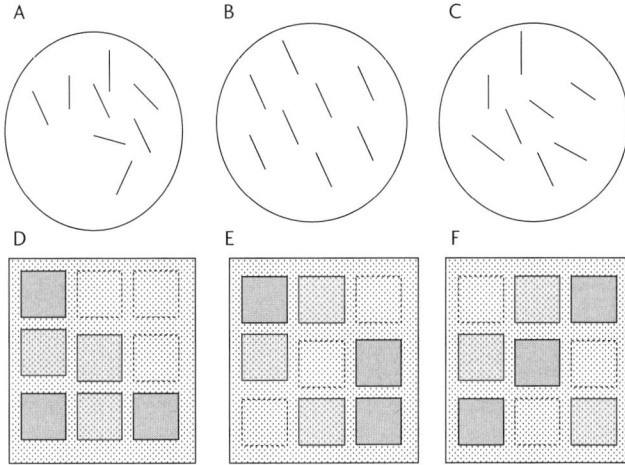

Fig. 3.6 Randomized and systematic arrangement of replicates in a mensurate (A–C) and manipulative/experimental (D–F) study. A–C. Placement and orientation of 100 m sweep net transects in a patch of grassland in a mensurate study. D–F. Experiment to assess the effect of shading (100%, 50%, and no shade) on grassland insect diversity (manipulative study). A, transects positioned and orientated randomly; B, transects positioned and orientated systematically; C, transects positioned systematically and orientated randomly; D, treatments (full shade, half shade, and unshaded) assigned randomly to grassland plots (inadequate placement); E, treatments also assigned randomly (improvement on D); and F, treatments assigned systematically to grassland plots (best option).

the investigator from making biased decisions (deliberately or unwittingly) about where the replicates should be placed, or which replicates to assign to a particular treatment (Hurlbert 1984) (Table 3.2).

Strangely, the potential downfall of randomization is that sample positions and assignments may by chance appear completely non-random. For example, in Figure 3.6A, it appears as though the south-west side of the patch was avoided when transects were placed, but this was in fact a random assignment of transects using randomly drawn x–y coordinates to place the centre of each transect (note that samples within a transect are by default not independent of each other). In this case, it may be better to place transects systematically (equal distances apart, where that distance is determined by the extent of the patch and the sample size) over the entire patch (Figures 3.5 and 3.6 B–C). If there is a known environmental gradient across the study area (e.g. shade, slope, temperature, soil quality) one may want to stratify the area, i.e. divide it into blocks perpendicular to the direction of the environmental gradient. Samples are then placed in each

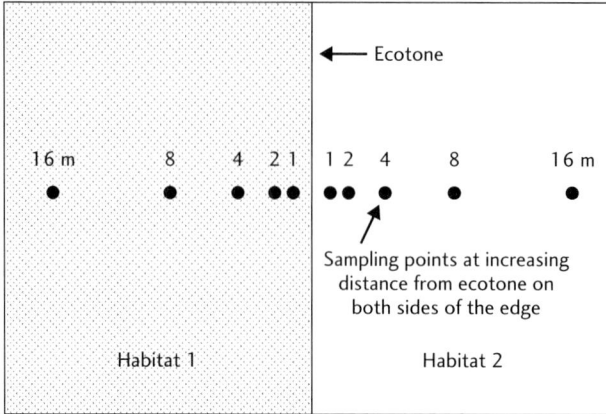

Fig. 3.7 Illustration of log$_2$ sample (or trap) placement approach to sampling across habitat edges or ecotones. The minimum distance used depends on the habitat structure, and realistically usually starts at 2 m (excluding the 1.0 m sampling point).

block (Figure 3.5). In Figure 3.6C, transect centres are systematically placed and compass directions randomly assigned to each transect. Placing transects systematically across the patch will achieve good coverage (and thus a result that is representative of the patch), whereas the random orientation of transects will avoid any bias as a result of unknown environmental gradients across the patch (Figure 3.6C). Therefore, these sample placement approaches may be used together to achieve unbiased, representative results for the extent sampled. Generally, where there is more than a single treatment (e.g. habitat type or disturbance level), the replicates of each treatment should also be *interspersed* (Hurlbert 1984) (Figure 3.3). Interspersion, along with isolation by geographic barrier or distance, is another way of avoiding non-independence (Figure 3.6).

As fragmentation is such a major issue in contemporary conservation, studies often involve close focus on the edge, interface, or *ecotone* between two vegetation types or land uses. The study may, for example, be about *contrast* (Wiens et al. 1993) between two habitat types, or penetration of the two habitat types by various species, i.e. *landscape variegation* (Ingham and Samways 1996). The point is that the focus is primarily on the edge, and secondarily on increasing distances away from the edge. The sampling is thus more intensive at the edge than away from it. Also, fragmentation studies to date have shown that edge effects often occur over short distances away from the edge (Ries et al. 2004). For insects, a suitable scaling for sample placement is m log$_2$, starting at the edge (Figure 3.7). Sampling is then done at 0, 1, 2, 4, 8, 16, 32, 64, …, n metres from the edge, as far into the habitat adjacent to the edge as necessary. In practice, 1.0 m may be simply too small a distance, so sampling is usually done at 0 m, 2 , 4 m, etc. A similar approach can be taken when studying road-zone effects, with sampling done in a $\log_2$ fashion from the edge of the road, path, or indeed any other linear corridor into adjacent habitat.

3.3.7 Sampling coverage and intensity

Sample unit size (grain) and sample area (extent) define the lower and upper spatial limits of any study of organisms sampled across space (Wiens 1989). These basic elements are combined to produce different *sampling coverages* and sampling intensities (Box 3.1). Grain and extent, in addition to the number of samples taken, are often interdependent and together may be expressed as either coverage or sampling intensity. The latter are seldom considered components of sampling design, but do have an impact on the outcome and interpretation of the results. For example, the larger the grain relative to the sample extent, the more widespread species and fewer localized species there are likely to be sampled when there is complete coverage of the sample extent (McGeoch and Gaston 2002). As sample extent increases, the sampling intensity of studies tends to decrease as a result of logistical constraints (Wiens 1989; Gaston 1996).

Although grain and sample number may be standardized as sample extent changes, the resultant change in sampling intensity may alter the way in which numbers of rare and common species are interpreted to change. A reduction in sampling intensity is likely to result in the loss of fine-scale details and an increase in differences between sample units (Wiens 1989). Low-intensity studies may therefore have apparently high predictability and patterns emerging from these will be far more readily generalized than from studies conducted at higher sampling intensities (Wiens 1989). At low sampling intensity, distances between samples may also be too great (or the grain too coarse) to detect smaller-scale environmental gradients.

3.3.8 Controls

The concept of *control* comes from manipulative experiments where one set of replicates is left 'untreated' or not manipulated. This control set of replicates serves as the baseline against which other treatments are compared (e.g. the unshaded plots in Figure 3.4). However, controls are also relevant to mensurate studies. For example, sampling the aquatic invertebrates in a river before a pollution event serves as the untreated control, or baseline condition, for aquatic invertebrate diversity in the absence of pollution. Similarly, simultaneously sampling several rivers, some of which are polluted and others not, provides treatment and control cases. Another form of control may be comparatively undisturbed or 'natural' habitat, in comparison with various states of anthropogenic disturbance.

Procedural controls are used in manipulative experiments, where the effect of the application of the treatment must be separated from the treatment effect itself. For example, the procedural effect of the application of a liquid insecticide may be the moisture effect on insect diversity, rather than the insecticidal content of the liquid. In this case, the procedural control would involve the application of the same volume of liquid as in the insecticide treatment, but without the active ingredient. Controls are also often necessary to control for changes over time in a system (for example, seasonal changes, rainfall events impacting the system). For example, in an experiment to evaluate the long-term impact of burning on grassland insect diversity, burned and unburned plots are monitored over time, such that the unburned plots serve not only as a burn control, but also as a control for separating the effects of burning from seasonal changes in the insect community.

3.4 Pilot studies

There is no substitute for really good planning, including clear formulation of the research/conservation question, and choosing appropriate statistics. Nevertheless, this is still only a first phase—the armchair phase—which should be developed into a concept note of at least a couple of pages, detailing all you intend to do. This concept note can then be used to bounce ideas off fellow researchers, site managers, and supervisors. Then, before embarking on a full-scale sampling programme based on this concept note, you need now to go out into the field and do some preliminary sampling, such as walking a transect, setting a few pitfall traps, collecting some focal insects, etc. Such a feasibility study or pilot study is invaluable, and enables you to go back to the concept note, refine and develop it into a research proposal, and then implement it, knowing, with confidence, that what you have planned to do is achievable in the time allotted to the project.

Pilot studies serve several useful functions. Once you have made preliminary decisions on the first ten questions in Table 3.1, it is usually wise to then conduct a pilot study as a preliminary attempt at implementing the study you have planned, in which the first set of data are collected relevant to the objective. Pilot studies are usually of short duration, involve few samples, and may be conducted over a smaller extent than planned for the final study. Pilot studies are used to:

- Test equipment and sample methods
- Assess if the sampling can logistically be completed in the time available

- Assess if sampling can logistically be conducted at the sites selected
- Practise the implementation of sampling techniques and methods
- Generate estimates of the variance for conducting power analysis
- Generate data for planning which statistics it will be necessary/most appropriate to use
- Identify unforeseen problems.

After the pilot study is complete, the original design may be modified to incorporate the lessons learnt. Pilot studies are almost always extremely useful, and most often indispensable to the success of a study. If the pilot study runs very smoothly, and no alterations to the initial design are made based on the outcome of the pilot study, then the data collected during the pilot phase may also in some instances be incorporated into the final analysis.

3.5 Coping with the unexpected

What happens when things go wrong? As ecologists, we should be used to fire, flood, drought, or other natural catastrophes interfering with our carefully planned sampling protocol, or at least expect it at some point in our career. The contents of pitfalls may be raided by hungry hornbills or vervet monkeys, intercept traps trampled by elephants, site markers eaten by termites, and Malaise traps disappear and turn up as kitchen curtains. What do you do if your study is delayed or biased, or you lose samples or replicates as a result of such events? Good pilot studies should help to eliminate many potentially nasty surprises from factors that it is possible to control or manage. Also, try to ensure in advance that the chosen sites will be secure from interference or planned disturbances for the duration of the study. In some conservation scenarios, site developments may be placed 'on hold', perhaps via a formal moratorium, pending environmental assessment. However, sadly, not all such events can be anticipated and the worse case scenario may involve a year's delay for the project, a start from scratch, or a move to a different study area. Such is the life of an ecological entomologist. However, always be on the look out for the silver lining, where disasters may be turned into valuable opportunities to improve understanding or management potential. For example, an unexpected fire across your study area halfway through sampling could turn a well-designed study into not only a comparison of habitats, but also a comparison of insect diversity before and after fire.

This raises the problem of missing samples. How many missing samples are acceptable? What does one do with missing data during analysis? Unfortunately, there is no single answer to these questions, and it depends on the sample design, and initial sample size, or number of replicates per treatment. Generally one or two missing data points per treatment will not severely affect the analysis or outcome, and the larger the number of replicates to begin with the better off you will be. In cases where, for example, one pitfall is lost from a set of 10 pitfalls that together form the grain of the study (i.e. all 10 pitfalls are used as a single sample), the entire set may have to be excluded from analysis. A set of 9 pitfalls is not comparable with one of 10, because the sample effort differs between them. However, instead of discarding the entire set you could use the average catch across available pitfalls (comparatively insensitive to small differences in pitfall number per set) in a set rather than the total catch (very sensitive to any differences in pitfall number between sets).

3.6 Data management

As soon as the objectives of a study start to be defined, the job of *data management* begins. Data management includes:

- Keeping track of which decisions were made and why during the design and planning phase
- Keeping record of data collection and data entry dates and error checking
- Collating records of observational comments and field notes made while the study was being conducted
- Reminders and updates of electronic and hard copy backups of data books and electronic data files; your institution may have formal requirements in place for this important data storage
- Keeping track of sample sorting and specimen identification, sample container and specimen labels (e.g. meaning of acronyms or nicknames used and housing of voucher collections and unsorted samples)
- Keeping track of exploratory data analysis and statistical analysis data files and results
- Keeping track of versions of the manuscripts produced
- Maintaining a *metadata file* that describes the date of creation, location, contents, updates and backup information of all of the above, i.e. of electronic and hard copy data files, samples, and voucher collections.

3.7 Summary

To ensure that the conclusions drawn from your research study are as reliable as possible, it is extremely important to plan the study well (Table 3.1). This includes a clear, detailed outline of the study objectives (including hypotheses to be tested) and formally designing a sampling protocol. During the design of the sampling protocol, decisions are made about, for example, *sample extent, grain and number* (Table 3.1). Samples must be well replicated and positioned, so that they are independent, to avoid *pseudoreplication*. This may involve measuring the spatial *autocorrelation* of variables of interest across the study area. In *manipulative*, and also sometimes *mensurate, studies*, the concept of having *controls* is important to ensure correct interpretation of treatment or disturbance effects. *Pilot studies* are used, amongst others, to iron out problems of sampling protocol before the main study is initiated. A well planned study not only ensures that the outcome provides valuable, accurate and novel information, but is reasonably robust to unexpected events while it is being conducted. Planning includes the development of a *data management* system that will be used throughout the duration of the research, and in many cases for posterity.

4

Collecting and recording insects

4.1 Introduction: What do you really need to collect?

Specimen collection (or recording) starts once the ecological or conservation question has been identified and once the study objectives have been set. The records or specimens collected will constitute your samples or voucher specimens from which to answer the questions posed at the start. The aim is always to collect insect specimens ethically, parsimoniously, and efficiently. In other words, time and effort is directed specifically to answering the question posed. This does not necessarily mean that you are so locked in that subsequent new ideas are precluded. However, any new set of records or samples must be planned from scratch, or should be directly comparable with results already gathered. This point re-emphasizes the importance of advance planning, as well as preliminary sampling, as part of a pilot study (see Chapter 3.4).

   Any single technique of collecting insects provides only one view on the world. An analogy is a set of windows around a building. Looking through one window (i.e. use one method in one way) provides one view, and when you look through a different window (use another method), you are likely to see a different picture of the outside world. Two adjacent views (two similar techniques) may give you similar but overlapping results, whereas a window on the opposite side of the building (two completely different techniques) will give you a totally different view (very different types of results). The outcome of this is that it is often worthwhile to use complementary sampling techniques. For example, when sampling moths in a wood, the noctuid and arctiid moths may be best sampled with a *light trap*, while the geometrid moths may be best sampled with a *light tower* (see below). Where the aim is to sample all the moth species in the area, it will be necessary to use additional techniques, such as *treacling* and *sleeving* (see below), as well as searches for larvae and pupae. This returns us to the point that you must be very clear on the aim of the project and on the amount of time and financial resources available. Should you focus on a restricted spectrum of insects, ranging from a single order to a single species, your methods can be tailored specifically for this. It is always important to routinely obtain as much background biological information as you can to streamline your approach and efforts. For example, if your insects are in the accessible adult stage during a particular part of the year only, there is no point in seeking them in other seasons. For example, much of the extensive published literature on the seasonality and biology of pest insects and sampling them in crops contains information invaluable for other contexts of sampling insects.

Another consideration is that different insect collecting techniques also operate or attract over different distances, which may also be different for different species. *Pitfall trapping* (see below), *sweep netting*, and hand-held *suction sampling* are basically point sampling, whereas *Malaise traps* will capture individuals over at least several metres. In turn, light traps and *pheromone traps* have an area of operation that can be enormous (attracting insects from up to several kilometres away). So, if you are comparing, say, different clumps of trees that are within sight of each other, light traps would be inappropriate.

Different collection techniques will be variably effective against different species, sometimes even for those in the same genus. Also, any one technique may be effective at certain times of the day or night (different moth species, for example, visit light traps at different times during the night, reflecting their different patterns of activity). Any particular species may be differentially sampled by any one technique according to the age of the component individuals. Most mature adult male dragonflies, for example, can be sampled along the river's edge, but the young adults of some species are usually not sampled because they move away from the water (even up to 20 km) to mature. There are also differences between the sexes in their response to a particular collection technique. An extreme case is among some moths, where the males are strong fliers while the females are flightless. Finally, bear in mind that for holometabolous insects (i.e. those where the young stages are morphologically and functionally very different from the adult) one is dealing with two functionally different organisms: a caterpillar is very different from a butterfly in both its habitat and its resource use. What this emphasizes is that the different sexes or life stages may require very different collection techniques. So it is important to bear in mind that when one collection technique is used, besides not necessarily capturing different species equally, it may also not capture different sexes, life stages and even, sometimes, different individuals equally. An assessment using a wide variety of techniques, and illustrating the complementarity of the different methods, is given by Clark and Samways (1997).

At the outset the study must be planned with a clear understanding of the spatial and temporal scale, and then enough sampling conducted to provide statistically sound results (see Chapter 3). Similarly, the collection technique to be used will hone that view. In other words, it is a question of *how* the sampling is done as well as *how much*, and deciding upon the equipment needed. While you do not want to take too few samples, you also do not want to take too many. Doing so may involve unnecessary and thus unethical collecting, to the extent that the samples cannot be sorted and analysed in the time available. Obvious as this may seem, it is still surprising to see how often research projects cause stress and even fail to materialize because of an undisciplined *sampling protocol* (see Chapter 3.3).

We now provide a broad overview of the collection techniques available. Inevitably in an introductory text such as this, not all techniques for every taxon and every concept can be accommodated. It is therefore essential to read around your specific topic, carefully scrutinizing recent research papers, to ensure that you are aware of the appropriate sampling options open to you at the outset of the study. This chapter should be read in conjunction with Chapters 2 and 3.

4.2 Named and nameless approaches to data portrayal

It is very useful to bear in mind when you start sampling that you are collecting data that can be analysed using two different broad approaches, which from a casual perspective may be described as *named* vs *nameless* approaches. To illustrate this point let us say that you sample a particular number of individuals of a moth species, which we shall call (fictitiously) *Mothus oneus*, in a particular place (where your trap is). At the same time, you also sample another species, *M. twous,* in the same spot, and a further species *M. threeus*, as well as other species to which you give names. You have thus collected three and more named entities in one place at one time (Table 4.1). In conceptual contrast, when you 'number crunch', plot and analyse these results simply as numbered entities (that may nevertheless have names), they effectively become nameless (as in Figure 4.1). In other words, the coordinates on a graph and the associated statistics are effectively faceless species, and without superscripts, there is no telling which point relates to which species (Figure 4.1). This production of graphs with faceless species has become a problem in recent years with journal space being so limited, and no space available (unless electronically) to list which species are included on any one graph. There are some wonderful numerical and graphical distillations of the data that provide crisp, deductive illustrations for the description of broad ecological or distributional patterns. Yet nameless approaches are

Table 4.1 *Named species and abundances of moths sampled using a particular trapping technique, e.g. a light trap. The 'scientific names' used here are of course fictitious, and are given simply for illustrative purposes*

Species	Abundance
Mothus oneus	21
M. twous	7
M. threeus	36
M. fourus	231
M. fiveus	333
Anothermothus primus	924 428
A. secundus	989
A. tertius	72
A. quartus	4
A. quintus	104
A. sextus	35 098
Trimothus alpha	667
T. beta	3 268
T. gamma	27 528
T. delta	7 402
T. epsilon	7 002
Fourmothus unus	73
F. duus	808

not always helpful, or sufficient, for on-the-ground conservation where particular species are the focus. The named species and associated quantitative data can be immensely useful, particularly for other researchers. These associated data, as with the voucher specimens, require careful and accessible storage (see Chapter 2). This can be done in the form of appendices in reports or theses, or as supplementary electronic data, and can be referenced in the scientific paper. This named vs nameless conceptualization is a special but important issue with insect and other biodiversity studies involving large numbers of species.

The above points need to be borne in mind when you choose collection techniques, because as with bycatch issues (see Chapter 2.1), you need to be mindful of other opportunities that may present themselves in the future. This can be very important, for example, when comparing biodiversity in an area from one time (one study) to the next (study). In summary, when planning the study, you need to think on the one hand how you will 'melt down' your data for the inductive, faceless statistics and graphics, and on the other hand, be aware of how and where you will include the named, semi-raw information. More

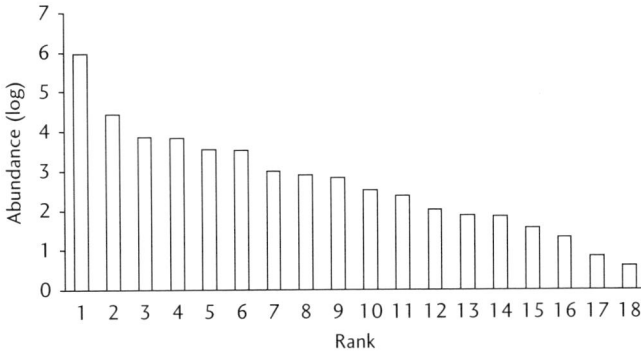

Fig. 4.1 An example of 'nameless' analysis (in this case in the form of a rank abundance distribution used for identifying and describing broad patterns in biodiversity) of the 'named' species data given in Table 4.

details on the inductive approach are given in Chapters 3 and 9, but the subject is raised again here to emphasize that you must plan and visualize what you are going to do with the specimens and data right at the start of the project (the software *Biota* is a very useful tool for biodiversity data management; see p. 373 for details). Because substantial effort (time, cost, and person-hours) goes into collecting insects, wherever possible the maximum information should be extracted from the specimens obtained.

We now describe some of the techniques and apparatus used to collect and sample insects. There are many variations on the details of these methods, and many types of apparatus are available commercially and can be found on the Internet. Should you wish to construct some of them yourself, you will find more details of measurements, at least of the more traditional items, in Martin (1977) and Uys and Urban (2006). But always bear in mind that any piece of apparatus may need to be adapted in one way or another for answering your particular research question. Another point to bear in mind is that insect ecology and conservation need not necessarily focus on specific taxa, but on, for example, the particular size of the component species (as large species tend to go locally extinct before smaller species in the same taxon), or on particular *functional groups* or *guilds*. Basset et al. (2004), using three complementary trapping techniques (Malaise, pitfall, and yellow *pan traps*) illustrated this by showing that indicator taxa (see McGeoch 2007) of forest disturbance belonging to selected chewer, parasitoid, and predator guilds had more discriminatory value than other guilds or the use of higher taxonomic levels (e.g. ordinal level). A brief overview of some of the techniques available, and the context of their application are given in Table 4.2.

Table 4.2 *Summary of some approaches and methods used to collect or sample insects in ecological studies. See text for additional details*

Target habitat	Some methods available	Comments
Forest canopy	Direct access, via walkways, cranes, platforms, ropes	Allows direct, close-range sweeping and foliage collection, but often expensive; safety issues are critical
	Fogging/misting using insecticides	Accessed from ground or by hoisting machinery to canopy; falling insects collected on trays/funnels on or near ground. Use only in calm, dry conditions; wide range of insects sampled
	Direct foliage samples	Obtains insects from twigs or branches by beating or using an aspirator; direct sampling of galls and leaf mines; collection of insects emerging from plant parts using sleeves; mostly targets sedentary taxa
Low canopy, shrubs	Sweep netting	Uses standard, coarse nets among foliage to directly capture free-living insects; fairly specific to certain taxa and life forms; done in calm, dry weather
	Beating	Uses standard horizontal canvas tray, with foliage jarred to dislodge insects; insects individually collected from tray using soft forceps or aspirator; done in calm, dry weather; easy replication
	Selective individual collection	For large, individually sought insects; collect by netting, forceps, insect tongs, with aspirator or with sticky rod; causes minimal disturbance
Low vegetation	Sweep netting	As above; collects wide range of insects that are dislodged relatively easily; can be standardized by certain lengths of sweeps in grasslands/crops
	Suction sampling	Uses suction devices, but must avoid clogging of machine; use only in dry conditions; both wide-necked, and narrow-necked (<15 cm diameter) models available, with latter (with greater airflow/nozzle area) preferred for insects near base of plants; easily available as conversions from garden leaf vacuums; standardization generally possible
	Direct examination	Vegetation may be placed in bags and taken to laboratory, to examine directly or place in emergence boxes, or twigs sleeved in the field or laboratory to obtain emerging insects
Ground/surface	Pitfall traps	Uses activity of surface-moving insects that are captured by falling in containers sunk to be level with ground surface, where preserved or held; many patterns and designs for this popular technique, with much literature on trapping variables, size, number, spatial arrangement, duration, preservatives, roofing, baits, etc.; can confuse insect activity with incidence, with some insects not caught representatively, although good for comparing different sites and sampling night-active insects

Table 4.2 *(Continued)*

Target habitat	Some methods available	Comments
	Soil and litter sifting	Area-based samples of dry leaf litter and debris sifted on white tray, with individuals retrieved individually by forceps or an aspirator; time-based examination can be used for standardization
	Tullgren–Berlese funnel extraction	Samples of litter placed in metal cylinders and heated from above by electric light, progressively inducing insects to move downwards and fall through a grid into a collecting jar with preservative; soil samples can be placed directly on finer mesh grid to prevent collapse; extract for several days; requires on-site electricity; readily standardized
	Winkler bag extraction	Similar principle to Berlese funnels, but dry litter/soil bagged and suspended without artificial heating to allow extraction of individuals; as not dependent on power supply, very useful for remote locations
Flying insects	Malaise traps	Large, mesh-based interception traps that can capture vast numbers of flying insects in preservative and over extended periods; indiscriminate and so needs careful use at sites with rare/threatened species; very effective for many flies and hymenopterans
	Window/interception traps	A variety of vertical interception panels have been designed to sample flying insects, that drop into collecting troughs; often effective for species that are rarely seen; may need some roofing to guard against flooding
	Suction traps	Relies on a powered fan to suck in insects from the surrounding air into a jar of preservative; good for very small insects; samples laborious to sort
	Light traps	Variety of designs available, relying on attractiveness of many nocturnal insects to light, particularly UV wavelengths; can retain catches alive for examination and then release; can also function by insects landing on a collection sheet; weather conditions influence catch; tend to have a wide area of sampling (many tens of metres and more) and therefore not suitable for point sampling

Table 4.2 (*Continued*)

Target habitat	Some methods available	Comments
	Pheromone/bait traps	For individual species in population studies, using specific pheromones or attractant baits; can be very effective for certain insect groups that are not easily collected by other methods, e.g. some powerful-flying canopy butterflies
	Pan traps/water traps	Coloured traps act as attractants to particular phytophagous and pollinating insects, and are useful in sampling and monitoring these species
Aquatic insects	Waterside vegetation	Sweeping or beating for adult insects on riparian or emergent vegetation; sample only in dry weather
	Dip netting	Retrieval of aquatic insets from submerged vegetation or the water column using a strong hand-held net; can be standardized in terms of lengths of sweeps of the net; nocturnal sampling can produce different species from day-time sampling
	Surber sampling	Area-based technique involving delimitation of substrate units in flowing water as basis for 'kick samples', disturbing the circumscribed area to release insects that are swept into a net suspended downstream; basis for much quantitative sampling of insect larvae
	Benthic grabs	Variety of mechanical grabs used to enclose and retrieve area-based samples of substrate from which to sift bottom-living invertebrates

4.3 High-canopy insects

Let's start at the top. Much insect life occurs way up in the tree canopy. The field of canopy insect research was partly pioneered in the Neotropics, and led to some remarkable discoveries. Among them is that perhaps half of all global insect diversity occurs on about 6% of the land surface, high up in the tropical forest canopy. Erwin (1982) was one of the pioneers, and his work on canopy insects stimulated a global debate estimating how many species may be living on the planet. Some have argued that such estimates per se do not really matter for conservation action, but the work stimulated much more focus on both the conservation of tropical forests in general, and on those parts of forests that should be conserved, as well as the extent to which disturbed forest still has conservation value. It also underscored the fact that much biodiversity is disappearing through *Centinelan extinctions* (these are extinctions of species before they are scientifically described, named after Centinela Ridge in Ecuador where botanists found that some tree species had disappeared before being scientifically described).

There are two approaches to *canopy sampling*:

- Direct sampling from the tree canopy once the upper reaches of the tree have been accessed
- Sampling of the tree canopy from the ground.

Access to the tree canopy has particular challenges, including the fact that you must not have a fear of heights! Personal injury insurance may impose very specific conditions on your activity. These access techniques come in two general forms: single-person access using ropes, and team efforts where almost permanent access is catered for by walkways, platforms, rafts, towers, or cranes (Figure 4.2). Such approaches are obviously expensive, and not usually within the province of an individual researcher. However, potential collaboration may be possible, sometimes with access fees to the facilities that can be put into a research proposal budget when arranged beforehand. Dial and Tobin (1994), Moffet and Lowman (1995), Mitchell et al. (2002), and Ozanne (2005) provide overviews of individual access techniques, while Mitchell et al. (2002) also give details of more permanent structures. Once in the canopy, techniques can be used as one would normally sample low vegetation on the ground. However, individual access using ropes limits use to techniques that are lightweight and can be physically carried by one person confined to ropes (usually, in fact,

Fig. 4.2 A gondola suspended from the canopy research crane erected in the Queensland rainforest, Australia (Courtesy Nigel Stork).

safety protocols may dictate two participants). Permanent and semi-permanent structures, in contrast, allow the use of heavier and more bulky equipment once in the canopy. One advantage of individual access is that the forest can be explored more freely, and samples taken far apart and on a regular basis (Stork, 2007).

For canopies more than 10 m in height, there is really only one on-the-ground technique, which is *chemical knockdown*. This has been remarkably successful in generating a huge amount of valuable data for diversity and conservation studies (Stork and Hammond 1997; Ozanne 2005). The principle involves blowing an insecticidal rapid knockdown compound into the tree canopy and collecting the falling insects near the ground. The insecticide may be a commercial formulation of a fog delivered for a few minutes from a commercial thermal fogging machine, or *fogger*. With this relatively passive technique, the machine may be operated from the ground, and the warm fog directed up to the canopy where it then rises up through the foliage. The insecticide fog surrounds the foliage, causing some of the insects to fall out the tree. The machine can also be hoisted into the tree, and triggered by a radio signal to start operating, to ensure delivery into the target region of the canopy (Figure 4.3). The technique is highly dependent on weather conditions and therefore must only be done on still, dry days, usually at dawn or dusk. A more active technique of chemical knockdown is to use a commercial *mistblower* that produces concentrated ultra-low-volume insecticidal droplets. Mistblowers are far more active than foggers but tend to be far more localized in their effect. They work by driving a fine insecticidal mist in and around the foliage; in this way they can actually physically dislodge insects.

Fig. 4.3 An insecticide fogger suspended high in the tree to deliver a knockdown insecticide into the tree canopy (Courtesy Nigel Stork).

Insecticides used in foggers and mistblowers come in commercial formulations that vary from one country to another. Natural pyrethrum, which comes from a chrysanthemum-type plant, has the advantage that it breaks down very rapidly once exposed to air and sunlight and thus leaves few residues in the tree. Also, many of the larger insects such as bush crickets become dislodged and stunned and fall to the ground where they can often be resuscitated should they be required alive. A more common commercial formulation is one of the synthetic pyrethroids, which are effective for knocking down most insects, and are easy to obtain.

Choice of fogger or mistblower depends very much on application, and we suggest that you contact a researcher who is currently active in the field to seek advice. Bear in mind though, that there is still a lot we do not know with regards to chemical knockdown from tree canopies, so comparing methodologies could be a research topic in its own right. Clearly however, using a mistblower with pyrethrum is very effective in comparison with some other techniques such as beating (see below) (Lowman et al. 1996). Of course, knockdown will only really sample species that are walking freely on the leaf or branch surface; it will not sample any species firmly attached to the vegetation, inside epiphytes (Yanoviak et al. 2003), or burrowing into the plant tissue. Epiphyte sampling requires special attention, and has been reviewed by Yanoviak and Fincke (2005). Sampling shoots, stems, and trunks has been reviewed by Speight (2005).

The dislodged insects need to be gathered as they fall down from the tree. Ozanne (2005) recommends vinyl hoops or funnels with a shiny surface and a collecting bottle at the bottom into which the knocked-down insects tumble. The collecting bottle can be partially filled with 70% ethanol as an instant

Fig. 4.4 Suspended funnels used to collect insects that fall from the forest canopy as a result of fogging with a knockdown insecticide (Courtesy Nigel Stork).

preservative. These hoops can be either hung in the tree or set on stands at pre-selected spots under the canopy (Figure 4.4). Besides being effective and efficient, these hoops also have the advantage that they have a standard surface area (0.5 m² or 1 m²) that allow some estimation of insect densities.

4.4 Low-canopy insects

Sampling inside bushes can be very difficult, because as soon as the bush is touched any wary insect such as a bush cricket soon jumps away. This problem is easily overcome using *insect tongs*, made from a pair of barbecue tongs to the end of which are attached two tea strainers facing each other (Figure 4.5). The tongs are carefully closed around the insect, which seems to freeze. It is then removed from the tongs by inserting the tea strainers in a clear plastic bag, opening and then withdrawing them, leaving the insect inside the bag. This is an excellent technique for catching unharmed, live insects for laboratory rearing.

Insects that readily fly away when approached can be collected from inside bushes or on tree trunks by the use of a 'sticky stick'. In this very effective technique a long cane or pole is smeared at its tip with a highly sticky substance, usually a polybutene-based product such as Plantex or Tanglefoot. The insect sticks to the end of the pole, which is then withdrawn. The specimen is peeled

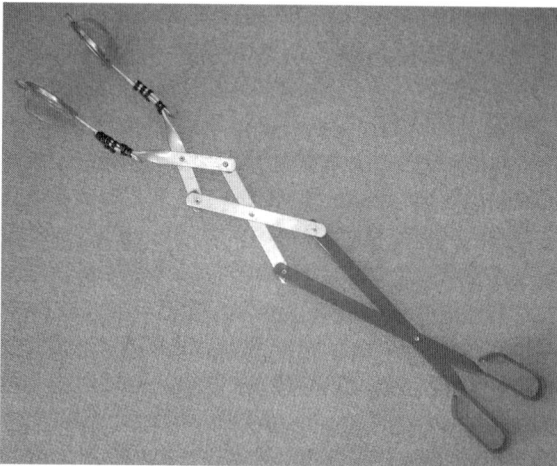

Fig. 4.5 Insect tongs made from barbecue tongs and two metal tea strainers. This contraption is excellent for catching insects in bushes, such as katydids (bush crickets), as the tongs can be closed around the insect causing only minimal alarm. The unharmed insect is removed from the tongs inside a clear plastic bag.

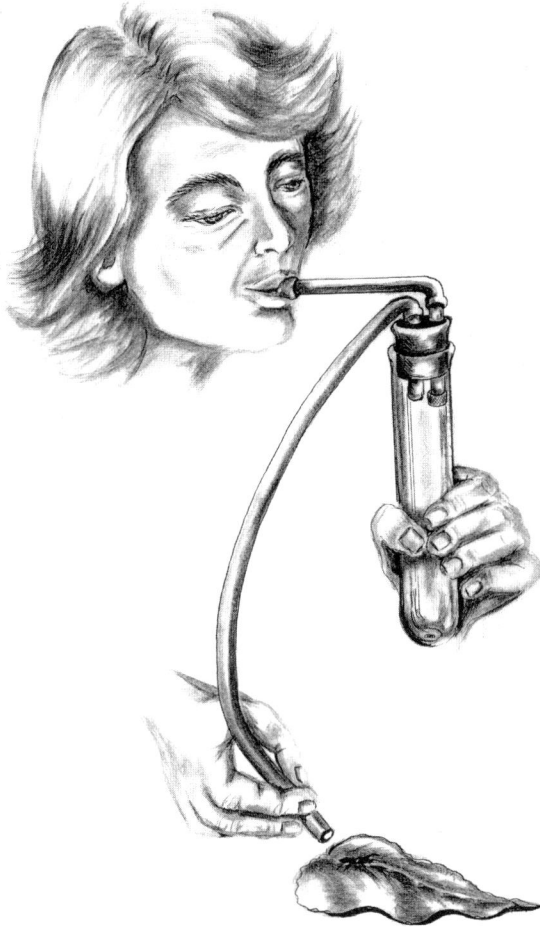

Fig. 4.6 Pooter (also called an aspirator) used for sucking up small insects from beating trays or directly from substrates.

off with a small, clean stick into a *killing jar*, or into benzene or xylene, which dissolve the polybutene.

An entomologically tried and tested technique for sampling individual small insects is the *pooter* (or *aspirator*) (Figure 4.6). It is a short, wide glass tube, stoppered at both ends, with an inlet tube through which small insects pass as they are sucked by inhalation through a tube at the other end. To avoid getting a mouthful of insects, it is essential that the inhalation tube is covered with a piece of gauze held in place by a rubber band. Where unhealthy or unsavoury habitats are being sampled (e.g. in caves, on certain fungi, around faeces) a gas-collecting rubber bulb can be used to suck the air through. This method is principally for

Fig. 4.7 Beating tray for collecting insects from bushes (Courtesy James Pryke).

extraction of specimens from a beating tray or sweep net (see below), although it can also be used for sampling bushes and flower heads direct.

Beating is one of the oldest and most effective techniques used in entomology. It simply involves using a tray or an 'upside down umbrella' and beating the bushes with a stout stick, causing the dislodged insects to fall into the tray (Figure 4.7). As with so many techniques it is highly selective in what it samples, but nevertheless produces a remarkable array of insect specimens. Bear in mind when sampling like this that not all insects are found just at the height at which it is convenient to beat a bush! Beating can also be used to sample high tree canopies from a walkway or a crane (see above). Also, very different results can be obtained depending on whether the bottom or top of the bush is sampled, and which compass quadrant of the tree (Samways 1985). Sometimes quite different insects can be found on the sunny side and the shady side of the tree. Sampling can be standardized when using the beating technique by, for example, beating a particular side of each tree at a particular height, the same number of times, with approximately equivalent force. Interestingly, Rohr et al. (2007) found that beating (in their case using a 1 m² sheet placed below the vegetation) was the most efficient of the 11 collecting methods that they used for monitoring the overall invertebrate assemblage in a national park.

Comparing different quadrants of the tree can also be tested using other techniques. These include branch *sleeving*, where a section of the branch is sleeved with a gauze (black is best, as it is easiest to see through) and a glass is attached at the upper and outer end to catch emerging or wandering insects (Figure 4.8). When carefully done, so that equal amounts of vegetation are included in each

Fig. 4.8 Sleeved branch for collecting emerging insects from the contained plant parts.

sleeve, it can be a very effective means of quantitatively comparing different trees or bushes. It can also be used to catch winged males emerging from sessile immatures, as in the case of scale insects. The method can also be employed to catch emerging parasitoids from insect hosts on the foliage.

A related principle to sleeving is the use of an *emergence box*. These are used to collect and quantitatively sample insects that emerge from leaves, twigs, flower heads, seeds, or fruit, including parasitoids and even hyperparasitoids of herbivorous insects. Vegetation samples are placed in a darkened box (and supplied with water where necessary) and a funnel and a collection vial placed at the side of the box. When the insects emerge, they go straight for the lighted area and are kept in the vial (Figure 4.9).

Sessile insects on trees or shrubs, or those with sessile life stages or conspicuous habitats or structures, readily lend themselves to hand collection, and can be quantified very effectively. They include, for example, scale insects, gall-forming, leaf-mining, and leaf-rolling insects, insect nests and pupae with conspicuous cocoons. These structures can be counted and measured in situ, and/or collected for rearing and identification of the adult insect.

4.5 Grass/herbaceous layer

The most common technique for sampling low vegetation is *sweep netting* (Figure 4.10). A sweep net is made from a stout frame, strong handle, and strong net bag that does not snag on sharp vegetation. The size of the net, in terms of diameter

Fig. 4.9 Emergence box used for extracting parasitoids from hosts contained in the box. The parasitoids emerge and move to the light, where they are contained in a glass tube. The closest wall is opened up here to show the inside, but would be closed when the box is in operation so that it is dark inside.

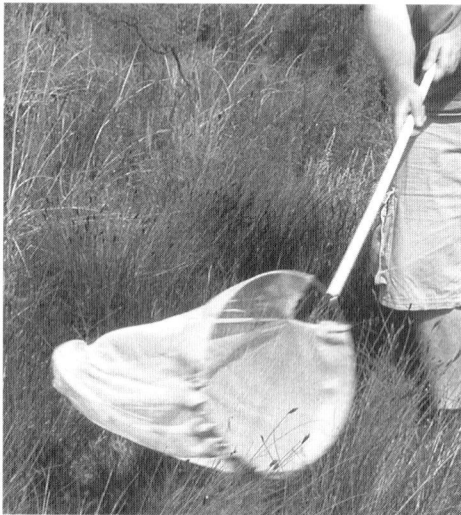

Fig. 4.10 Sweep netting involves using a very strong net that can cope with the resistance of being moved swiftly through low-growing vegetation (Courtesy Sven Vrdoljak).

or width, is a matter of comfort for handling, i.e. how strong the arms of the collector are, and the robustness of the vegetation. The net can be used quantitatively when the vegetation being sampled is more or less of the same structure and resistance. The net can be dragged through the vegetation for a set distance, speed, and height. The length of transect and numbers of transects depend on what the aim of the project is, and, in particular, how much material can be

Fig. 4.11 Suction sampler in action, being used to sample insects from low-growing vegetation (Courtesy James Pryke).

sorted. Preliminary sampling, or a pilot study (Chapter 3) is thus essential. One advantage of sweep netting is that the collected specimens can if necessary be removed from the net selectively according to focal taxa, size class or whatever, using soft forceps, a fine sable-hair paintbrush, or a pooter (see above).

Another technique used for low vegetation is use of a *suction sampler*. These machines are available commercially and are widely used (Figure 4.11). Also useful, but of course far less powerful, are battery-operated vacuum cleaners. Vacuuming techniques can be used quantitatively so long as due recognition is given to any differences in the vegetation structure.

4.6 Sampling flying insects

When you require voucher specimens of larger, flying insects you may net them with a standard butterfly net. The size, weight, and even colour of netting depend on personal preference. A good overall compromise is a circular or kite-shaped net (Figure 4.12), with a stout pole, about 1.5 m long (to double up as a walking staff in difficult terrain or for support in streams), with black gauze (so that the insects can be easily seen through it). For high-flying insects, telescopic aluminum poles are available from fishing tackle shops, i.e. those normally used for fish landing nets, or hardware stores (as window cleaning accessories). These can be used to sample as much as 8 m high along tree margins (Figure 4.13) but become very ungainly to use for most other purposes. When an insect is netted,

Fig. 4.12 Aerial kite net on a stout pole. Besides being a useful collection tool in its own right, it also doubles up as a walking staff and a support in difficult terrain.

it is immediately confined by rapidly twisting the net so that the bag is closed against the net rim.

Some very large insects such as large orthopterans can be shot using standard dust shot in a shotgun. Naturally of course this can be very disturbing to people and wildlife alike, and so must be used only where appropriate and where permission has been obtained.

Dusk-flying predacious insects such as aeshnid dragonflies, generally fly too high to net but can be caught using a *buri* (a Japanese word), which is a piece of strong, dark cotton thread, about 25 cm long to each of three ends of which is attached a small stone or lead shot. The buri is thrown in the air in the flight path of the dragonfly, which goes to seize it as prey and then becomes tangled in the thread and is thus brought down (Figure 4.14).

Fig. 4.13 A very long-handled aerial net can be a useful collection tool for some high-flying insects.

The techniques outlined so far are *active* as the operator has to actively operate the trap. Other techniques are *passive*, and involve a trap being put in place which insects are attracted to or encounter by chance. Various traps, for example, are designed for catching insects in flight. These may be divided into those that actively attract the insect, such as *light traps*, and those that the insects fly into by chance, and known as *interception traps*. All have enormous biases and the results should be treated with appropriate caution when the study objective is to quantify insect diversity. Light traps, which are light sources operated at night, come in many designs, some of which are commercially available (Figures 4.15 and 4.16). Fry and Waring (1996) and Muirhead-Thomson (1991) provide details on various light trap designs. These traps are very effective for many, but not all, moths, and may also be used to catch some beetles, bugs, and ant

Fig. 4.14 Buri. This is a three-stranded missile (like bolas) with a small weight tied to the end of each string. It is an effective piece of equipment for collecting dusk-flying dragonflies. It is thrown into the air to intercept the insect's flight path. The dragonfly chases it as if it were prey, and in doing so, gets entangled and brought down.

Fig. 4.15 Simple, portable light trap, made of a card cylinder inserted into a jar with pieces of an egg box in the bottom on which the captured insects can sit.

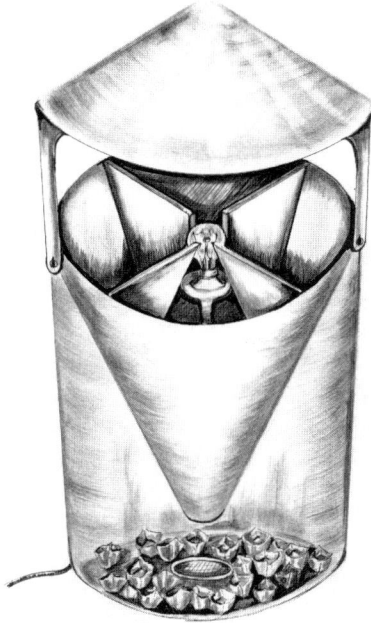

Fig. 4.16 Light trap with a roof to keep off rain and a mesh-covered plug hole to drain out any water that may enter the trap. Insects enter the funnel and are 'knocked down' by the vertical veins. Pieces of egg box at the bottom of the trap give the insects something on which to sit. Although shown here as transparent, the trap is normally made of a material such as aluminium.

lions among many other taxa. Some interesting new dragonfly records have come from light traps being operated by lepidopterists. Normally these traps are operated to catch the living insects, which can be selectively kept or released, so having minimal impact on the population.

Light traps work on the basis that insects mistake artificial light for the moon, and fly with a constant angle to the light. In the case of the moon, which is effectively at an infinite distance, the constant angle is maintained as the insect moves across the landscape. However, as a light source is nearby, the insect in trying to fly while maintaining a constant angle, goes round and round in a spiral all the while closing in on the light source. Some traps use mercury vapour light bulbs, and others use fluorescent tubes. Fox et al. (2006) point out that the attractiveness of moth trap lights depends partly upon the level of background lighting, as well as the type of bulb used. Increased light pollution in the UK over recent decades has decreased the efficiency of light traps using regular incandescent bulbs, emphasizing the value of mercury vapour bulbs as being generally more attractive to moths than regular bulbs.

A variant of the normal light trap is the *light tower*, where a light source is placed inside a frame about 2 m high and 1 m in diameter, clothed in a thin white sheet (Figure 4.17). The insects are collected as they land on the outside of this tower. In remote tropical locations a white sheet hung up and lighted by a paraffin lamp can pull in an enormous number of insects. They can be collected from the sheet with soft forceps, or tapped into a killing jar, but it is best to use a head lamp for extra vision and to have both hands free.

Light traps can be used to provide good quantitative data. Nevertheless, results vary according to species and weather conditions, and to flight periods during the night and also relative to the flight behaviour of the subject insects (Hirao et al. 2008). It is important to bear in mind that not all moths, for example, are attracted to light. Also, different species fly at different times during the night, and flight behaviour can be affected by factors such as lunar cycle, cloudiness, wind, humidity, and rain. Also, light traps can attract insects from many metres and more, and therefore do not provide point samples of species. Certainly, moths captured in a light trap may not necessarily be from that immediate habitat and may only have been passing through. A practical consideration is that

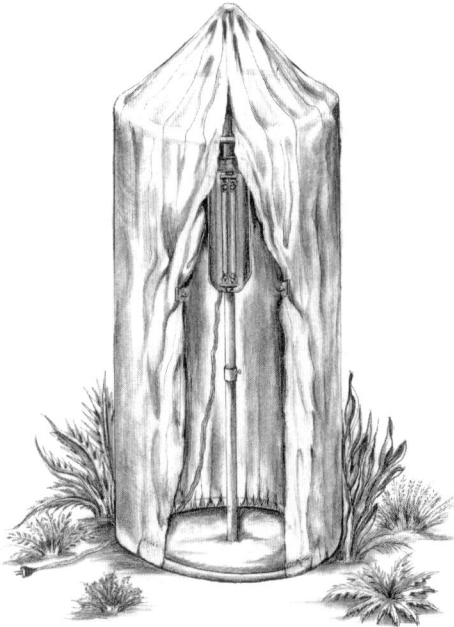

Fig. 4.17 Light tower. This is a tent made of a thin white sheet over a frame, with a light inside. The tent is closed during operation. Insects attracted to the tower rest on the outside and are collected by hand.

light traps generally need an electrical generator that can be noisy, heavy, and vulnerable to theft. However, mercury vapour fluorescent tubes (black light) are available that can be run from 12 V vehicle batteries, and solar-charged batteries are available for use in more remote areas.

Another attractant used principally for moths is *treacling*. This is shrouded by myth and curiosity, with moth collectors in the past often having their 'secret formula' (Figure 4.18). Basically, the method of making the attractant consists of placing black treacle in a saucepan and warming it, then introducing brown sugar, dark beer and some rum, sherry or brandy. The treacle is then brushed on tree trunks at dusk and the site revisited during the night. Sometimes a ring of moths can be seen sitting around the treacle smear, imbibing the artificial nectar. The moths become tipsy on the alcohol and can be gently tapped with a paintbrush into a collecting vial or killing jar.

There are almost as many flight interception traps as there are research projects, but the principle is always the same. In the case of *window traps*, a glass or Perspex sheet is erected into which insects fly and are knocked off

Fig. 4.18 Treacling is the daubing of a tree trunk with a mixture of treacle or molasses with a little beer or brandy. It can be very effective in attracting some species of insects, especially moths, with different species attracted to different recipes and different heights of the treacle daub on the tree trunk.

balance (Figure 4.19). A tray of water with a little detergent to reduce water tension is placed at the base of the sheet, so that the insects tumble into it. Alternatively, glass sheets can be covered with transparent kitchen plastic film (cling-film, Glad wrap, Saran wrap) over which is smeared Plantex or Tanglefoot or a similar transparent sticky substance. The specimens become trapped on the sticky surface, and can be stored by covering the original sticky surface with another sheet of film to seal in the sticky surface, insects and all. The sheets are then stored in a file separated by sheets of paper. Other types of interception traps involve use of very fine black netting suspended by ropes across paths or

Fig. 4.19 (Upper panel) Two flight interception traps set at right angles to each other and anchored with guy ropes to survive high winds. (Lower panel) Looking at the catch inside the trough at the base of the 'window' of the flight interception trap (Courtesy of James Pryke).

forest clearings. Again, a tray or trough with liquid is placed at the base of the trap to retain the intercepted insects.

One of the well-tried and certainly favourite flight-interception techniques of dipterists is the *Malaise trap* (Figure 4.20). Again, there are many designs but each is effectively an open-sided, dark-coloured tent with a central vertical interception sheet and a funnel mechanism leading to a collecting bottle. Many types of insect are attracted to the overall dark shape of the trap, move into it, and are then funnelled into the collecting bottle. Innovations include incorporating colour panels: blue, white, or yellow panels can be effective for attracting pollinators (Campbell and Hanula 2007). Fraser et al. (2008) point out that while a small number of Malaise traps can yield useful information on insect assemblages, they are likely to seriously underestimate total species richness. To achieve extensive species coverage, they recommend that sampling should continue over several weeks with widely separated traps sampling both core and edge habitats.

Malaise trapping is a relatively effective *point technique*. What this means is that it is capturing insects within the immediate vicinity of the trap (although of course this increases as the landscape opens up, with Malaise traps operating over many tens of metres in semi-desert). Malaise traps are far more of a point technique than light traps (which cannot be used for fine-scale spatial studies

Fig. 4.20 Malaise trap. This catches insects both actively, by attracting certain insects, especially flies, by its dark form, and passively, by some insects inadvertently flying into it. On hitting the black gauze, the insects fly upwards and are guided by the tent cover into a bottle at the highest point containing a preservative, such as 70% ethanol (Courtesy James Pryke).

over just a few metres), but they are not such a point technique as pitfall traps, for example (see below). Indeed, Malaise traps can produce a great deal of data and are highly complementary to other techniques, particularly as they are so effective for many species of flies that may not be captured using other techniques. Although traditionally used as ground traps, as in Figure 4.20, they can also be built around a frame that can be hoisted up into the tree canopy for sampling insects in forests. Sampling at different heights in the forest will produce more representative results than just sampling on the ground. This may also be the case in open habitats.

Økland (1996) tested the efficiency of various traps for sampling saproxylic beetles, and found that *trunk window traps* (consisting of a 'window' placed almost horizontally against the tree trunk and a collecting funnel underneath) are suitable for comparing different substrates within the same forest environment, whereas window traps are more suitable for comparing different forest environments.

Another technique used for sampling flying insects is the use of *pan traps*, which are simply coloured bowls placed out in the open and containing water with a little detergent to reduce surface tension (Figure 4.21). There is still much to explore in terms of which colours to use for which taxa (Kirk 1984; Campbell and Hanula 2007), and so we recommend some preliminary sampling to see how effective different colours are relative to the goals of the project. Fluorescent yellow-green is good for insects attracted to shoot tips, such as some thrips, aphids, leafhoppers, and psyllids. Blue, yellow, or white can be effective for

Fig. 4.21 Pan trap (yellow in this case for catching pollinators). It contains water with a little detergent to reduce surface tension (Courtesy Sven Vrdoljak).

pollinators, while some flies are attracted to dark colours. Little is known about combinations of colours such as blue and yellow stripes, so this could be a field of investigation. The bowls or pans can be spray-painted and placed on stands placed in the field at half a metre or so in height. One bias to watch out for is that catches are dependent on, for example, the number and colour of flowering plants in the vicinity.

A variation on the theme of coloured traps is to use coloured *sticky traps* (Figure 4.22). The base can be a galvanized iron or Perspex sheet (25 cm × 25 cm) painted or covered with coloured sticky tape hung in the field, say, from a branch. The trap can be covered with clear film (cling-film, Glad wrap, Saran wrap) and smeared with sticky polybutene such as Tanglefoot or Plantex. As with the window traps mentioned above, the sticky surface can be covered with another sheet of clear film to seal in the sticky surface and the sampled insects with it. A disadvantage of this technique is that individual insects are not always readily extracted from the surface of the trap and need to be identified and counted *in situ*.

Butterflies, such as some nymphaline and charaxine species, as well as chafer beetles and many flies can be attracted to a *baited trap* hung in a tree and baited with fermenting fruit such as banana with a little rum (Figure 4.23). The trap consists of a landing platform on which the bait is placed. Above a gap of up to 5 cm is a gauze tower, around 1 m high and 30 cm diameter. When butterflies are being sampled, the traps should be cleared out regularly as the butterflies tend to flutter and damage themselves. Alternative baits such as rotting fish or meat, or even mammalian faeces, can be use as baits for certain species such as dung beetles.

Fig. 4.22 Coloured trap, in this case fluorescent yellow for capturing a range of insects from thrips and psyllids to hymenopteran parasitoids. The coloured surface is covered with transparent kitchen wrap and smeared with a sticky polybutene.

Fig. 4.23 Baited trap. This has a landing platform in the centre of which is placed a bait, such as rotting fruit. The trap consists of a net cylinder above the bait. Insects that land on the platform and head for the bait then tend to go upwards, and are trapped in the net.

Butterflies, being conspicuous and day-flying, can be quantitatively sampled simply by visual surveys along *transects* of variable length (Pollard and Yates found that 3 km was optimum for British butterflies) but consistently 5 m wide. The transect is walked slowly and all the butterflies in it recorded on a data sheet. It is a very robust and tested method that has yielded an enormous amount of data over the years (see Chapter 6.5.7 for more details). It can be used in sub-tropical and tropical areas, provided that there is a clear view of the area being sampled. In these species-rich areas, it is usually necessary to carry a pair of *close-focus binoculars* to aid identification. Many of the well-known makes now have the option to focus down to 2 m and some to as little as 1 m. These very close-focus binoculars are a very useful tool indeed (Figure 4.24). When selecting them for purchase, see whether you can put the toe of your shoe into sharp focus. If you can, then you have a very useful and versatile instrument. Opticron also make a lightweight *monocular* that is very compact and useful (Figure 4.24).

Fig. 4.24 The insect conservationists' two best friends: close-focus binoculars and a monocular. When in the shop choosing one of these, do make sure that you can focus on your toes while standing. Very few binoculars can focus this closely, so you will probably need to test several pairs. The ones shown here are lightweight 8×32 Kamakura, and have given years of service often under gruelling conditions. The monocular is an Opticron 8×20 Very Near Focus, and handy to have in your pocket for emergency use when you spot an interesting-looking insect while out on a family picnic.

Of course, you need to know your butterflies well and be able to identify them from a distance and on the wing.

Butterflies are not the only group of flying insects that can be sampled visually. Dragonflies and damselflies can also be sampled by walking a set distance along the water's edge and numbers of individuals recorded per set distance, such as 100 m of bank and sampling 2 m into the water and 2 m onto the bank (Figure 4.24). For these odonates, it is essential to use close-focus binoculars or a monocular.

Certain insects can be trapped using a *pheromone trap*. Pheromones are substances secreted into the air by the insect for communication, particularly to attract members of the opposite sex. A recently emerged female moth can be placed in an assembly trap to attract males. This type of pheromone trap is a darkened box, inside which a female is caged. The trap has a funnel-shaped opening into which the males fly and come to rest inside the box. These traps, being species-specific, are of course principally for monitoring changes in population levels over time.

Fig. 4.25 Commercially manufactured tent trap for monitoring Red scale *Aonidiella aurantii*. A rubber capsule, impregnated with the female sex pheromone that attracts males, hangs by a wire inside the tent (which is made of card). The undersurfaces of the tent are smeared with a thin layer of sticky polybutene. This trap also attracts parasitoids of the Red scale. The delta trap works on the same principle as the tent trap but simply has a floor that is the sticky surface.

Commercially available traps, such as tent traps (Figure 4.25) and delta traps, support a rubber capsule impregnated with an artificially synthesized phero-mone. Usually it is the female sex pheromone, which attracts the males. Traps are available for a range of insects, from scale insects to moths, but generally only those of pest status, although attractants are now being made available for certain threatened species (Box 4.1).

A little-explored but very valuable side-effect of these pheromone traps is that they attract many parasitoids, and even hyperparasitoids, that are specific to that particular host. Pheromone traps can be used, for example, to monitor these parasitoids in relative proportions to the males of the host (Samways 1988b).

Box 4.1 Monitoring rare and threatened insects with pheromone attractants

Since the first chemical identification of an insect sex pheromone several decades ago, pheromones have been applied mainly for monitoring and control of pest insects. Virtually no effort has been directed towards pheromone-mediated moni-toring of rare and threatened insects, in spite of an ever-growing interest in bio-diversity and conservation. We attribute this to tradition rather than to a lack of potential applications. The legendary attractiveness of many insect pheromones would be ideally suited for monitoring elusive species even at very low population densities.

To our knowledge, the first pheromone identified expressly for conservation purposes was the male-produced pheromone of the saproxylic scarab beetle *Osmoderma eremita*. This was the first insect to be assigned special protection within the European Union. The pheromone has been used for parallel monitoring of *O. eremita* and the predatory saproxylic click beetle *Elater ferrugineus*, for which the *O. eremita* pheromone apparently functions as a prey or habitat cue (Larsson et al. 2003; Svensson et al. 2004). Pheromone monitoring of these two species has yielded improved population estimates and information regarding population fluctuations and dispersal biology (Larsson et al. 2008). Air sampling of its pheromone has also been used to determine occupancy of *O. eremita* in individual tree hollows (Svensson et al. 2003). Possibilities for monitoring of *E. ferrugineus* have been further extended with the recent identification of the female-produced sex pheromone of this species (Tolasch et al. 2007).

For better or worse, most pheromone systems are specific for a single species or for limited groups of species. This has the advantage that investigations could be directed selectively towards target species and would not require extensive entomological skills. On the other hand, identification of a pheromone represents a significant effort, which may limit the number of species that could be targeted. The rarity of many species may also constitute a bottleneck in the identification process. Occasionally, whole groups of related species may share a very limited number of pheromone components, enabling trial-and-error screening for attractants without a full-scale identification process. Examples of this are moths of the families Sesiidae (clearwing moths) and Tineidae (clothes moths and relatives), which are often trapped by collectors and taxonomists on generalized pheromone lures (Priesner et al. 1989; Buda et al. 1993). Sesiid and tineid moths are frequently xylophagous or saproxylic with a potential to be useful indicators of forest biodiversity (Larsson et al. 2008).

Pheromone attractants have the potential to vastly improve precise monitoring of many insect species in space and time. Their applications include determination of presence or absence with high accuracy, and standardized methods for relative population estimates that could be used to track population fluctuations over time and evaluate how individual species are affected by landscape and habitat management or climate change. Monitoring with typical female-produced sex pheromones are usually non-invasive from a population perspective, as they target only males and would seldom out-compete the natural pheromone released by females. Pheromone-mediated monitoring could likely be applied to most flying or highly mobile insects that use chemical communication over some distance. This excludes many visually conspicuous, diurnal insects like butterflies or dragonflies (which are likely over-represented as model organisms), but should include just about every moth, a large fraction of beetle species from different families, and a variable number of dipteran and hymenopteran species among the major insect orders.

(Left) *Osmoderma eremita* females dispersing between host trees can be intercepted by traps baited with synthetic pheromone. (Right) A collection jar from a pheromone trap containing two trapped females, one of which has been captured before. The proportion of marked individuals recaptured in pheromone traps can be used to estimate absolute population density.

Mattias C. Larsson, Swedish University of Agricultural Sciences
Glenn P. Svensson, Lund University
Nils Ryrholm, University of Gävle

4.7 Sampling insects on the ground

Insects that move around on the surface of the ground are known as *epigaeic fauna*. However, there is often a circadian rhythm of movement, with insects and other arthropods going deep into the litter, and even into the soil during the day, only to emerge on to the soil surface during the night. Furthermore, many species are active nearer the surface during wet periods, going deeper in dry spells.

Although various techniques are available for sampling insects on and in the litter and soil, one should not overlook the possibility of direct visual searches for some species, particularly in and around rock crevices, under stones, and among logs. Indeed, this is often the preferred technique for looking for non-insect arthropods such as millipedes, centipedes, and velvet worms (Onychophora). Specimens can be simply picked up with soft forceps and placed directly in 70% ethanol. Smaller individuals can be sucked up using a pooter (see above). A trowel or a garden fork can be used to look for moth pupae or cockroaches, for example under tree canopies.

The most popular technique for sampling epigaeic fauna is by using *pitfall traps*, as reviewed in detail by Woodcock (2005). In principle, a pitfall trap consists of a large test tube, bottle, or plastic cup sunk into the ground and partially filled with a preservation fluid, such as seven parts 75% ethanol and three parts glycerol (or, if for DNA analysis, >90% ethanol). The glycerol reduces evaporation. Ethylene glycol (antifreeze) is another popular preservative for pitfall traps, although it is toxic to humans, and of course, is not readily available in tropical countries. Make sure that the preservative used does not repel the insect group of interest.

The traps are dug into the ground with minimal disturbance to the substrate and the rim left flush with the ground surface. It is generally a good idea to leave the traps closed for a few days before setting, to compensate for the *digging-in* (or disturbance) *effect*. Pitfall traps are normally left open for a certain amount of time, usually 3–5 days, although Borgelt and New (2006) provide strong evidence that 7 days is the minimum suitable period for some applications, but it depends also on weather conditions. Pitfall traps deployed in the rainy season may well fill up with water within a day or two, and thus require much more regular emptying, or providing each trap with a roof or shelter as noted below. At the end of this trapping period, the traps are emptied into collection bottles and labelled with pencil on paper inside the bottle. It is generally better to rest the traps by closing them up between the sampling periods, to minimize disturbance effects and to provide time for the researcher to sort, or at least properly store, the captured specimens (sometimes several hundreds or more). After the rest period, trapping can be resumed. This may be done throughout the year if appropriate, or at set intervals: for example, four 7-day trapping periods at four discrete times of the year, representing spring, summer, autumn, and winter. The choice of trapping frequency is dependent on the study objective (Chapter 3). At high latitudes it may of course not be appropriate to sample during the winter. When traps are closed between sampling periods it is essential to mark where they are, because they become overgrown and difficult to relocate. This can be done directly, by marker stakes for example, and/or using fine GPS coordinates.

There has been considerable debate as to which type of pitfall trap is best, although choice of trap depends in part on the types of focal insects (or other arthropods). The *Majer-type pitfall trap* (Figure 4.26) is excellent for ants (although other types are also effective for ants; Agosti et al. 2000). It is essentially a boiler tube, about 18 mm in diameter. It is inserted into a sleeve (made from irrigation pipe or similar) that is sunk permanently in the ground. When the boiler tube is removed, the sleeve is corked up. The sleeve needs to be banged into the ground with a custom-made, very tight-fitting awl. Some researchers working in firm soils have found that the sleeve is unnecessary and the tube can be sunk directly into the ground, but traps may then become very difficult to extract easily.

For larger insects such as ground beetles, larger traps need to be used (Figure 4.27). The diameter of the mouth of the pitfall trap has a large effect on the

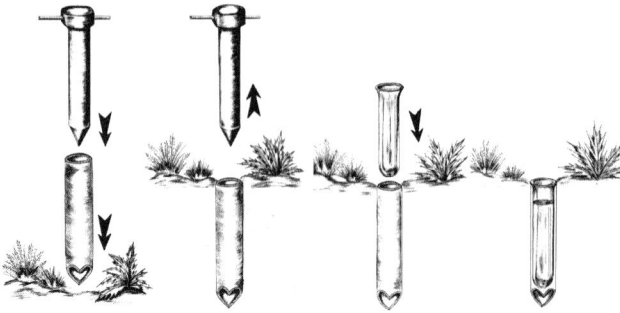

Fig. 4.26 Majer-type pitfall trap for sampling surface-foraging ants. This trap is effective for most ground surface-living ants, and excludes large insects and other bycatch. The sleeve consists of a chamfered strong plastic tube (such as irrigation piping) that fits snugly over a purpose-made steel peg. The two are driven into the ground and the peg extracted. The tube is then the housing for a test-tube or boiler tube that is chosen to fit fairly tightly into the sleeve. The trap is then partially filled with a preservative (e.g. 2/3 70% ethanol and 1/3 glycerol) and left operational for a few days, after which the contents are poured out. When the trap is not in use, the sleeve is stoppered. A shelter for the trap, consisting of a roof on four stilts to keep out rain, is an additional option.

Fig. 4.27 Large-diameter pitfall trap that catches a wide variety of insects and other surface-foraging arthropods, both small and large. One plastic cup sits inside another, with the inside one being removed to empty the contents. Various preservatives can be used, with one of the most versatile being 80% ethanol and glycerol (Courtesy James Pryke).

diversity and body-size spectrum of insects trapped. Disposable plastic coffee cups have been very popular because they are cheap, easy to obtain, and light to carry. One cup can be sunk as the sleeve and a second one inserted inside the first as the trap. These larger traps are very prone to flooding during rains, and so they may require a shelter, such as a plastic Petri dish on wire legs or a galvanized nail holding the shelter a few cm above the trap. These shelters of course are very vulnerable to mammalian disturbance, including that by humans.

Pitfall traps can also be baited. The bait is placed in a separate container, covered with gauze cloth to prevent the attracted insects remaining on the bait, inside the pitfall traps and on a pedestal so that it is above the preservative. Choice of bait depends on the target taxa. Rotting fish or meat can be used to attract carrion beetles, while chafers can be attracted to fermenting fruit. Coprophagous beetles can be attracted to traps baited with mammalian faeces. The quantity of bait per trap should be standardized.

As many authors have emphasized, the catch from pitfall traps depends on the activity patterns of the local species, and also on habitat structure, and so the proportionate number of captured individuals may not necessarily reflect the true numbers of individuals from one species to the next. Because habitat structure can have a significant effect on pitfall catches, it is advisable not to use pitfalls to compare habitats with very different structures. But this largely goes for most trapping techniques. Pitfall traps are excellent for point sampling and for comparative sampling in different habitats. However, it is essential that results are made comparable by the traps being open for exactly the same amount of time (e.g. if a 3 day rotation is decided upon, then it must be adhered to). This may require some preliminary sampling to ascertain the best length of time for the trap to be active. Also, all the traps must be flush with the ground so that their trapping efficiency is equal.

Trap layout depends on the aims of the project, although our advice is that individual traps be replicated in one spot to improve efficiency (see also discussion in Chapter 3). One way to do this is to place them in a square: say 1 m² for the Majer trap and 2 m² for cup traps. These would not, however, be replicates in a statistical sense, but would be considered as one 'trap set'. Each set would then be a replicate for analysis. The number of traps in a set will vary, but some workers suggest 10–15 traps is adequate.

Some authors have combined pitfall traps with *barrier traps*, to deflect the wandering insects into the pitfall trap. There are almost as many designs as there are studies using them, with a generalized design given in Figure 4.28. The barrier or fence can be a wooden plank or plastic corrugation, about 2 m long, at each end of which is placed a pitfall trap. These traps are particularly effective for large, fast-walking insects such ground beetles and tiger beetles. As they scuttle

Fig. 4.28 Polythene barrier trap, 50 m long (above), at either end of which is a pitfall trap (below), being used here to catch large savanna arthropods (Courtesy James Pryke).

across the surface of the ground, they bump into the barrier and then follow it along, only to fall into one of the pitfall traps at either end.

Cockroaches and other insects that are shy of light can be captured by placing dark-coloured cardboard or wooden *condos* into which they retreat as night time ends. These condos should be only about 1 cm high and have an entrance on either end that is funnel-shaped to encourage them in. Traps can also be baited, either with a carbohydrate such as wheatgerm or with something like meatloaf, depending on the target taxa. Traps like this are very effective for insects living on the floor of caves.

Finally, in the case of large insects such as grasshoppers in savanna, or tiger beetles, simply the use of a good pair of close-focus binoculars can provide an enormous amount of good quantitative data. Transects, usually 50 m or 100 m long and 4 m or 5 m wide, are walked slowly and all target taxa seen (and verified with the binoculars) recorded on a data sheet. A *voucher collection* (see Chapter 2.5.3) is maintained, against which you can compare the species you see in the field. Any new species is captured and added to the voucher collection. As it is usually difficult to remember a long list of scientific names, the observed species can be given vivid nicknames that are descriptive in the field. These nicknames are later cross-referenced with the voucher collection to provide the scientific names. This technique only works for taxa (species or genera) that are readily told apart on sight. Even then, capture of individual voucher specimens may be wise, to guard against any mistakes in identification.

Watts et al. (2008) developed a method for tracking (without having to collect) rare, elusive, and nocturnal giant weta in New Zealand using tracking tunnels. The tunnels are baited (with peanut butter in the case of weta) to attract the insects inside them, but as they enter the tunnel they are obliged to walk through an ink 'footbath'. As they progress to the bait in the centre of the trap, they leave inked footprints behind on the white card base (Watts et al. 2008). The method has been so effective for weta that it is worth trying out on other large, ground-living insects.

4.8 Sampling insects in the leaf litter and soil

Sampling insects below the surface of the litter or soil generally goes hand in hand with sampling other arthropods and even other invertebrates (New 1998). Together, they all constitute the litter or soil fauna. Insects living in logs (saproxylic species) can often be classed with litter insects as they can be similar in taxonomy and sampled in the same way. The most commonly employed technique is that of the *Berlese–Tullgren funnel* (Figure 4.29). It can be used for litter, soil, log, and bark samples. The principle is that a sample is placed over a coarse gauze sheet (about 5 mm mesh size) and a 60 W light bulb used to provide both light and heat to drive the fauna down into a funnel, at the bottom of which is a collecting bottle with 70% ethanol. It is usually necessary to place a small gauze mat with a smaller mesh size (about 1 mm) in the centre of the coarse gauze to prevent litter and soil samples from falling into the funnel and dirtying the collecting bottle.

Fig. 4.29 (Upper panel) Bank of Berlese–Tullgren funnels with a series of six 60 W light bulbs above each funnel. (Lower panel) Bank of funnels with the canopy holding the light bulbs removed to show the funnels and the collecting bottles at the base containing 70% ethanol underneath (Courtesy James Pryke).

The samples can be placed in the funnel according to a set mass or volume. This makes quantitative comparisons between sites possible. However, it is essential to bear in mind that the results can be highly biased if the samples are sometimes taken after rain and at another times after a dry spell. In the case of log samples, the woody material can be broken up by hand and then placed inside the funnel. As one usually works with replicated samples the funnels are normally constructed in rows (Figure 4.29). This also means that extraction of the insects and other fauna from several samples can take place at the same time.

Working on the same principle as the Berlese–Tullgren funnel, and where electricity is not available, an alternative technique is the *Winkler bag*. This is simply a gauze-covered frame shaped into a funnel at the bottom and at the end of which is placed a collecting tube with 70% ethanol. Inside the frame is placed a bag made of 4 mm mesh that acts as the retaining sack for the litter sample. The whole of the Winkler bag is suspended from a tree or a frame. As the litter dries out, the insects and other invertebrates move down and move through the mesh of the inner sack and tumble down into the collecting bottle at the bottom. Quantitative data can be obtained by placing a set weight or set volume of litter or soil in the inner sack. The whole bag is lightweight and can be folded up very easily, making it ideal for expeditions to remote areas.

A *ground emergence trap* can be used to sample emerging insects from the litter layer and soil. It is particularly useful at the start of the summer season and after rains in dry areas. The trap consists of a tent made of black gauze pegged to the ground with a collecting bottle placed in the top corner to collect the insects, especially flies, emerging from the ground.

A similar principle can be applied to logs using *trunk emergence traps*. These are constructed by wrapping the entire circumference of a random 30 cm long section of the log (although never closer than 50 cm to the end of the log) and the trunk of snags with polypropylene weed barrier cloth, which permits moisture and air to pass through, but not light. Pieces of strong wire are inserted into holes in the log to keep the cloth away from the log, and the cloth sleeve wired at either end to confine emerging beetles. Foam or carpet underlay is placed between the cloth and the trunk, under the wire seals, to accommodate the rugosity of the bark. The insects are collected in a 250 ml clear plastic bottle attached by a screw thread to a lid at the top end of the sleeve, and half-filled with 50% propylene glycol. Beetles emerging from the log fly towards the light and are caught in the bottle (Johansson et al. 2006).

Finally, it is necessary to mention termites as a special case (the same principles also often apply to ants). They are eusocial insects and require special collection techniques (Jones et al. 2005 provides an overview). Of course not all termites live in the soil; some are arboreal or live inside rotting timber. With termites it may not be necessary (or appropriate) to sample individuals. Instead their mounds can be recorded, including mound size, density, and dispersion. Termite mounds can also be dug up, but bear in mind that searching for the queen can involve a very deep dig indeed, as she may be several metres below the soil surface. Jones et al. (2005) recommend that a quick way to make a rough estimate of termite population size is to take a sample of known volume of the mound, extract and count the termites, and multiply up to the volume of the mound. The problem is that mounds can be irregular in shape. This is partly overcome by the mounds being assumed to be regular shapes such as hemispheres or cones (or more accurately using photography and image analysis software). The initial sample must be taken quickly, before the alarmed termites retreat further inside the nest. They can be sorted from the nest material by hand searching or by floating the individuals out in water in a flotation tray. Caution is required because termite occupation of the upper part of the mound depends on the time of day, temperature, and whether it is a dry or wet period.

4.9 Sampling aquatic insects

Insects are abundant in the aquatic environment, particularly where the water is shallow and with an abundance of water weed. Sampling under these conditions requires a very strong *dip net* (Figure 4.30). The most versatile water net is triangular, about 30 cm across at the far end, with a frame made of stainless steel. The bag is best made from stainless steel with 1 mm gauge mesh. Obviously the handle must also be very robust. Such a net can be pushed vigorously into water weed to dislodge insects.

Many insect larvae and other invertebrates live on the bottom of water bodies. Sampling of these is well summarized by Southwood and Henderson (2000). They point out that there are five principal variables that affect the choice of sampler for bottom fauna: (1) the type of animal being sampled; (2) the nature

Fig. 4.30 Dip net used for catching aquatic arthropods: it must have a sturdy frame and tough gauze (Courtesy Ricky Taylor).

of the bottom substrate, whether soft or hard; (3) the speed of the current; (4) the depth of the water; and (5) the objective of the study.

From a conservation perspective, rivers and wetlands can be assessed for ecosystem health based on the composition of the benthic macroinvertebrates (Hauer and Lamberti 2007). Each taxon (usually family in the case of insects, but a higher taxon for most of the other invertebrates) is assigned a sensitivity score (see Gerber and Gabriel 2002 for an example) and the total score reflects the condition of the water body. A much more robust and meaningful comparative score is the *average score per taxon (ASPT)*, which is simply the total of the sensitivity scores at any one site divided by the number of taxa recorded. Smith et al. (2007) provide a worked example and application of this approach.

To obtain the macroinvertebrate score for river assessment, the dip net mentioned above is used as a *kick net*, where the net is held upright, and then, immediately upstream of the net, the stream bottom is stirred up by moving your foot around, until all the bottom is stirred up in a standardized set of movements. The net is then searched to remove excessive debris and large insects removed with soft forceps. The contents can also be tipped into a white tray of clear water. Specimens can be sorted in the tray using a fine sable-hair brush and specimens removed with soft forceps and, after being counted, returned to the water, or, where needed for later identification, placed in vials with 80% ethanol. This technique, when employed over a set time period (usually 15 min) or over a particular area of water, provides much quantitative and comparative data. When a set time period is used, all microhabitats at the site being assessed are included in the 15 min period. As species-level assessments provide much more sensitivity than the use of higher-taxon surrogates, and for measures of ecological integrity, selected individuals of insect larvae can be reared through to adulthood in specially adapted aquaria. This method enables matching of any particular larva to an adult, and is particularly useful in the case of dragonflies, details of which are given by Rice (2008) and Samways (2008).

Emerging aquatic insects can be sampled, even quantitatively, using an *aquatic emergence trap* placed over the surface of the water (Freitag 2004). As with many terrestrial traps, such a trap is sampling continuously once it is in place, and has the advantage of collecting aquatic insects that have emerged from the water as adults and might otherwise have dispersed. It is essentially a fabric tent that is placed over the surface of the water, usually adjacent to the bank. There are various designs depending on intended use (see Freitag 2004), with the simplest one being a frame of plastic tubing and a nylon tent that is floated at the shore to sample emerging dragonflies (Figure 4.31). The advantage of this system is that the larval 'skins' (*exuviae*) are kept on the inside of the tent and can be matched to the adult that is also captured in the tent.

Fig. 4.31 Aquatic emergence trap, being used here to sample emerging adult insects. As the insects leave the water to become adults, they fly up to the top of the tent where they are directed by a Perspex funnel into a chamber made out of two plastic jars facing each other, with a hole cut into the side of the top one. If the insects need to be kept alive, no preservative is placed in the chamber. Otherwise, a preservative such as 70% ethanol is introduced. (Courtesy John Simaika).

Various other methods are also available for sampling bottom fauna, including removal of set amounts of substrate using, for example, a grab, which is a piece of apparatus which acts as a pincer to literally grab a standardized amount of bottom substrate. Various types of trawls and dredges are also available for sampling along set distances across the bed of still water (Southwood and Henderson 2000). Many of the vast number of techniques along these lines have been developed for specific purposes. Some, such as trawls and drags, can be very disturbing, even destructive, of the habitat, and must therefore be used with caution in a conservation context. Some may only be partially effective, with certain species escaping from the haul.

A very useful method for sampling in still and clear water, which is very non-intrusive (and therefore very effective for determining exactly where in the water body and where among plants certain insect species are living), is by diving using underwater scuba gear. Sampling can be done in various ways, with the lead originally coming from coral reef assessments. Sampling can be along set transects under the water (e.g. a 50 m tape measure can be laid out and individuals at each 10 cm can be recorded; or, individuals can be collected using the insect tongs mentioned in section 4.4 and individual positions recorded). This method can generate much detailed and accurate information relatively quickly. The deployment of scuba gear was used, for example, to assess where trout food was distributed around a reservoir—as it turned out, this was in the 0.5 m weedy margins (Samways et al. 1996), leading to trout fishermen immediately changing where they cast their flies!

4.10 Sound recording

Many insects produce sound. This can be very useful for various conservation assessments, especially as many orthopteran species produce high-intensity sounds, and, as a group, they tend to be sensitive to landscape change (see Samways 1997a for a review). Recognition of the specific nature of orthopteran song and the development of lightweight, tight-focus recording techniques has led to incorporation of sound recordings of these insects into taxonomic analyses and evaluations. Flowing from this, the use of insect song can be developed into assessments of habitat quality. Table 4.3 gives an initial categorization of song types. The relative proportions of these, for example, can give an indication of how the landscape has changed in favour or not of the various species. Methods are currently being developed using digital spectrograms to bring much more detailed focus to these acoustic assessment methods.

When recording insects, it is important to bear in mind that their song, with its high-frequency components, rapidly attenuates in vegetation, especially the broad-leaf variety. This means that one of the golden rules of recording insect song is to make sure that the microphone is as close to the singing insect as possible. There is a trade-off here, because the closer you get the more likely

Table 4.3 *Song characteristics for major taxonomic groups of Orthoptera (from Riede 1998)*

Taxon	Spectrum	Frequency range (kHz)	Infrastructure (human ear terms)
Tettigonioidea	Wide	5–101	Any
	Narrow	1–101	Note and chirp
Grylloidea	Narrow	1–11	Trill
Acridoidea			
Gomphocerinae	Wide	2–40	Zip, buzz, and rattle
Oedipodinae	Wide	2–40	Tick and rattle
Acridinae	Wide	2–40	Tick and rattle
Hyalopterix	Narrow	3–5	Trill
Romaleinae	Wide	2–40	Rattle
Pneumoridae	Wide	1–8	Rattle and trill

the insect is to be disturbed and stop singing. One way to overcome this is to suspend the microphone from the tip of a fishing rod and lower it over the singing insect. The insect is far less disturbed when a small object descends from above than when a bulk approaches from the side. Be careful not to break or crunch any vegetation underfoot, as this gives out a sharp, transient sound to which orthopterans are very sensitive. For nocturnal singing insects, and after recording their song, you may wish to retain the specimen, in which case you can search for the insect in the vegetation with a head torch and catch it with the insect tongs mentioned in section 4.4.

Use of a parabola, and especially the immensely useful bat detector, for capturing insect songs is generally acceptable for most ecological and conservation applications, especially as most of the insect's energy is put into producing sound around the carrier frequency, which includes the audible range and, with many bush crickets (katydids), also or even exclusively in the ultrasonic range (Hung and Prestwich 2004). Diwakar et al. (2007) have even shown that an experienced listener can accurately record the various songs of audible crickets for conservation assessment purposes.

Simple chirp counts may be used at low densities of grasshoppers, but may cause confusion when large numbers are present, because of difficulties in distinguishing the songs of individuals or species. However, recent developments in automatic recognition bioacoustic techniques, by which parameters such as chirp length and pulse length can be extracted from recordings (Schwenker et al. 2003), have led to improved possibilities for species identification and enumeration. Work on British grasshoppers (Chesmore 2004) showed that species identifications can approach 100% accuracy under low noise conditions, using identifications based on artificial neural networks. Gardiner et al. (2005) considered that acoustic techniques have been under-exploited for monitoring Orthoptera, and that these have important roles in identifying the presence of (1) cryptic species, (2) non-macropterous individuals (which may not be amenable to flushing), (3) species that are very rare and otherwise likely to be overlooked, and also in habitats where more traditional sampling methods are difficult to implement. Advances continue to be made rapidly, with moves toward automated identification systems incorporating abundance estimates. Riede (1998) and Riede et al. (2006) have given much relevant background to the subject, emphasizing the values of acoustic parameters in inventory, as a firm estimate of species richness that helps to overcome gaps in formal descriptive taxonomy, and as tools in monitoring habitat use (and therefore habitat quality) by Orthoptera.

4.11 Summary

It is important to bear in mind that any one technique will only collect a particular subset of fauna. Thus, depending on the research question being posed and the specific aim of the project, it is often wise to use a variety of techniques for different suites of taxa. For comparable catches and therefore comparable data, techniques must be *standardized*, e.g. for placement, design, duration, bait type and quantity used. Individual traps or technique applications should generally not interfere with each other (e.g. be placed too close together or repeated to soon after each other). Care must be taken to construct, install and apply each trap or piece of equipment so that there is minimal sampling error and disturbance to the environment. Care should be taken when interpreting the meaning of number of individuals of each species caught, for example, as either abundance or density. In most cases trap-based estimates are density estimates (number of insects per unit area or volume), rather than actual abundance estimates, because most trapping techniques have a fixed area over which they operate. Time of day, season, and weather conditions generally have a significant impact on catch, as does habitat structure and landscape topography, and these must be standardized. More than a single technique may be necessary to optimize inventories (called *sampling sets*). *Pilot studies* are important for testing the efficacy of a technique, unless it is a standard, well-known, tried and tested method. And lastly, do bear in mind that replication must be done in a way to avoid *pseudoreplication*, as mentioned in Chapter 3.3.4.

As a corollary, we should mention that there are a multitude of techniques, variations, and minor modifications of specific techniques for collecting insects. Most of the traditional ones are discussed by Southwood and Henderson (2000). However, such techniques are often modified by individual researchers and for individual studies, to suit the specific purpose and context of that study. In this chapter we have provided only a broad overview of a range of techniques available and commonly used examples of each.

5

Measuring environmental variables

5.1 Introduction: selection of variables

Insects respond to a range of conditions around them. Many insects have precise requirements for food (such as a particular host plant or dung type) and less readily quantified environmental requirements, such as suitable *microclimate*. Their incidence, abundance and distributions are governed strongly by such features. These conditions (broadly, *environmental variables*, EVs) may be determined by other organisms (*biotic variables*, such as interactions with plants and natural enemies), the physical world (*abiotic variables*, such as temperature or stream flow), or a combination of both (such as litter depth). Abiotic and biotic variables may combine to strongly influence the seasonal activity patterns of insects, and favourable conditions dictate the rate and timing of such development. Less predictable events—such as fires or floods—may be superimposed upon these patterns and provide opportunities for insects to exploit novel conditions and expand their activities or, conversely, cause direct mortality. The importance of timing of weather events is clearly seen, for example, in the desert, when flowers appear in profusion only some time after the rain, making conditions appropriate for pollinators that then appear rapidly and in unexpectedly high numbers.

The focus of this chapter is on measuring EVs that affect insect life (including activity patterns, survival, natality, population dynamics, and interactions with other species), both to its benefit and to its detriment. It is the balance of these two that is often critical for conservation, where of course we wish to create and sustain conditions that are advantageous to the species in question. Indeed, a starting point in conservation is that physical and biological conditions must be optimal for survival and success of the species. In turn, for long-term conservation, environmental conditions must be managed so that they remain optimal. This may involve interventions such as deliberate burning, selective grazing by mammalian herbivores, or the removal of alien plants. A note of caution is perhaps needed, because many insects of conservation interest are ecological specialists, and as a consequence are rare and potentially vulnerable. Many species are known from few, even single, sites that are presumed to be remnants of a previously more widespread range. Because we have no other model, we assume that the environmental conditions of the occupied site are optimal for the focal insect. Management then seeks to define and replicate these conditions to provide optimal habitat for the species. However, the insect may be only just surviving in the last remaining but impoverished habitat, rather than thriving (Davies et al. 2006). Subtle appreciation is needed to understand this, as otherwise we run the risk of replicating suboptimal habitat conditions for species.

As stated by Wikelski and Cooke (2006), successful species conservation relies on an understanding of the physiological responses of species to their changing environment. Achieving such understanding will involve both measuring the environment and understanding how environmental change affects the physiology of the species. The effect of EVs on the physiology of the insect (other than direct mortality) will then translate into changes in, for example, individual behaviour, fecundity, and generation time, that will in turn translate into changes in population size, distribution, and species assemblage structure. Understanding how insects respond to their abiotic environments has become particularly important in the face of the current rapid rate of climate change (Chown and Nicolson 2004).

The role of EVs can be illustrated using the reintroduction of insects into a habitat after the local population has become extinct. This procedure is predicated on the causes of earlier decline being known and countered so that the loss will not recur. Elsewhere the species may be surviving, and hopefully thriving, and it is there that we can assume that conditions are at least suitable, if not optimal. These suitable conditions are provided by a combination of site parameters, such as adequate patch (habitat) size, good patch quality, and minimal patch isolation. This would be a *source habitat*, where the species is breeding and from which individuals are able to disperse to establish in adjacent patches. This contrasts with a *sink habitat*, where conditions are suboptimal and the local population is as a consequence less viable, with few to no individuals being produced and surviving in the patch. The existence of source habitats is especially important in the long term, because if conditions in at least some patches are not optimal, the species may well go extinct. Site features, such as those above, are one category of EV. Resource supply is another category, and 'habitat quality' may be regarded in terms of the coincidence of the critical resources and a climate regime suitable for the growth, maintenance, and reproduction of the insect species.

The identification of source habitats is an important first step to understanding and being able to measure what are the suitable or optimal environmental conditions for a species. It is also here that as many as possible of the important EVs should be recorded. These data then provide a yardstick against which other habitats (suitable or unsuitable for the focal species) can be compared. It will most likely be necessary to measure these conditions over a whole year, and/or at critical times, such as at high water levels or during droughts in marshland.

When comparing conditions between one patch and another (e.g. a source habitat and a sink habitat), or trying to understand what drives seasonal and longer-term population dynamics of a species, the key question becomes, which EVs should you measure? In turn, if the goal is to maintain habitat quality in good condition over the long-term, the question becomes, which, when, and how often, should critical EVs be measured? The measurement of EVs is also common in assemblage and community studies, where the question of interest has to do with which EVs are responsible for patterns of species richness, abundance, or community structure.

The choice of which EVs to measure, when, and at what interval, will depend on the precise ecological question you are asking or the conservation approach being taken. It is all too easy to attempt to measure as many EVs as possible, without understanding the rationale for doing so. Species are likely to vary in their requirements, and more than a single, often interacting, EV is usually important to them. Indeed, EVs often interact with each other in a non-linear fashion. For example, temperature and relative humidity in a microhabitat may increase together until a point is reached where the grassland habitat dries out, at which time temperature may still continue to increase but humidity drops. Different species respond to these changing conditions in different ways. In the case of drying grass, one species of bush cricket, for example, may move to neighbouring bushes while another may move to longer grass. This is precisely why it is so much more valuable to use species rather than higher taxonomic levels when monitoring environmental conditions. A range of species will display a range of sensitivities to the various and changing conditions surrounding them. Where your study involves whole assemblages or communities, or even particular food webs, you will need to ascertain which EVs shape such groupings, i.e. which EVs are associated with high and low species richness, different assemblage structures, abundances, and interactions.

Elevation, latitude, and longitude are often measured as relational (independent) variables, especially in studies conducted over broad spatial scales. However, these are not true EVs. Rather they are used as surrogates (or proxies) for a suite of abiotic EVs that are well known to vary in fairly predictable ways across gradients in elevation, latitude, or longitude. Similarly, slope and aspect (which can be measured with a clinometer and compass respectively) may be surrogates for drainage conditions, wind exposure, timing of exposure to direct sunlight, and amount of sunshine received daily. Wherever possible it is better to measure EVs directly than to rely on surrogate measures. For example, elevation is a generally good surrogate for temperature. However, fine-scale variations in temperature (microclimates) across an elevational gradient are often not well predicted by a measure of evelation.

Deciding on which EVs to measure, and when, is the subject of this chapter. More broadly, you may need to choose those EVs that characterize, for example, 'high-quality indigenous grassland' or 'high-quality ponds', rather than necessarily considering the representative *focal species* living there, to define a broad suite of umbrella values that must be sustained.

Much of need for selection of EVs will initially be site-focused, and the nature and characteristics of any particular site may dictate the features of interest, e.g. closeness and size of rocks on a hill in a savanna landscape. It is often worthwhile taking a series of digital photographs of the features of likely interest (not least to act as a memory aid in the final selection of which particular features to investigate) as a record of the site, and as a foundation for assessing changes in the future.

5.2 Conceptualizing environmental variable data

Before discussing the 'what', 'how', and 'when' of measuring environmental data, let us plan ahead and visualize where you are going to enter these data in the data matrix relative to the insect data itself. The important point here is that the insect abundance or diversity data (see Chapter 9) are often collected at the same time and, of course, in the same place as the various EVs. EVs can therefore be considered *relational data* because they relate to the insect abundance or diversity data, and usually during analysis one relates the insect data to the EV data in the hope of finding a relationship or correlation between them. Ludwig and Reynolds (1988) provide an excellent and highly recommended introduction based on a *data matrix* view (see also Chapter 9.3 on the assemblage matrix and relational data). In principle, the analysis of insect data is based on measurements of abundance of species across *sampling units* (SUs) (see Chapter 3.3). A SU is a discrete, sampled part of a habitat. It may be the number of insect individuals in a particular quadrat, transect, unit of soil, volume of log, volume of water, a particular number of fruits, or so many leaves, etc. The SU is your basic working unit in the matrix. This matrix is arranged with SUs listed across (columns) and species listed down (rows) (a simplified example is given in Table 5.1; see also Chapter 9). The important point here is that at the EV matrix is constructed with the same columns as the species/assemblage matrix (for convenience, usually at the bottom of the data matrix; see Ludwig and Reynolds 1988) so that analyses can be conducted by making comparisons of rows and columns across the two matrices (Table 5.1).

You may also wish to have a seasonal perspective, with samples from spring, summer, autumn, and sometimes winter, when influential EVs also vary considerably. This is strongly advised as a minimum basis for understanding the life histories of such short-lived animals. In this case, prepare four data matrices, one for each sampling period. If time allows (because remember that you have to sort all the samples and enter the data into the matrix), you may wish to sample twice during each of these quarterly periods. In practice, a good compromise is to sample six times a year, i.e. every two months. Whatever time period you choose to make your species assessments, the same must be done for the EVs. Importantly, for sound statistical analysis, every SU should as far as possible have its own independent measure of each EV.

Depending on the ecological or conservation question, it may be appropriate to focus on certain times of the year: for example, sampling intensively in summer and perhaps not at all in the winter. Because of the fine-grain nature of insect distributions within their overall habitat, tempting surrogates (such as using temperature data from the nearest weather station, or even several metres away) can sometimes be of little value.

To achieve broad spatial representation, many one-off samples of species and EVs are often used in preference to repeated sampling at fewer localities (an optimal compromise may in some cases be to sample a few sites several times and the reminder only once). Again, the approach adopted must be driven by the conservation or ecological question that you have decided to focus on (see Chapter 3). And this is where you must be very astute with your environmental data. Do not just measure EVs for the sake of it, but rather measure those that are likely to be meaningful. Of course, when the final analysis of these data is done, some of the EVs will turn out to be important whereas others may not.

5.3 Choosing the environmental variables

Deciding which EVs to measure is rarely easy, and the difficulty of choice is rarely mentioned in textbooks. One of the problems is of course that in some cases you do not know which EVs are indeed important until you start measuring, gathering, and analysing the data. In some cases one may be guided by

Table 5.1 *A subset of a data matrix in which species and environmental data are entered relative to the sampling unit (SU) in which they occur. In this example, there are 18 SUs. Each of these SUs may, for example, be a quadrat (or the pooled data from say, four quadrats or four pitfall traps). In this matrix there are six SUs (six replicates) in each of the three different habitats (A–C) that are being compared. These habitat types in turn would have to be replicated (to avoid pseudoreplication), say three to six times, for statistical analysis (leading to a larger matrix). In this example only five insect species were sampled, but of course it could have been any number (n). Species 2 is clearly rare (N = 9 individuals) in the sampled area and it was not sampled at all in habitat B. Four environmental variables (EVs) were measured here. EV1 is percentage shade caused by tree canopies over the focal SU at midday. EV2 percentage bare ground, and inversely proportional to this is EV3, percentage ground cover. Also related is EV4, which in this case is maximum height of ground cover. This data matrix is the starting point for all subsequent hypothesis testing and analysis, whether simply descriptive, statistically univariate, bivariate, or multivariate. N = the total number of individuals per species across SUs and habitats*

	Habitat A						Habitat B						Habitat C						N
Sampling Units (SUs)																			
Species	1	2	3	4	5	6	7	8	9	10	11	12	13	14	15	16	17	18 (...n)	
1	14	8	21	4	16	9	1	0	0	1	2	0	47	40	22	63	71	108	427
2	0	0	1	0	0	0	0	0	0	0	0	0	0	2	2	2	1	1	9
3	7	14	7	6	7	2	28	24	32	17	31	27	6	4	4	3	2	7	228
4	2	2	0	3	4	2	9	7	11	14	9	7	0	0	0	0	0	0	70
5	107	99	78	111	123	166	105	115	121	97	91	142	14	12	12	14	10	18	1435
S = number of species	4	4	4	4	4	4	4	3	3	4	4	3	3	4	4	4	4	4	
Environmental variables (EVs) 1	70	75	80	75	75	75	60	65	60	55	60	60	80	80	80	85	75	80	
2	58	62	61	57	40	67	28	24	27	27	26	26	84	77	81	80	81	82	
3	42	38	39	43	60	33	72	76	73	73	74	74	16	13	13	19	20	18	
4	7	8	7	8	7	7	12	22	19	21	17	13	5	5	4	5	4	6	
...																			
n																			

literature on related taxa and sometimes even in similar climatic areas or disturbance types. Dung beetle assemblages, for example, are well known to be sensitive to shade and a study on dung beetles may consider canopy cover or light intensity as an EV to be measured. Soil temperature just below the ground surface has been shown, for example, to be an excellent predictor of activity patterns of the invasive Argentine ant (Human et al. 1998).

EVs are chosen according to the question you are asking. Say, for example, you are studying the effect of woody invasive alien plants on ground invertebrates. One EV that you would measure is percentage shade cover. In contrast, wind speed at the particular time you are visiting the site may be largely irrelevant to ground-dwelling invertebrates. In contrast, where you aim to measure flight patterns of butterflies, moths, or flies, it becomes vitally important to measure prevailing wind, cloud cover, and temperature conditions. You would therefore need to measure flight activity of the butterflies and at the same time and place measure wind speed and other EVs. Australia's Golden sun moth *Synemon plana*, for example, has very finely timed flight activity. Males fly only in calm conditions, in bright sunlight, with temperatures above about 20 °C, without rain, and for a few hours in the middle of the day (Gibson and New 2007).

To emphasize this important point, let us take another, rather different, example. Imagine that you have sunk *pitfall traps* into the ground. Should you measure soil or air temperature, and if so when and how often? Start by thinking it through. What relevance would a point measurement at any one moment in time be to the current pitfall catch? This question applies to both soil and air temperature. Insects and other invertebrates caught by pitfalls may be dependent on the temperatures last night or yesterday afternoon, and in fact perhaps even 2 weeks ago. It may have rained in the night, perhaps stimulating high activity. Present insect activity is generally the net result of the various influences and impacts of both previous weeks and previous seasons.

It is critical that you work through which EVs are likely to be important and whether measuring them the way you are is appropriate and will provide a sensible, meaningful answer. This 'thinking through' is of course also part of hypothesis formulation (Chapter 3). Furthermore, bear in mind what is critical for one species may not be so for another. A salient reminder of this is the tropical African ground beetle *Africobatus harpaloides* that tolerates only a 0.9 °C change in daily temperature before dormancy is induced (Greenwood 1987). This illustrates just how critical temperature is for some species. Unfortunately this sort of physiological information is not yet available for the vast majority of insect species (see Chown and Nicolson 2004 for an overview of insect physiological ecology).

It helps considerably to understand how the insect results relate to EVs after entering your EV data into the data matrix. In Table 5.1, EVs 2 and 3 are the inverse of each other simply because where there is no vegetation cover, there is bare ground. Other EVs may not be so directly equivalent. Air temperature 50 cm above the ground may be related to height of the grass sward, yet the relationship is unlikely to be linear. For example, the temperature will be different, but not necessarily proportional, when the grass is, say, 5, 10, 20 or 50 cm tall. This is because the height of the sward itself may affect the *microclimate* on the ground, and influence the vitality and activity of *epigaeic* insects such as certain beetles and ants.

Selecting EVs is done *a priori*. Only when the analyses have later been completed will it become evident which EVs are statistically significant. A pilot study can be very helpful for deciding which EVs to measure in the full study and which can be left out. Rarely is one EV alone the sole driver of insect presence/absence or abundance. Nevertheless, one EV may be a key factor, as may two or three EVs combined. It is important for conservation to determine which of the EVs are significant (Figure 5.1). EVs that are positively and significantly related to insect population size, total abundance, or species richness would be encouraged in management plans, whereas EVs that are negatively related to insect vitality factors would have to be removed or reduced. Identifying a key threat and removing it can lead to an immediate recovery of some insect species.

Fig. 5.1 Before it was rediscovered in 2003 the Ceres streamjack damselfly *Metacnemis angusta* was feared extinct, having last been seen in 1920. To this day, it is still only known from one pool cleared of invasive alien riparian trees, in a river braid, with the water plant *Aponogeton* and cleared of invasive alien riparian trees. The point here is that the key threat was identified: invasive alien plants completely shade out the habitat. By addressing the key threat, and restoring appropriate environmental variables, especially sunlight and growth of indigenous vegetation, the species recovered.

This was shown categorically when invasive alien woody plants were removed and highly threatened damselflies immediately returned to the site (Samways and Taylor 2004).

Sometimes, when you measure specific EVs in the immediate vicinity of a large, mobile insect (such as shade cover, air temperature, and water flow rate close to a perching dragonfly, or when there is a key impact such as road dust on parasitoids; Figure 5.2) you may find that your EVs account for as much as three-quarters of the variance in the insect data. However, for many insect–EV interactions, such high explanatory power (strong correlations between the insect data and EVs) is seldom obtained. Nonetheless, weak correlations may be significant, suggesting that important EVs remain unmeasured. Selecting various EVs or certain combinations of them for statistical testing of significance will often single out which are the really important ones. However, do bear the following three issues in mind: (1) The EVs chosen may not be the only important ones, and other important EVs, both abiotic and biotic, may

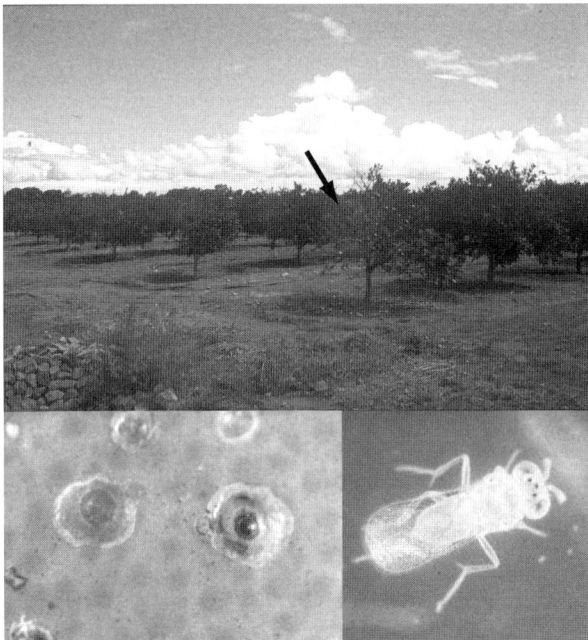

Fig. 5.2 The line of citrus trees adjacent to the dust road suffered defoliation (above, arrowed) as a result of a severe Red scale *Aonidiella aurantii* (bottom left) infestation. Dust is a key environmental variable inhibiting the searching behaviour of the parasitoid *Aphytis africanus* that normally keeps the Red scale under control.

not have been measured. (2) The abundance or diversity of your focal population or assemblage may be influenced by, say, a pathogen or competition (biotic interactions), that may, in turn, be responding to different EVs than the EVs to which the focal subject is directly responding. (3) Finally, be very aware of autocorrelation, where different EVs are inevitably spatially or temporally associated. Temperature and shade, for example, usually fluctuate together over both space and time. This means that these two EVs are not independent of each other, and during analysis must be dealt with carefully and usually separately (refer to a general statistics text on the subject of colinearity, or non-independence, among relational variables). Similarly, because different EVs are measured in a range of, often very different, units (e.g. percentage cloud cover, mm of rainfall, degrees Celsius temperature), EVs may need some form of transformation before inclusion in a statistical test (as in fact do insect abundance and richness data) (see Crawley 1993 and Quinn and Keough 2002).

A list of possible main and commonly measured EVs to consider including in your study, where relevant, is given in Table 5.2. As a general rule, where possible try to obtain quantitative measures of the EV of interest rather than using a qualitative measure. Sometimes, qualitative estimates, or categories of condition, are the only way to measure the EV of interest. However in many cases quantitative alternatives are available. For example, if given the choice between measuring light intensity in a forest patch vs estimated percentage canopy cover (Figure 5.3), choose light intensity (Figure 5.4) because it will provide a more robust (and often more accurate and sensitive) measure that is easier to deal with statistically when it comes to analysing the data. Similarly, for example, weigh standardized area-based samples of leaf litter to establish litter density, rather than estimating litter cover as, say, 'high', 'medium', or 'low' (the latter provides a categorical EV and the former a continuous one). Basically, wherever possible avoid human subjectivity when measuring EVs.

Some general principles for measuring EVs include:

- It is better to choose a wide range of EVs, several of which may be complementary. Indeed, measuring a wide range of EVs is often not actually a major practical issue because so many can be measured simultaneously with increasingly sophisticated and affordable electronic instruments.
- It is essential to replicate your measurements (*replication*) as far as possible in the same way (and at the same spatial and temporal sampling points) as the actual insect samples. Several measurements can also be taken in the same *microclimate* or habitat patch, and the average of these measurements can be used in the analysis. In this case the multiple measures are not

Table 5.2 *Some essential environmental variables (EVs) to consider measuring when undertaking a study in insect ecology and conservation*

Category	Comments
Abiotic environmental variables	
Soil	
pH	Can be determined with a pH meter
Temperature and moisture	Can be measured with an appropriate probe; relevance to the study must be considered very carefully as the insects and other arthropods present may be present as a result of past temperature and moisture events, e.g. the point of sampling may have gone through a climatic bottleneck in the past such as a high temperatures and drought, or inundation
Compaction	Can be measured with a penetrometer
Physical and chemical composition	Soil samples must be fresh; essential to consult an expert; for introductions to techniques: see Tan (2005) and Carter and Gregorich (2007)
Water	
pH, dissolved oxygen, and conductivity	Can be measured with a commercially-available water analyser, which must be carefully calibrated; for background information see Hauer and Lamberti (2007)
Turbidity	Traditionally measured with a Secchi disk; water samples can be assessed using a portable turbidimeter;
Pollution	A specialized field; for an introduction to techniques, see Nollet (2007)
Above ground	
Temperature and humidity	Temperature probes are available, but it may be more meaningful to have a continuous set of data from a data logger or even a local weather station; remember that the current insect assemblage will be strongly determined by past temperatures, especially adverse conditions. Similar issues arise with relative humidity, which can be measured with electronic hygrometers. Insect distributions can be defined by a combination of temperature and humidity; for example, high temperatures and low humidity can be a lethal combination, and measured as the Saturation Deficit Index (see text). Microclimate is generally highly relevant to insects, i.e. the climate within the specific habitats or substrates in which they are found

Table 5.2 *(continued)*

Category	Comments
Dew	Heavy morning dews can be a massive mortality factor for small insects and mites in exposed situations (Samways (1979)
Wind	Wind meters are available for measuring current wind speed, which is only meaningful for behavioural studies, and for standardizing sampling, e.g. sampling is only done on windless days ; for ecological studies, it may be more important to have long-term records from a nearby weather station, or, where extent of exposure is required, it is necessary to install data loggers which monitor wind continuously; wind can interact with rainfall, and together, they can have storm impact
Cloud cover	Measurement of cloud cover is very important when sampling flying insects, and it is important to monitoring under standardized conditions, e.g.<10% cloud cover; long-term cloud cover data can be obtained from local weather stations; related to cloud cover is insolation, which is for example the number of sunlight hours a particular microhabitat experiences. Insolation can also be measured using an infrared thermometer, which records the heat being emitted from the surface of the ground
Pollution	Some important variables to consider are particulate deposits, toxic substances (including insecticides), smog index, dust along road verges or on leaf surfaces
Geographic	
Elevation	Often a critical measurement for many insect studies, and is usually expressed as metres above sea level (m a.s.l.); NB 'altitude' refers to height in the air, whereas 'elevation' is height of the land.
Aspect	Very different results can be obtained from different sides of e.g. a mountain or side of a railway cutting, and where relevant this should be stated
Slope	The angle of a hillside is intimately associated with the vegetation vis-à-vis soil or rock cover and may be very influential for insect habitat and flight path. Slope may be measured with a clinometer
Disturbance	
Fire	In some ecosystems, especially savanna, fire frequency and intensity strongly influences insect assemblages; also consider the long-term fire history, which strongly influences the vegetation structure and composition, as well as more recent fire events (i.e. over last few years)

Table 5.2 *(continued)*

Category	Comments
Other disturbances	Vehicle impact, human foot traffic impacts, effect of fence lines, excessive mowing, boat impacts on water bodies, building of dams, channelization of rivers, as well as the larger impacts of agriculture and urbanization and their changing of microclimatic EVs are among the many considerations
Seasonal	
Time of year	This is critical to many insect ecology and conservation studies, and especially the Red Listing process (Samways and Grant 2006, see Chapter 6). Normally, spatial issues are considered first, then temporal issues. Do not slavishly repeat sampling every calendar month at the expense of good spatial sampling. Unless it is a fine-tuned monitoring project, researching, for example, timing of eclosion of a particular species, most insect assemblage studies can be addressed by repeat sampling six times per year, or even four times, one for each season
Historical	
Past events	This can be a very important variable, with coupling and decoupling of plant/insect assemblages over time (Ponel et al. 2003). The current sampled assemblage is not a fixed entity but one of continual change. Species are responding to past natural (Hewitt 2003) and anthropogenic impacts (Williams 2002)

Primarily biotic environmental variables

Plant architecture or form

Plant height and density	This is an important, although complex, set of EVs. We strongly recommend consulting a good text (e.g. Kent and Coker 1992) and a plant ecologist, so that you measure plant characteristics in accordance with the aims of your study. See text. Examples are given in Pollard and Yates (1993)
Canopy cover	The amount of sunlight reaching the ground is a very important determinant for many insect species. Sunflecks and tree gaps in forest allow light to penetrate to the lower storey. Tree canopy cover can be assessed using a 0.5 m^2 quadrat with 10 cm-spaced cross-wires held at arm's length and repeated, or with a camera with a wide-angle lens pointing upwards (Pollard and Yates 1993). In the field of insect conservation, plant traits can be very important, for example for some insects only certain plant forms of a particular host plant species are suitable for oviposition (Samways and Lu 2007)

Table 5.2 *(continued)*

Category	Comments
Plant composition	
Species composition	This is intimately associated with plant architecture or plant form. For herbivorous insects the plant species present are an important variable that should always be considered. Unless the study is of particular species with particular associations with certain plant species, it is recommended that only the dominant species (making up 50–90% of the cover) are identified. We strongly recommend careful consideration of the aim of the study and discuss the measuring of plant variables with a plant ecologist. We also strongly recommend that all alien plants, particularly when invasive of natural ecosystems, be recorded, as invasive alien plants are a major threat to many insect species worldwide. Hopkins et al. (2002) make the interesting point that host plants with restricted geographical ranges can also have rare insects
Leaf and twig litter	
Depth	This can be a very important variable for a whole range of invertebrates and should be measured with a rod with a round point such as a knitting needle
Relative humidity	The relative humidity of leaf litter is a very important determinant of the invertebrate community. Under dry conditions, the individuals go deeper into the litter, sometimes even into the soil below. It can be measured using an electronic hygrometer with a probe
Alien plants	
Species, structure, invasiveness	Invasion of habitats by alien plants is a major threatening process that often requires measurement. Besides measuring the effects of such invasives, it is also important to consider how they might be spreading and therefore of increasing impact. There is also the converse to consider: how suppression or removal, when feasible, of such plants might improve the habitat for conservation
Logs	
Age (decay class)	Age of the log is important, as logs of different species of trees tend to converge with respect to their invertebrate fauna as they age
Tree species, size and position (standing or lying)	Different-sized logs, and logs in different positions, can have different combinations of invertebrates. Size is also associated with tree species, as can be texture, e.g. coconut logs have a hard exterior and soft interior whereas sympatric *Pisonia* logs are even throughout (Kelly and Samways 2003).

Table 5.2 *(continued)*

Category	Comments
Relative humidity and exposure	Logs can vary in their moisture levels, which can be measured with a hygrometric probe. Log moisture in turn depends on the exposure of the log to sunshine vis-à-vis canopy shade
Macrophytes	
Cover	Macrophyte cover can be measured with a pin frame or a quadrat frame (see Williams 1987; Kent and Coker 1992)
Composition	Macrophyte species composition can be very important for some species of aquatic insects, and should be recorded
Submerged plants	
Cover	This can be measured under the water with a line transect while snorkelling. Where the water poses a health risk e.g. bilharzia, waders must be used and the hands covered with waterproof gloves. The floating macrophyte cover is removed and the measurements taken just under the surface. It is not necessary to go too deep as most aquatic insects are in the top 0.5 m
Composition	Plant species composition should be recorded
Disturbance	
Megaherbivore impact	This can be very important, via for example manuring, soil compaction and grazing intensity (see Samways and Kreutzinger 2001). It may also have been important in the past, before indigenous large mammals were hunted out; current domestic livestock grazing can also be a significant variable

Notes: Many of these EVs are interrelated, and may have an integrated affect upon the focal insects of your study. Not all these variables need of course be measured in any one study. These are just some of the EVs to consider. Depending on the specific aim of your study, you may wish to measure some very specific EVs, such as boulder size, presence of large mammals or fish, or rates of human visitation to your site

replicates, but merely repeated measures to ensure accuracy of the mean measure that is eventually used.

- Environmental conditions are inherently variable across the landscape and even within a habitat patch, giving rise to a range of microclimates. When determining the value of the EVs across a habitat, it is essential to take EV measurements in a range of microclimates across that habitat. These measured EVs may be averaged when the aim is to characterize the habitat, although you should **always** keep the raw data values separate, as these may turn out to provide a vital clue in the conservation of an insect species, e.g. specific temperature range required for oviposition.

- Before going into the field and using instruments to collect data, make sure the instruments are calibrated. This is especially important for instruments used for

Fig. 5.3 Different levels of tree canopy cover recorded with a wide-angle lens.

measuring water conditions, as they tend to be very sensitive. Details are usually given with the manufacturers' instructions, and these should be read carefully.

• Equally obvious, be familiar with the operating procedures of instruments before going in the field, and of course carry any essential spare parts, especially extra batteries.

5.4 Measuring air and soil environmental variables

Various aspects of the aerial and soil conditions at any one point location are interrelated, with the plant structure and composition the primary producer linking the two media. Temperature directly affects metabolic processes, growth, and reproduction in insects, and there is an optimal range of temperatures at which these processes and activities take place. There are also lethal temperatures above and below which the insect may die (see Chown and Nicolson 2004). For many insects, lethal temperatures and humidity are closely associated

Fig. 5.4 Young female Blue basker dragonfly *Urothemis edwardsii* obelisking to keep cool in the hot midday sunshine. The key environmental variable to measure is the temperature and humidity of the air (with a shaded probe) in the immediate microclimate.

and can be combined into a measure of climatic severity known as the *saturation deficit index* (SDI). SDI is saturation vapour pressure (at maximum temperature) minus minimum relative humidity, multiplied by saturation vapour pressure (at maximum temperature). In practice, the SDI can be approximated with a high degree of accuracy using the maximum temperature and minimum relative humidity (RH) during each week (Samways 1987):

$$\text{max temp (°C)} - \text{min RH (\%)} \times \text{max temp (°C)}$$

Insects are subject to temperature effects at various spatial scales. The smallest scale is the *microclimate*, which includes conditions immediately around the insect, and is strongly affected by vegetation type and microtopography. Microclimate can vary enormously, both over time and over space and in ways very different from those observed from weather station data. For example, microtemperatures may be buffered, or more extreme, and also change at different rates compared with temperatures recorded in standard Stevenson screens 1.5 m above the ground (used to measure meso- and macroclimate; for a good treatment of measuring climate refer to Bonan 2002). When interested in insect behaviour, distribution, or physiology always try to measure temperature and humidity at a scale closely matched to the scale of the microhabitat that the insect occupies. Insects often move readily in response to these variations, benefiting greatly from their high flight mobility. They may even show extreme behavioural activities, such as *obelisking* in dragonflies to keep cool while sometimes also still maintaining territories on perches at the water's edge (Figure 5.5).

A scale insect on the upper surface of a leaf in the savanna may experience an amazing 60 °C, while one on the underside of the same leaf may only be at 40 °C, which is nevertheless still high. Meanwhile a moth larva inside the crown of the same tree or an ant foraging up the trunk may only be experiencing 32 °C. Yet on the other side of the globe a boreid scorpion fly might be foraging underneath a snow layer, where it is several degrees warmer than on the surface of the snow, especially when the wind blows.

Such point microclimatic temperatures can be measured with a digital thermometer probe, where only the tip records the temperature. Some pointed probes can be inserted into the soil or leaf tissue to ascertain temperatures where cryptic insects may be residing. For most entomological applications, it is only necessary to choose a probe with a temperature range of 0 °C to +40 °C. The development of inexpensive, easy to use, and highly portable miniature temperature and humidity loggers has made the measurement of microclimate temperature and humidity much easier, e.g. see temperature loggers in Fig. 5.4. Dozens of these microloggers can be scattered across your sampling extent and used to generate spatial temperature and humidity profiles. Each individual micrologger is programmed electronically in advance to take measurements at set intervals (from seconds to hours) for days or even months. Once removed from the field the data from each micrologger are readily downloaded using system-specific hardware and software.

Soil temperatures can vary enormously with depth and position around grass tussocks, for example, and if the grass has been burned or not. These microclimates in turn are influenced by mesoclimatic differences, up and down hills, and on one side vs the other (Figure 5.6).

The next largest scale is the *mesoclimate* (Yoshino 1975), which covers several tens to several hundreds of metres and may, for example, relate to conditions on either side of a hill that can vary enormously and dictate the local distribution of insects (Figure 5.6). Finally, there is the large scale, which is over kilometres or larger, through regional to global. Obviously the microclimate is determined in large part by changes at greater spatial scales, as witnessed by the effects of global climate change on insect distributions and *phenology*.

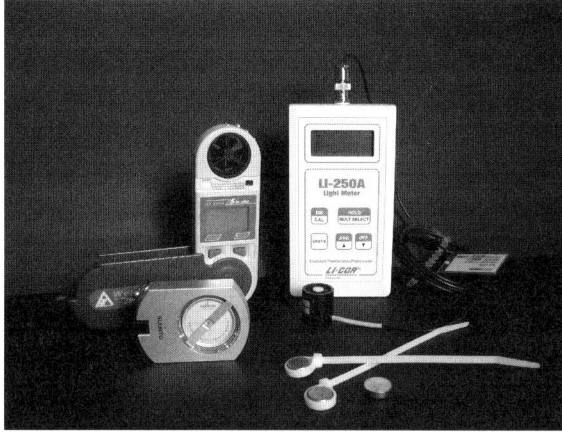

Fig. 5.5 A selection of instruments used by ecologists to record abiotic environmental conditions. From front in a clockwise direction: temperature loggers (plastic cable ties aid their location in the soil (Thermochron iButtons, manufactured by Dallas Semiconductors; clinometer (Suunto); hand-held wind meter, with built-in temperature and humidity sensors (AZ Instruments); light meter with sensor for measuring photosynthetically active radiation or light intensity (Li-Cor) (Courtesy Peter le Roux).

In some cases it is not possible to obtain an independent EV measurement for each SU. For example, you may have 5 independent sampling grids (or SUs) for each of 10 sites, but only one Stevenson screen series of temperatures available to characterize ambient air temperature at each of the 10 sites. Each SU thus does not have an independent temperature measure, and temperature will either have to be excluded from the analysis or the insect sample results pooled across the 5 sample grids for analysis.

For measuring mesoclimatic differences, and to measure temperature changes over time, it may be necessary to install semipermanent temperature (and also perhaps humidity) probes (Figure 5.4). These are either linked to a data logger for electronic capture in the field, or in some models the data are directly down-loadable from the instrument itself. Often the most useful way to illustrate such results in temperature fluctuation in a way that is meaningful for insect ecology is to display the *moving average*, where the original data is smoothed. Davis (2002) gives an excellent background to this. However, such an analysis must be inter-preted with caution because high peaks and low troughs, that may be biologically significant, are lost with such smoothing (see, for example Le Roux and McGeoch 2008). More sophisticated series analytical techniques, such as Fourier analysis, are also available, and again described in detail by Davis (2002).

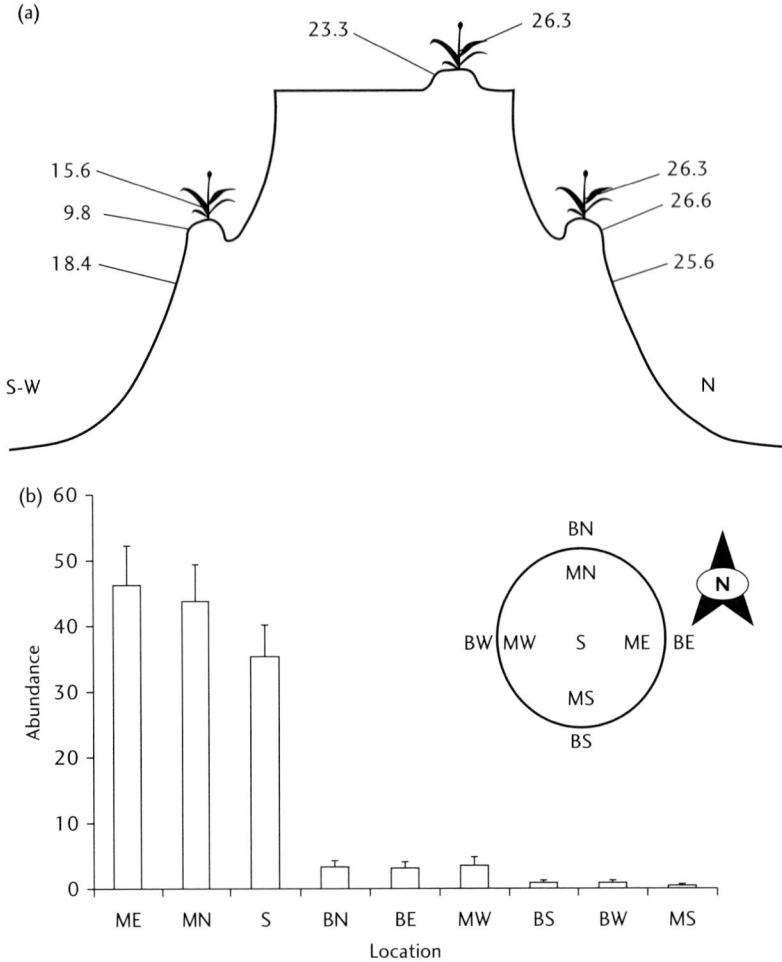

Fig. 5.6 (a) Soil temperatures (in °C, 10 mm below the surface) at about midday when the sun was shining from the north during one winter day on the summit of a hill in the South African Drakensberg. This diagram emphasizes the enormous differences which can occur in the various microclimates across this relatively small landscape, which are of major significance for the small-bodied insects living there. Note: the bumps on the slopes are schematic exaggerations of the perennial root system of grass tussocks. (b) Mean grasshopper abundance was very sensitive to these temperature differences. The locations on the hill were ME, midslope east; MN, midslope north; S, summit; BN, base north; BE, base east; MW, midslope west; BS, base south; BW, base west; MS, midslope west (from Samways 1990).

Wind speed is often interrelated with other aspects of microclimate, including temperature and relative humidity. A breeze can have a strong cooling effect, and it can also affect flight behaviour. This must be borne in mind when trapping insects using the various techniques outlined in Chapter 4. In turn, breeziness may be associated with cloud cover, all of which can dramatically alter an insect's flight behaviour. On Easter Island, the only dragonfly inhabitant, the Globe Skimmer *Pantala flavescens*, reacts immediately when a cloud passes over and drops into the grass. It does not do this on the African mainland, where it will even fly in the warm rain. It seems that there has been strong selection on Easter Island, the world's most remote island, to land safely as soon as breezy and cloudy conditions arise. To be blown out to sea would mean almost certain death. Indeed, many insects have evolved flightlessness as a possible counter to this risk associated with island life.

General light levels can be approximated for insects using standard light meters. Some caution in interpretation is required because insect eyesight is biased away from the red end of the spectrum towards the ultraviolet. It is also important to bear in mind that many insects can see *polarized light*, so they can effectively navigate even on overcast days. Cloud cover is relatively easily estimated in terms of percentage of sky covered (to the nearest 10%). Shade, which is mostly a function of vegetation density, particularly canopy density, is also very important for many insects. It can be estimated using a 0.5 m² wire quadrat with cross-wires spaced 10 cm apart and held at arm's length, and the procedure repeated across the horizon. Digital photography can help with the estimation of both cloud cover and tree canopy, using a fisheye or wide-angle lens pointing vertically upwards (Figure 5.3). Using an image analyser, the area of gaps in the canopy can be accurately quantified.

Soil characteristics are another important group of EVs to measure, particularly but not only for ground-dwelling insects and those that spend some or most of their life in the soil. For example, some Orthoptera will only oviposit in soils of a specific type, humidity, and temperature. The litter layer on top of the soil is also important. It is usually composed of decomposing plant material, such as leaves and sticks. The depth of the litter layer (for comparative purposes across sites within a study) can easily be measured using a bluntly pointed rod, like a large knitting needle. The rod is applied with equal pressure (e.g. under its own weight or with a set weight added) at each SU and the depth of penetration measured.

Soil litter per 0.5 m² can also be removed and thoroughly air dried to obtain an estimate of dry biomass per area. Soil compaction can also be estimated with commercial apparatus, such as a radioactive density gauge (e.g. Troxler 3411-B)

Fig. 5.7 Soil compaction assessment instrument (Troxler 3411-B). Sensing steel rod (a), hammer used to create a hole in the ground for the sensing rod (b), hammer stand used to ensure stable hole on the ground (c), base plate used to stabilize the apparatus during sampling (d) (Courtesy Rembu Magoba).

that measures compaction in terms of kg/m³ from dry density (Figure 5.7). Such a gauge also measures soil moisture. Without this apparatus moisture content can be estimated gravimetrically by measuring a fresh sample and then heating it at just over 100 °C until a constant mass has been obtained. The difference in the weights of the 'before' and 'after' samples represents the moisture content of the soil, and is normally expressed as a percentage of the initial sample mass. All these approaches to soil characterization require repeated sampling to obtain a mean value and an indication of variation about the mean. The number of replicates should at a minimum be the same as the SUs, although it is often better is to take several measurements per SU and insert the average of these measurements per SU into the data matrix.

The nutrient status of soils can provide valuable background information because it influences the type of plants present, which, in turn, affects the insect community. Simple probes for measuring pH are available from garden and horticultural suppliers. Similarly, instrumentation is available for measuring significant nutrients such as phosphorus, potassium, and nitrogen, as well as in some cases rarer minerals and compounds. Soil samples can often be sent to a commercial soil-analysis laboratory. Where the aim is to assess its nutrient status, it is important to consider the treatment of soil post-sampling and pre-analysis. For example, drying soil at too high a temperature can alter its nutrient quality. Also, soil left in warm, humid conditions will encourage biological activity (such as fungal growth) that will also change soil nutrient status. Soil coarseness (the distribution of particle sizes in the soil) and the percentage clay content are often important determinants of insect activity and behaviour. For example, lepidopteran caterpillars will often only pupate at suitable soil sites, locusts and grasshoppers are often particular about the soil in which they will oviposit, and soil type is also a well-known determinant of local dung beetle diversity.

5.5 Measuring vegetation characteristics

Plants are an intrinsic component of the habitat of most insects, and can be viewed not only as physical habitat but also as providing other critical resources (such as food and protection from natural enemies). It is generally necessary to characterize the physical place where insects live, more properly called the *biotope*. This means recording the plant species and plant structures present. A very good introductory text is Williams (1987) and excellent, more advanced texts are Kent and Coker (1992) and Stohlgren (2007).

When undertaking a single-species study, it may be very important to identify the particular plant species or the several species with which that insect is associated (Cappuccino and Price 1995). For more advanced studies on exact relationships between insects and plants, it may also be necessary to record the functional traits of the host plants. This was illustrated in the case of the Karkloof blue butterfly *Orachrysops ariadne*, which will lay its eggs only on the prostrate variety and not the upright variety of its leguminous host plant. Likewise, Eltham copper caterpillars in Australia prefer stunted specimens of their plant host to individuals that to us appear much more healthy.

When undertaking food web studies, it is also essential to identify the plant species that are the basis of the focal food web. This contrasts with a study, say, on the whole insect assemblage, where the general characteristics of the plant assemblage (as given in Table 5.2) are normally sufficient. However, such a coarse study may well lead to a more detailed study on certain focal species. Should you be recording water beetles in a quadrat in a pool, there is little point in recording every single plant species and plant form present. Yet recording which are the dominant species and the general structure in terms of proportion may be very important. A SU with 80% bulrushes, 10% lilies, and 10% open water is a very different habitat from one with 10% bulrushes, 90% lilies, and 0% open water. A rule of thumb is only to record plant species that account for 5% (i.e. 1/20) or more of the quadrat. Where there are various rare plants, whether aquatic or terrestrial, they can sometimes collectively be classed as 'other plants'. Finer detail than this may be meaningless for a general study (but not of course for a species-specific study or one where the objective is to examine the relationship between plant and insect species diversity).

As with measuring climate and soil, when measuring plant EVs it is essential that the measurements are taken within the areas where the insects are being sampled. The proximity and spatial scale of measurement of the plant variables must be comparable with that of the insects. Plants can be surveyed using either *quadrats* or *intercept methods* (*line transects*). Quadrats are usually square frames with four cross-wires (such as taut fishing line) at 90°, giving 25 grid squares (Figure 5.8). The size of the quadrat can be 0.5 × 0.5 m, 1 × 1 m, or 2 × 2 m, depending on the structure of the plant community and density of vegetation. Normally, percentage cover is estimated by scoring presence or absence of the dominant plant species in each grid square, or percentage bare ground).

Intercept methods are an alternative to quadrats, and a *pin frame* can be used. The plant species (or bare ground in some cases) is scored whenever the pin touches it (Figure 5.9). A practical modification of this method is to use a length of rope with knots along it at equal distances (Figure 5.9). The plant species (or bare ground or rock), and its height, nearest to each knot is recorded or measured. Length of rope and distance between knots depends on the terrain and size of the plants being sampled. Long pieces of rope (50 or 100 m) with knots at 1 m or 2 m intervals are convenient in grassland, savanna, shrubland, and open woodlands. This method is very effective for habitats that are thorny, bushy, or rocky, where placement of quadrats or pin frames is not really feasible.

There are many other aspects of the plant environment that may also be important, including plant density, age structure, plant architecture, and even plant condition. Often these features can only be scored on a scale of 1–10, or 1–5, ranging from low to high, young to old, etc. Additionally, there is sometimes merit in making an estimate of canopy complexity relative to canopy height. This variable may be related to ground cover (also called understorey vegetation), for example, where the canopy shades the ground. An example may include the following scores: score 1 = low canopy only (90% of vegetation <0.5 m in height); score 2 = low vegetation and bushes; score 3 = mostly bushes; score 4 = bushes and small trees (<20 m in height); score 5 = dominated by tall trees (>20 m in height). Recording such complexity depends on the ecological or conservation question. As outlined earlier, scoring, or categorical measures such as this, while convenient, are often less amenable to statistical analysis (particularly where multiple EVs are involved) than quantitative, continuous EVs, such as mean vegetation height across 10 sampling points per SU.

Tree height can be measured with an electronic clinometer, available from forestry instrument suppliers (Figure 5.4). Should you be caught out in the field

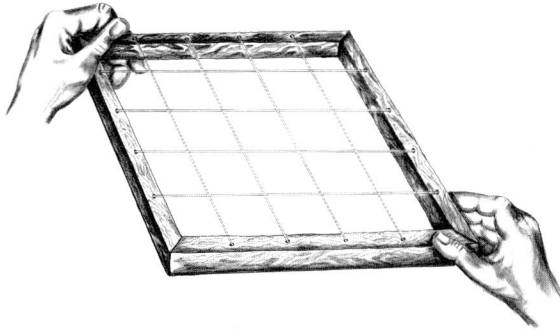

Fig. 5.8 Plant sampling quadrat (size 0.5 × 0.5 m).

Fig. 5.9 Line intercept methods for measuring plant cover: pin frame (above), and a graduated rope (below).

without such an instrument, this is one EVs that can be measured without electronic gadgetry. A simple mechanical clinometer can be made and pointed at the top of the tree, with the base given as an angle (Figure 5.10). The value of the tangent of the angle of elevation = tan A. Height of the tree is then the distance from the foot of the tree to the observer multiplied by tan A. This value, plus the height of the observer's eyes from the ground, is the height of the tree.

For insect conservation, there is one other issue that is conceptually important. One of the reasons why insect species are threatened and their populations in decline, especially in disturbed habitats, is that alien plants are invading and

Fig. 5.10 Measuring tree height. x, tree height; y, distance from observer to base of tree; z, height of observer; A, angle of elevation. Although electronic clinometers are commercially available, height of an object such as that of a tree can also be measured by simple mechanical means. A clinometer is pointed at the top of the tree, then lowered to the height of the observer, and the angle recorded. Distances y and z are measured, and the value of the tangent of the angle of elevation (tan A) determined from tables. Height of tree $(x) = y \tan A + z$.

causing severe declines in habitat quality. Thus it may be very important to record percentage cover of alien plant species relative to the abundance of all other plants in the SU.

5.6 Aquatic environmental variables

Many aquatic insects are highly sensitive to the condition of the water around them. Indeed, the assemblage structure of these and other macroinvertebrates often can be used to determine the quality of the water body. There are several *key environmental factors* that determine the presence/absence of a species and its abundance in a particular area of fresh water. These key factors are discussed in detail by Hauer and Lamberti (2007). In the case of streams and rivers, the speed and depth of flow can be very important, although it is essential to be aware that the flow may well vary through the year, and also varies according to precise microhabitat. Small streams, for example, may have riffles, glides, and deposition zones (pools), each of which has its own particular physical and biotic characteristics, which may be as important as the overall conditions of the body

Fig. 5.11 Any one site provides a range of conditions for a whole range of insect species living in different microhabitats. This stream, on top of Table Mountain (Table Mountain National Park, Cape Town, South Africa) is very rich in point endemic species, each of which lives in a particular part of the stream and/or among particular vegetation on the bank. When sampling 'the stream' it is essential to sample the range of microhabitats to have a truly representative sample.

of water (Figure 5.11). Abiotic EVs must thus be measured where the insects are sampled, alongside, where appropriate, the features of the plant assemblage in that microhabitat.

Flow rate can be measured with a dedicated gauge, but for most practical purposes a piece of wood moving on the surface and timed between two points that are a known distance apart is suitable ('Pooh sticks'). Flow rate, like vegetation characteristics, can be quantitatively measured by placing them into one of 5–10 categories, according the range of flow rates being recorded.

Water depth is a critical feature for many aquatic insects, and easily measured using a long ruler or a graduated plumb line. Related to flow and depth is turbidity, which is the extent to which suspended solids in the water absorb light rays. Spectrophotometry can be used to give accurate measurements of the amount of suspended material in water samples, but for most practical purposes in insect ecology and conservation a *Secchi disc* can be used (Figure 5.12). This is a 20 cm diameter metal disc painted with alternate black and white quadrants. It is lowered into the water on a graduated line to a point where it just disappears, and then carefully raised again to the depth where it reappears. This gives a comparative depth measure, Secchi disc transparency. By way of a guideline, the

Fig. 5.12 Secchi disc: still one of the most robust and simple means of determining the particle content (turbidity) of water.

eutrophic zone (where there is enough light to enable photosynthesis) is about 2.5 times the Secchi depth.

Salinity is an important EV for many aquatic insects, with most species being very intolerant of even slightly salty conditions. Conductivity can provide a practical index of salinity, where sea salt alone is the dominant ionic component. However, where there is agricultural runoff or effluent input into a water body, conductivity may reflect general ion content of the water from other sources. If salinity is not an issue then conductivity can reflect anthropogenic disturbance of the water system. Insects are also susceptible to changing conditions of acidity and alkalinity, and thus pH should be measured alongside conductivity. Both can be measured with carefully calibrated commercial apparatus.

The oxygen content of the water is also an important determinant of where certain insect species can or cannot live. It is often correlated with other EVs such as flow rate, water temperature, depth, and suspended organic matter (and bottom organic matter where the water is shallow). Temperature is measured with an electronic thermometer, while dissolved oxygen is measured with a commercial water analyser. Where water is polluted with organic matter, oxygen may be removed by decomposition faster than it is replaced. This leads to a biochemical oxygen demand (BOD), which is measured by keeping a water sample at 20 °C for 5 days and determining the oxygen used during that time. Organic pollution can also cause *eutrophication*, resulting from increased inputs of nitrates and phosphates, usually from agricultural activity adjacent to the water body. The impact on insects may be indirect through stimulation of algal blooms, which, on death, can take out critical amounts of oxygen from the water. For eutrophication and pollution studies, the water must be analysed using spectrophotometry, details of which are given in Hauer and Lamberti (2007).

5.7 Summary

Selection of which *environmental variables* (EVs) you are going to measure is an intrinsic part of project planning. EVs and their measurement should not be an afterthought. Indeed, it is difficult to envisage a research or conservation question that will not require the measurement of at least a few EVs. A good way to visualize this part of the study is to prepare a *data matrix* where the EV values for each *sampling unit* (SU) are listed as a block underneath that of the species (see Table 5.1). EVs are subsequently measured in or close to each SU, being the basic replicate on which analysis and statistics will be performed. Which EVs to incorporate into the study depends on the precise ecological or conservation question. It also depends on what exactly you want the EVs to tell you, and this requires some careful thought and common sense. Some of the EVs will have an interactive effect, that gives you an opportunity to cross-check and see whether combinations of EVs are having an impact. Other EVs will be correlated with each other and, if so, require careful consideration when they are included jointly in an analysis. This is clearly seen in aquatic studies where features such as stream flow, water depth, oxygen content, temperature, turbidity, and eutrophication can be interrelated. You should select a wide range of EVs to measure in each of your SUs, bearing in mind that the total of these, if they have been chosen well, may account for up to 50–75% of the variation in the insect data.

Estimating population size and condition

6.1 Introduction: the relevance of populations

The *population* is often the unit of primary interest in insect conservation studies. Declines in population size, and in the number and geographical extent of populations, are often the major harbinger of conservation need. Indeed, the two major conservation paradigms both concern species populations. The *declining population paradigm* implies losses due to threats, and when these threats are defined, the paradigm provides a basis for averting the decline. The *small population paradigm* emphasizes the environmental risks faced by species with small populations, as well as genetic problems arising simply from a population being small (Caughley 1994). It follows that an understanding of population structure and dynamics underpins much practical conservation at the species level, and can help us assess when conservation is needed, and how to proceed with it.

Much species-focused conservation is initiated as *crisis management*, responding to the decline or loss of populations. On the positive side, the rapid development and short generation times of many insects renders them important tools in environmental assessment, because they often respond to imposed changes much more rapidly than many other organisms.

Declines in abundance and distribution are often very difficult to quantify. Differentiating changes in populations due to normal fluctuations in time and space from those that are a consequence of environmental change (e.g. habitat destruction, pollution, climate change) is a central theme in insect conservation. It is also a basis for informed management, i.e. opportunity to draw on quantitative or semi-quantitative information rather than having to guess at likely drivers of population change. However, the situation is not simply to do with fluctuations about mean population size. Continuing landscape modification and global climate change, and the various adverse *synergisms* associated with this change, have actually caused some species to increase in overall mean abundance (often generalist and invasive species) whereas others have declined (Woiwod 2003) (Figure 6.1).

Categorizing *conservation status* (i.e. conservation 'need') depends on knowledge of insect population sizes and dynamics, from which their vulnerability to extinction is assessed. The number of populations known, the geographical distribution and geographical range of these, and their size, are all parameters frequently used in such assessments. For many very rare or highly localized species, only single populations or single occupied sites are known. In the case of

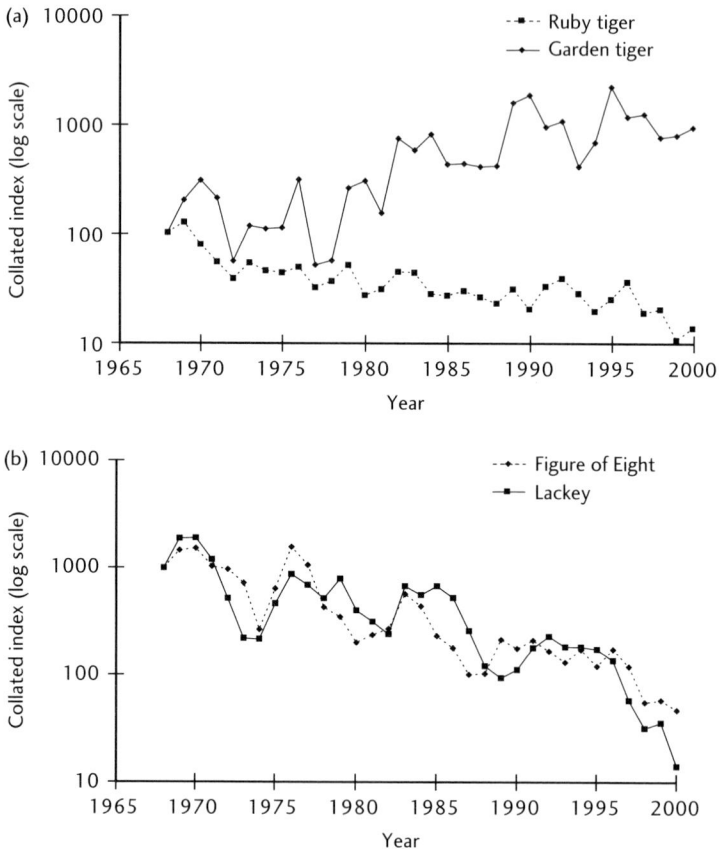

Fig. 6.1 Population fluctuations and trends in British moth abundances. (a) The Ruby tiger moth *Phragmatobia fuliginosa* has increased in abundance over the years, while the Garden tiger *Arctia caja* has declined steadily. (b) The Lackey *Malacosoma neustria* and Figure of Eight *Diloba caeruleocephala* moths have declined especially severely in recent years (from Woiwod 2003).

these *point endemics*, the population is then essentially equivalent to the species for conservation purposes.

In this chapter, we review aspects of single-species conservation—the so-called *fine filter approach*, in contrast to the *coarse filter approach* of conserving the landscape; Chapter 10). While population size and condition are underlying considerations in species conservation, it is important to recognize that practical conservation work on populations must always recognize the ecological and evolutionary background of the characteristics, condition, and health of the host habitat, and how these have changed, and continue to change, over time.

6.2 Assessing changes in population levels

Populations are continually dynamic. Historically, most conservation interest has been placed on population size, i.e. the numbers of individuals present. However, there is a functional distinction between *effective population size* (the number of individuals contributing to the next generation) and *census population size* (the total number of individuals). Populations increase in size through births and immigration, and decrease through deaths and emigration, so that knowledge of these processes informs and helps us to interpret numerical change. Most insects are *apparent* for only parts of each year, i.e. present in some relatively easily visible and assessable stage. This means that understanding patterns of seasonal development is an integral part of quantifying population size. For example, most studies of butterfly populations have involved census of the conspicuous adult stage, which may be present (although reasonably predictably) for only a few weeks in each year. The remainder of the year is passed in less conspicuous, sometimes diapausing or otherwise inactive or cryptic, immature stages (egg, larva, or pupa).

Many of the most intensive studies on insect populations have been on abundant pest species, aimed at improving pest management through more effectively orchestrated control measures. For these insects, their very high abundance has enabled development of the *life table* approach, by which mortality factors are quantified for each major growth stage—eggs, different larval instars, pupa (if present), adult—or period to indicate the key factors causing loss (Varley et al. 1973). This appraisal is then augmented by information on migrations and their causes. However, such detailed measurements are not possible for most insects of conservation interest, and non-pest and threatened species have seldom been assessed in as much detail. Nevertheless, the general principles of field population dynamics were illustrated in large part by studies on a few abundant but innocuous temperate region taxa, such as British grasshoppers (Richards and Waloff 1954). These studies revealed the considerable dangers of extrapolating from one population to another. They also emphasized the importance of inter-site and inter-population mortality effects, reflecting such factors as local incidence of natural enemies. Such studies also underline the importance of studying several populations of a species to determine what level of generality is reasonable. Following on from this, it is wise to consider separately each population of a species of conservation concern, because each population may have a unique combination of factors influencing its well-being, either favourably or detrimentally.

Population size and changes in abundance are most commonly quantified for a single life-stage across generations, rather than by using the more comprehensive life table approach (Price 1997). For rarer species, any such information may be difficult to obtain, because only very low numbers are recorded even at the highest abundance levels, and simply determining presence of the species at a site may be the more viable option. When determining the significance of perceived changes in abundance, an allied dilemma relates to determining what form the focal population takes. For example, many historical studies of butterfly populations compare the structure and dynamics of *open* and *closed populations*:

- An *open population* is one whose geographical boundaries are not well circumscribed. It is subject to immigration and emigration, so that changes in numbers are not due wholly to internal demographics, namely births and deaths. Such populations may experience sudden changes due to migrations, and determining the arena (study extent, see Chapter 3) for their study is intrinsically difficult.
- A *closed population*, by contrast, is geographically discrete (perhaps confined to a patch of suitable habitat in a landscape), and sufficiently isolated for migration to play only a minor role in its usual dynamics. Changes in abundance mainly reflect internal processes, with certain changes in the environment interacting with these. The localized nature of such populations defines the spatial extent of possible study, but if this (or the population itself) is very small, the possible harmful effects of the study, such as trampling the site, may need to be considered. Many such populations of insects that arouse conservation interest occupy such very small areas, with a hectare or less apparently enough to sustain some insect species indefinitely, as long as conditions remain suitable. Small range size may, of course, predispose a population to vulnerability. This is especially in the case of point endemics that by twist of fate happen to be 'unlucky' and live in an area firmly within the human footprint, vis-à-vis another point endemic that 'by luck' is missed by that footprint and is able to survive by default in a small area, as long as it is not disturbed (Figure 6.2).

Fig. 6.2 The giant flightless cockroach *Aptera fusca* is a 'lucky endemic'. While being a highly localized or point endemic, it has the good fortune to live on top of Table Mountain (Cape Town, South Africa) and has escaped the urban human footprint. The lowland point endemics have not been as lucky (Samways 2006a).

Reasons for studying a particular population may extend well beyond delineation of numerical status or change, and may be orientated more towards quantifying features of the species' ecology and development, as an *autecological study*. Such aims must be formulated clearly at the onset of a study. Thus, the only known population of the South African Brenton blue butterfly *Orachrysops niobe* was the subject of an ecological study with five major aims (Silberbauer and Britton 1999):

1) To establish the conservation status of the butterfly
2) To determine the distribution and behaviour of adults and immature stages
3) To assess site attributes, particularly the distribution of the sole larval food plant
4) To determine the existence of any symbiotic relationship between the butterfly and ants
5) To survey other areas to determine whether the population was indeed the only one present in the area, and to find sites that could be suitable for possible translocations in the future.

Such aims clearly dictate the procedures to be undertaken. Sometimes, however, there is little information to go on once the species has become extremely rare. The last known wild individuals of the Lord Howe Island stick insect were living among and feeding on guano-covered indigenous *Melaleuca* plants that were stunted from prevailing sea conditions. It was not known whether this was a specialized population of what was a slightly more generalist species (Figure 6.3). Under captive conditions for *ex situ* conservation (see Chapter 8), the endemic *Melaleuca howeana* was confirmed as a suitable and sole host for long-term rearing, although the species has now been reared through one generation on blackberry *Rubus fruticosus* (Honan 2008).

6.3 Significance of surveys

Several distinct activities are involved in population studies, and should be clearly distinguished. These were formulated by Hellawell (1991), and their place in butterfly recording activities discussed by Harding et al. (1995):

- **Survey.** Observations, ranging from qualitative to quantitative but usually undertaken to a defined procedure and within a restricted period, and without anticipating what the outcomes may be.
- **Surveillance.** Repeated surveys that provide a time series to indicate changes or variability, again without presumptions of the outcome.
- **Monitoring.** Intermittent surveillance to estimate trends or extent of variation from an established base line data set (see McGeoch 2007 for discussion on insect monitoring).

The more qualitative aspects of a survey may simply include *presence/absence data*, either for a focal species or for a suite of species, which leads to the

Fig. 6.3 Lord Howe Island stick insect *Dryococelus australis* on its preferred and natural food plant *Melaleuca howeana* (Courtesy Patrick Honan).

compilation of a *species list* for a site or habitat. Such a multispecies list can be used as a tool in wider biodiversity evaluation, or as part of a more extensive geographical study of a particular species. The major objectives of the most comprehensive insect recording scheme (for British butterflies) are as follows (Harding et al. 1995):

- To provide knowledge of the distribution and rarity status of the butterfly fauna (in this case, of the United Kingdom)
- To identify sites of importance for butterflies and their conservation
- To monitor changes in the abundance and ranges of species, both at important sites and in the wider landscape
- To assess the factors that may be causing such changes
- To provide a reliable and quantified basis from which to advance conservation decision-making, and to advise planning and legislation at local, national, and international levels.

Studies on changes in abundance and distribution of the most popular insect groups to the extent that they have occurred in the UK (Asher et al. 2001) or the Netherlands (van Swaay 1995) can be undertaken only exceptionally, and with the aid of large numbers of (usually voluntary) participants and effective coordination. An important difference in focus is whether a single species population is to be studied alone, or whether those of a number of coexisting species are to be monitored simultaneously, as in the butterfly schemes noted above. The approaches may be very similar, and each depends on adequate field methods and the ability to identify the focal taxon or taxa. Thus, *transect walks* (below; and Chapter 4.6) were used as one technique to monitor butterflies in different habitats in Costa Rica (Sparrow et al.1994), in conjunction with timed observations (15 min) at fruit-bait traps. Some parameters relevant to such surveys are noted in Table 6.1.

6.4 Evaluating population structure

The concept of a *metapopulation* (Hanski 1999), which is a population structure that is widespread among insects, has modified our thinking on the simple open–closed dichotomy mentioned in section 6.2. Various forms of metapopulations have been described, and their maintenance depends essentially on assuring continued *connectivity* (or accessibility on the part of the insect) between the various possible *habitat patches* in the landscape (Chapter 7) (Figure 6.4). Thomas (1995) characterized metapopulations by three features: (1) occasional movements between local populations, (2) colonization and extinction of habitat patches (with patches often fragmented within the wider landscape), and (3) local populations occurring in groups rather than as single isolates. A metapopulation thus comprises a series of more or less independent demographic units, each of which may be deemed an independent closed population over the short term because it occupies a discrete area and migration is not evident. However, over the wider landscape, a metapopulation is maintained through rolling colonization–extinction–recolonization within and across populations. Responses to declining habitat condition manifest as *local extinctions*, i.e. extinction of a local population within the broader network of the surviving metapopulation. In contrast to the more conventional conservation focus, such local extinctions may be of little concern, as they are part of the normal population dynamics of the species. Thus, if a species has a metapopulation structure, the loss of a local population may not necessarily be of as much conservation concern as the loss of a fully closed population.

Table 6.1 *Some guidelines for design of a monitoring programme for tropical butterflies (after Sparrow et al. 1994).*

Conservation issues

Focus on specific land management issues, both for rapid estimates and continued monitoring to compare management effects

Include both natural and disturbed habitat types, to evaluate differences in occurrence and relate these to specific vegetation and geographic features

Determine criteria for selecting species to be monitored

Identify subset of taxa amenable to study, e.g. by being conspicuous, trap-prone or otherwise easily sampled

Concentrate on relatively common, habitat-specific species (note difference from single species monitoring, often rare species) to assess habitat condition through wider representation

Distinguish between common species typical of the study area and cosmopolitan species tolerant of disturbance and that are likely to be abundant in any highly altered area

Biological knowledge of the study taxa is invaluable, and needed if long-term monitoring is to help design of conservation strategy

Assess biophysical aspects of potential study sites

Control for elevation in sites – many tropical butterflies stratify along elevational gradients

Pay attention to geophysical aspects, such as slope, aspect, vegetation cover, etc.

Control for light-gap size, as one of the most critical influential variables in forest butterfly assemblages

Evaluate sampling methods

Combine sampling methods to maximize productivity. Standardised trapping, netting and visual sampling are all useful approaches

Determine proper sampling frequency

Frequency should be determined by monitoring needs, logistic constraints and seasonal changes

Concentrate most effort during season of peak emergence or abundance, enabling more effective comparison of annual shifts in abundance and incidence

---

The uneven distribution of individuals across space, such as apparent aggregations or localized *colonies* across a landscape, is a useful clue to population structure. An isolated discrete population without close neighbours is associated with the traditional concept of a closed population, while a group of such entities within a larger area (although each discrete) is more likely to represent a metapopulation. Delimiting a metapopulation involves close examination of the relationships between such spatial clusters of individuals, and has considerable

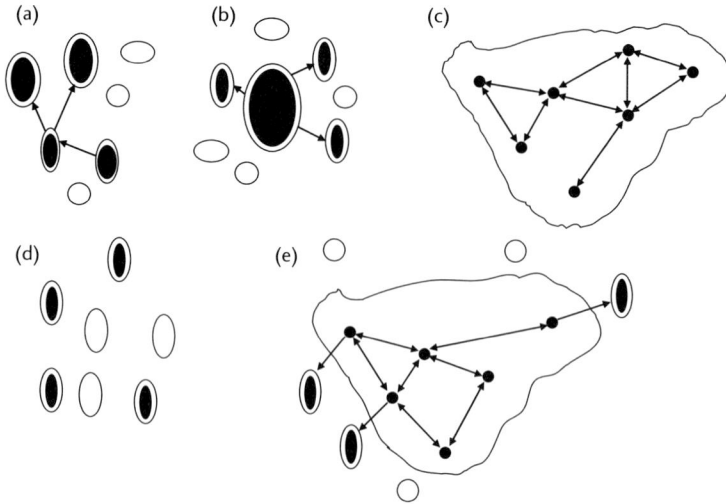

Fig. 6.4 Metapopulation structures and dynamics: (a) 'classical', (b) mainland–island, (c) patchy, (d) non-equilibrium, (e) intermediate case combining features of a–d. Filled symbols denote occupied habitat patches; unfilled symbols vacant habitat patches; arrows indicate dispersal. Outer black lines encompassing filled symbols (either groups or individual symbols) denote boundaries of local populations (from Harrison and Hastings 1996).

importance for managing the species involved. The three possible approaches for determining a metapopulation involve (1) evaluating the frequency of migration among local populations, (2) analysing the spatial pattern of populations, and (3) analysing patterns of genetic differentiation. The last has traditionally been regarded as the most significant of these, but is perhaps the least convincing. This is because local aggregations in mobile species may give the appearance of local populations, but lack direct evidence that they are independent (or largely independent) demographic units. Measurements of dispersal can directly demonstrate frequency of movements between habitat patches, and genetic studies can integrate levels of gene flow and help to record even rare dispersal events.

In the case of the Silver-studded blue butterfly *Plebejus argus* in the UK, the evidence from the three approaches (as mentioned above) was complementary in revealing a metapopulation structure (Lewis et al. 1997). These researchers measured migration and dispersal by *mark–release–recapture* (MRR, Chapter 7) in seven habitat patches and by releasing butterflies into unoccupied habitats. They examined (1) colonization of newly established habitat patches over 7 years, and (2) the genetic differentiation (through allozymes) of local populations within a supposed metapopulation. Although the butterfly is relatively sedentary, local migration was detected at six of seven sites by MRR. However, it seemed that even the most isolated patches received migrants regularly. Patch isolation was not a major barrier to colonization, but dispersal between patches separated by more than about 5 km was unlikely, so that past fragmentation may already have defined the limit of the population to networks of patches closer than this. The high local population densities of butterflies were thought to be enough to prevent much genetic differentiation through *genetic drift*, in spite of low levels of migration. The value of using several methods, as in this study, is that none would independently have provided such a clear picture in suggesting that the population dynamics of *P. argus* closely matches the classical concept of a meta-population. The precise spatial population structure (Figure 6.4) of any given species may be important to determine, because optimal management should ideally incorporate such knowledge in attempting to assure that particularly important populations are sustained.

6.5 Measuring population size

6.5.1 Some underlying principles

A variety of methods are available for estimating insect population size, each with advantages and shortcomings, making it sometimes very difficult to decide which might be the optimal technique to use. There have been few critical reviews comparing the effectiveness of different assessment techniques for a particular set of insect species. However, Gardiner et al. (2005) made a comparison for grassland grasshoppers, and reviewed 112 publications on Orthoptera sampling. Sweep netting (Chapter 4.5) was by far the most frequently used technique (51 papers), followed by transects (19) and open quadrats (15), with a further five methods (pitfall traps, ring counts, box quadrats, timed counts, night traps) less popular. But popularity does not necessarily mean that this is automatically the best technique for your study. You should review or try the various methods and see which one suits best your goals in your particular geographical setting. Orthoptera may also be subject to acoustic surveys, using sound recordings to identify species that are sometimes morphologically very similar (see Chapter 4.10).

Selecting any particular method involves careful consideration of the needs of a study. Southwood and Henderson (2000), drawing on a classification of population estimates (Morris 1960, a review that is still highly relevant), distinguished between *absolute population estimates* and *relative population estimates*. An absolute population estimate is the number of insects per unit area, and knowing this is fundamental to calculating mortality and other details of population dynamics. A relative population estimate is less precise, and does not allow for such detailed interpretation of *population density* or *population intensity* (the number of insects per unit of habitat), but is applied in a wide range of conservation contexts to compare and rank numbers in space and time. Wherever possible, take absolute population estimates rather than relative estimates. Direct counts of all individuals in an insect population are rarely possible, except for very small and circumscribed closed populations such as the Brenton blue, above. A further example is the Lord Howe Island stick insect (*Dryococelus australis*) (Figure 6.3): the sole small population of this large and conspicuous insect could be counted by torchlight in the few square metres over which it extended (Priddel et al. 2003).

Estimates of trends in population size are almost invariably based on some form of sample. Ideally you should aim to take an absolute population estimate and plot how it changes across seasons to give an indication of *phenology* and apparency of the different life stages. Determining when and how to attempt such measurements is difficult, not least because of logistic constraints limiting the amount of time or funding allocated to any assessment. In many studies in insect conservation, only a single-occasion survey each season or each generation may be feasible. You must decide what data you need to accurately answer the conservation question, and weigh this against resources available to carry out the project.

Unlike pest species, for which knowledge of population sizes and changes forms the basis for pest management, there is not usually any economic incentive to estimating population sizes for conservation. Population studies on pests often involve intensive studies (repeated every few days, for example) and often extended for the whole life of the crop, to measure when the population has reached the economic threshold and when to apply control measures. Because conservation population studies are generally not motivated by economic demands, populations of threatened species are seldom monitored as intensely or as frequently as pest species. Nonetheless, well-developed strategies for sampling insect pest species are often very useful for application to species of conservation concern. It may therefore be helpful to consider the insect pest management literature when thinking about designing an insect population study.

Most population size estimates for insects are based simply on counts of individuals that are seen or sampled in a series of units of habitat. At times, this may involve destructive sampling, such as removal of tree bark to count bark-dwelling insects, or splitting of stems to find stem-borers, so that habitat destruction becomes an important limitation to the study. More commonly, observations based on area or transect counts in a more open habitat are feasible, either with capture of the insects involved (sweeping for grasshoppers or leafhoppers in herbage or grassland) or without (counting butterflies).

6.5.2 Temporal considerations

Recording at different times of the day (Chapter 7), or under different weather conditions, can produce highly unreliable comparisons, and you would need to specify a series of threshold conditions and times to produce valid sampling for comparison across occasions or across sites. As an extreme example, many insects are nocturnal, others diurnal, and still others crepuscular. Many others have more subtle patterns of activity related to time of day and influenced heavily by temperature, cloud cover, rain, and wind. You may need to undertake preliminary observations to determine when samples might be most valid or informative. Some Australian diurnal sun moths (Castniidae) are a case in point. The males typically fly in grassland for only 2–4 hours in the middle of the day, in bright sunlight, and cease flying immediately if the wind rises above a gentle breeze or the temperature falls markedly; and females fly very rarely under any conditions (Gibson and New 2007). At least part of the rationale of considering most sun moths as threatened is simply that they are difficult to detect or monitor under even slightly suboptimal conditions. In other words, specify the set of conditions or thresholds under which you consider the sampling to be valid. Stick within these boundaries: for example, do not sample one population on a day with 0% cloud cover and another on a day with 100% cloud cover. Rather, set a standard, such as only sampling on days with 10% or less cloud cover, in which case, you may sample on one day when there is 5% cloud cover and another day when there is 8% cover.

Seasonal variation may also be superimposed on diurnal patterns. Many butterflies are *protandrous*, with males emerging well before the females, or the two sexes may have different longevity. Thus, the flight season (representing the window of opportunity for sampling) may be broadly predictable each generation or year, but precise appearance time may be modified by local weather conditions and will influence daily activity and the actual dates of first emergence. However, superimposed upon this is the fact that insects in some areas are emerging earlier in the year as a result of global climate change (Roy and Sparks 2000). Reliance on single-occasion sampling is unwise if any form of population size quantification is needed. This is also a very critical issue when Red Listing an insect species (see section 6.6.2), which is only realistically possible once the species' *phenology* has been worked out. Samways and Grant (2006) discuss this in detail (Figure 6.5).

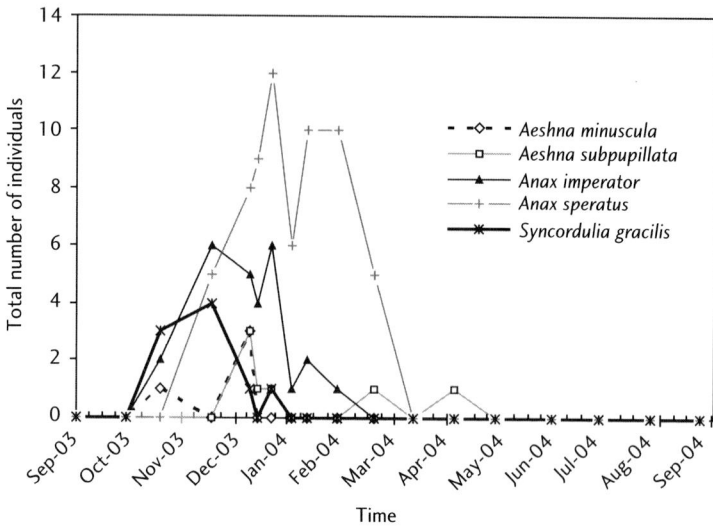

Fig. 6.5 Phenology of five dragonfly species in the Cape Floristic Region, South Africa. The Yellow presba dragonfly *Syncordulia gracilis* is a Red Listed species and this particular site is a stronghold for it. Clearly, it is critical to assess this species only in late November or early December, the only time window when the population level is high. For much of the year, this species is simply not apparent. Ideally such sampling should be repeated for more than a single year because inter-annual variability may result from differences in annual weather conditions (from Samways and Grant 2006).

There has been considerable debate over the extent to which diversity in ecosystems begets stability (Ives 2007). This is important in the realm of insect ecology and conservation, as insects are the most diverse of all organisms. Furthermore, many insect populations are among the most variable of all organisms. Local populations, even of common species, can go extinct, to reappear at some later time through metapopulation dynamics (Dempster 1989). This population variability over time at any one location has important implications for conservation as too much variation, perhaps induced by habitat disturbance, can lead to population oscillations and finally a population crash. Where landscape fragmentation thwarts recolonization of a patch, this eventually leads, patch by patch, to range retraction (Gaston 2003). Thus it becomes important to monitor changes in population levels over time to ascertain whether or not the population is oscillating within certain 'safe' limits.

This temporal approach to single populations translates into the diversity/stability debate of biological communities. Tilman (1996), for example, pointed out that diverse communities may maintain their function when disturbed, doing so through changes, sometimes large ones, in species abundances. The outcome is that particular species may fluctuate greatly in more diverse communities yet the community as a whole fluctuates less. However, this is a complex field fraught with semantic differences, as discussed by Ives (2007). Indeed, there is much that insect ecologists and conservationists can contribute to this debate (see for example Engen et al. 2002).

Measures of variability of populations and of species richness largely employ the same mensurate techniques (Figure 6.6), but on the understanding that some numerical features are more mathematical outcomes than biological realities. A comparative analysis of the performance of some of the available measures of

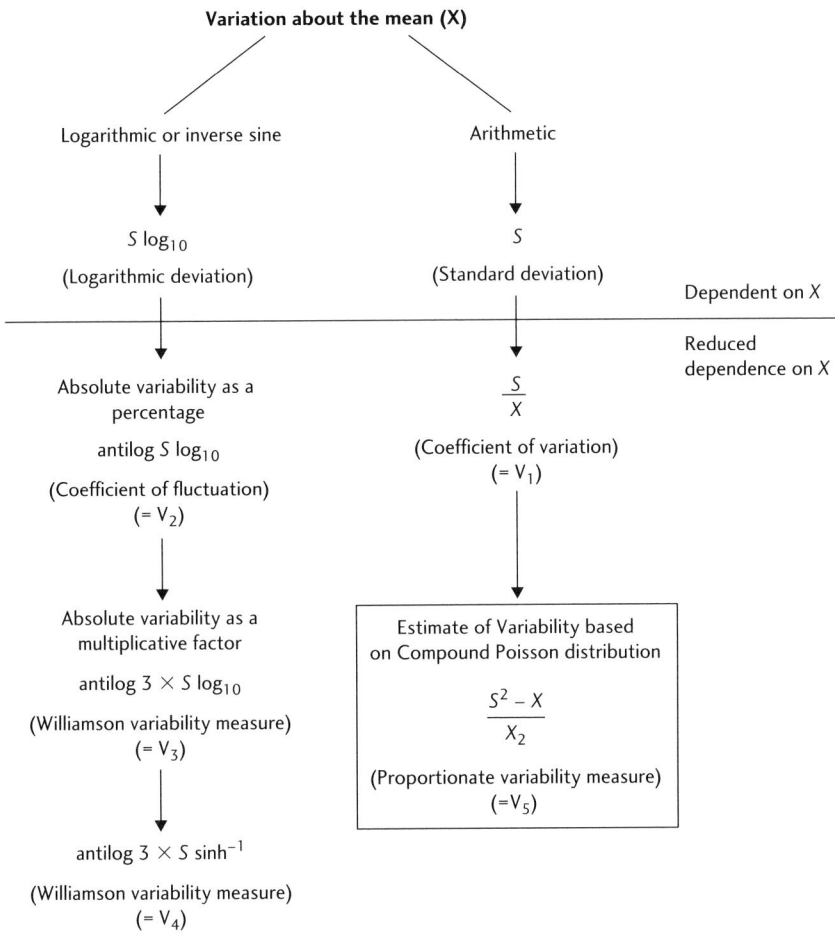

Fig. 6.6 Relatedness of five temporal variability measures (from Samways 1990), where X is the mean.

temporal variability is given in Box 6.1. The results to date seem to indicate that the rare species tend to stay consistently rare, while common species tend to be rather more volatile, at least for ants. The conservation message, with our current state of knowledge, is that populations with a consistently low coefficient of variation (V_1) or proportionate variability (V_5) are possibly more secure. Deviations

---

Box 6.1 Choosing a temporal variability measure

Temporal variability applies both to single-species population level changes over time and to changes in species richness at the same location, with the measures for both essentially the same. Five such variability measures have been compared using insect data (Samways 1990) and we make suggestions here arising from that study. These five measures are related as shown in Figure 6.6.

Standard deviation $\log_{10}$ population size has been widely used in community stability studies. However, it is not recommended here for comparative variability studies as it is not independent of the mean, especially for widely differing population densities.

- V_1 is the familiar coefficient of variation (S/x), where S is the standard deviation and x the mean population density.
- V_2 is the coefficient of fluctuation (antilog $S \log_{10}$) (Whittaker 1975).
- V_3 is the Williamson measure based on $\log (x + 1)$ transformed data (Williamson 1984).
- V_4 is the same measure but based on the $\sinh^{-1}$ transformation (Williamson 1981), and is the antilogarithm of 3 × standard deviation of transformed raw values. The justification is that with normalized data, about 90% of the records fall within 1.5 standard deviations of the mean. The antilogarithm of this range is the multiplicative factor from the expected minimum to expected maximum. The advantage of this measure, as with V_1 and V_2, is that it makes conceptualization and comparison of temporal data straightforward. For example, species A varies by a factor of, say, 3 times, whereas species B varies by 8 times. The $\sinh^{-1}(x)$ transformation has the advantage that it is more accurate than logarithmic transformations at low population densities, which arguably are those of more conservation concern.
- V_5 measures proportionate variability, after account has been taken of the mean. $V_5 = (S^2 - x)/x$, and is based on the compound Poisson distribution. It gives an estimate of variability in expected number of individuals after removal of sampling error. This, of course, is only an estimate, as sampling is not necessarily independent for all species.

From a conservation perspective, it is important to note that none of these variability measures is entirely satisfactory when population densities are low and when zeros are involved (which give artificially low variabilities). As Whittaker (1975) has pointed out, this is biological nonsense as zero densities imply that the species is absolutely unstable and has become extinct.

from the historical levels are likely to be indicative of some sort of stress, with the overarching effect of climate among them (Figure 6.6). However, much more research is required in the field of the significance of variability, and the comments we make here are more open to challenge than being definitive.

6.5.3 Spatial considerations and differences between sexes

Even deciding where to study an insect population may require substantial thought and preliminary investigation. The two sexes of the same species, for example, may frequent different habitats, comprising breeding and non-breeding areas. Hine's emerald dragonfly *Somatochlora hineana* in North America is one such example. Males primarily occupy wetland habitats, whereas females use dry meadows and marginal wetland areas, moving to wetlands only to mate and oviposit (Foster and Soluk 2006). The frequently used strategy of sampling dragonflies only in wetland areas may thus be inadequate to interpret the population dynamics and sex ratio of this particular species. Other species are migratory or change habitats locally during their adult life, making it imperative that you thoroughly investigate aspects of the biology and behaviour of a species before conservation action can be recommended.

In some cases, elusiveness of one sex may dictate that surveys are confined to the more conspicuous or accessible sex. Brenton blue butterfly *Orachrysops niobe* males are active and fly conspicuously, whereas females are cryptic. Census surveys of the Brenton blue were thus taken of males only (Silberbauer and Britton 1999). In a detailed survey of the Australian Turquoise jewel butterfly *Hypochrysops halyaetus*, Dover and Rowlinson (2005) found the male data were spatially autocorrelated (see Chapter 3.3.5), whereas female data were not. Males were positively correlated with the proportion of bare ground, probably reflecting preferred perching sites from which to pursue mates; females were influenced by host plant density and negatively by dense ground cover of other plant species. These points emphasize that doing good ecology, and certainly doing good conservation, depends on behavioural insight of the focal species.

6.5.4 Accuracy of estimates relative to feasibility

Mark–release–recapture (MRR) studies (methods of which are given in Chapter 7), although relatively laborious to undertake, may provide much more reliable information on population size than simple field counts, although the method requires at least two or more visits to any site. In addition to estimating population sizes, the method can be very useful for quantifying population structure, as individuals moving between the population sites sampled can be detected. In general, reliance on single-occasion size estimates is unwise if reliable quantitative or semi-quantitative, rather than purely relative, data are needed. However, studies of critical species on sites that are remote or difficult to access may preclude additional visits.

One very practical decision you need to make is whether some broad estimate of abundance is needed (such as 'thousands', 'hundreds', 'tens') or more precise estimates that are inevitably more difficult (and costly) to obtain. In the former case, for example with the aim of ranking sites or populations within a metapopulation, relatively simple inspections may be adequate, as long as they are conducted under the right environmental conditions and at the right time of year. More precise estimates are commonly associated with detailed *autecological* studies that may extend over several generations and may provide detailed information needed for conservation management. However, you should be aware that the natural vagaries in the size of many insect populations, whereby numbers may differ several-fold in successive generations, can confound interpretation of abundance trends. Some insects even undergo natural 'boom-and-bust' cycles reflecting phenomena such as variable diapause leading to a pattern of high abundance followed by very low numbers or apparent absence. *Population release* is a phenomenon found in many insect species, where occasionally ideal combinations of resource availability and quality, weather, and low natural enemy impacts coincide to cause unusually large populations (Price 1997). Excessive or unrealistic demand for precision in population size measurements can lead to much effort for little practical return.

6.5.5 Special considerations for threatened species

Population sizes of conservation-significant or threatened taxa are usually assessed non-destructively. If possible, you should not harm any individual or remove any from the population. Also, you should aim not to disturb the habitat. *Subcortical* insects (normally collected or sampled by removing bark from trees) or those found in dead wood (sampled by breaking up dead fallen or standing timber) need particular attention, because habitat loss is fundamental to sampling them, and is usually a prerequisite for estimating density and abundance. The need for such destruction must be considered carefully, and the damage minimized. The purpose of the exercise must be very clear from the outset, with the realization that it is very rarely possible to obtain total census accounts of any insect population.

Many of the populations you need to sample will be small or highly restricted and isolated, and presumed to be closed populations. The various small urban habitat patches supporting populations of the Eltham copper butterfly *Paralucia pyrodiscus lucida* near Melbourne, Australia, for example, are isolated from each other by housing developments. There is no evidence that the butterflies can cross such patently inhospitable terrain.

For many rare species, simply detecting the presence of a population, with no more than a very general estimate of size, may be your most important need. The two major reasons for requiring just a general estimate are (1) that conservation status (below) may be determined on the number of populations, rather than the number of individuals, of a species, and (2) that the difficulty of observing very rare species often precludes any reliable quantification, with simply their presence being the significant point. In contrast to this, demonstrating a genuine absence can be very difficult. New and Britton (1997) observed only five hilltopping individuals of the Australian small ant-blue butterfly *Acrodipsas mymecophila* at the only known occupied site in Victoria during three summers of observation. Such species are inherently difficult to detect, but their presence may be very important in conservation. In this case, the population condition (simply recorded as *extant*) is more informative than its size (inferred as 'small'), but difficulties then occur in learning enough about the species to facilitate informed management. In such cases, it is always worth trawling published literature for details, or seeking expert knowledge, on any closely related species as a possible guide for action. However, always be cautious with such an approach because the species that is threatened may differ from a common sibling in one apparently small key trait that makes all the difference for survival, as was shown for the Karkloof blue butterfly *O. ariadne* and a common sympatric sibling (Samways and Lu 2007).

Three categories of approach have been used to estimate insect population sizes in the field (1) direct counting methods, (2) MRR methods, and (3) removal methods. The last is usually not relevant for species conservation, because the main aim in conservation is usually not to deplete the focal population, other than by very temporary holding of specimens. Again for the Brenton blue butterfly, Silberbauer and Britton (1999) held captured males in a cooler container as they were taken (to avoid double counting of patrolling insects) and released them after the relatively short sampling period, once no further males were evident. Each approach makes various assumptions about the population and about the accessibility and activity of the individuals. Furthermore, many of the assumptions are difficult to validate.

6.5.6 Direct counts

Direct counts of individual insects (or insect structures such as galls, cocoons or mines) are usually based on some form of area-based or time-based survey, or a combination of these. In some contexts total counts can be taken across an entire site, but logistics usually preclude this. For dragonflies, for example, it is sometimes possible to assess recruitment into a highly seasonal adult population by counting the *exuviae* (the last moulted skins of the larva) left attached to emergent or waterside vegetation as the adults eclose. See Box 6.2, Ott (2007) for a good example of this in practice in the case of the threatened Orange-spotted emerald dragonfly *Oxygastra curtisii* in Germany. Such counts around a particular length of stream or the perimeter of a pond may provide the best possible population size estimate. Other approaches have been less frequently used, but can include counts of butterflies or other insects attracted to baits of various kinds. Zonneveld (1991) counted nymphalid butterflies feeding on flowering *Buddleia* shrubs in the Netherlands, under specified weather conditions.

Box 6.2 Using dragonfly exuviae in ecological investigations and population assessment

---

When dragonflies and damselflies emerge as adults from the aquatic larval stage, they leave behind their cast 'skins' or exuviae (singular: exuvia). These are usually fixed to the substrate from which they emerged. Such exuviae illustrate that the species has bred in that specific area and can also be used to estimate local population levels.

How to find, collect, and preserve exuviae

Exuviae are found close to the water where the larvae live. Each species has a typical orientation (horizontal or vertical) and substrate on which they emerge.

Exuvia of the Orange-spotted emerald dragonfly *Oxygastra curtisii*.

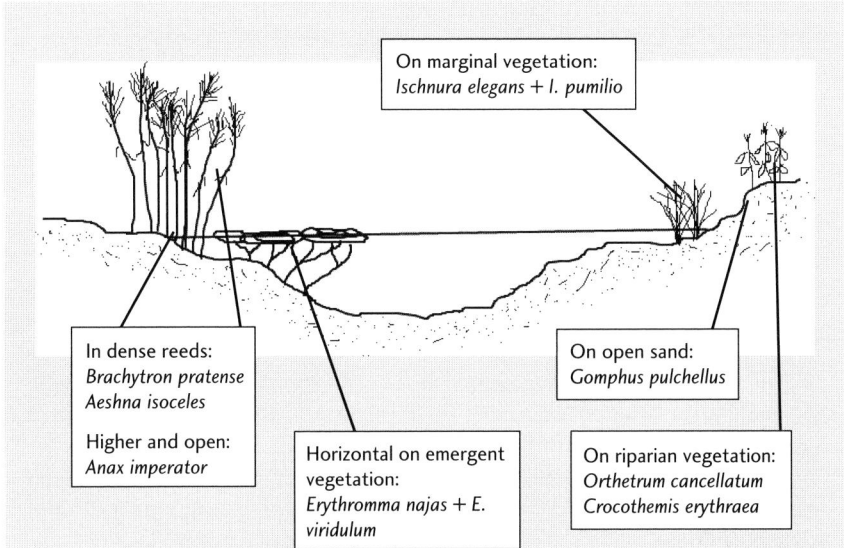

On marginal vegetation:
Ischnura elegans + *I. pumilio*

In dense reeds:
Brachytron pratense
Aeshna isoceles

Higher and open:
Anax imperator

Horizontal on emergent
vegetation:
Erythromma najas + *E.
viridulum*

On open sand:
Gomphus pulchellus

On riparian vegetation:
Orthetrum cancellatum
Crocothemis erythraea

Different dragonfly species emerge in different microhabitats

For example, most clubtail dragonflies (Gomphidae) emerge horizontally on sandy or stony river banks close to the water line, while damselflies such as *Ischnura pumilio* and *Coenagrion hastulatum* emerge on to vertical structures and very close to the water surface, often with their terminal parts still touching the water surface. *Erythromma viridulum* emerges horizontally on to floating macrophytes or algae. *Aeshna isoceles* is well hidden among reeds, while *Anax imperator* can be found up to 1.5 m above the water surface on bulrush stems or on other vertical vegetation. However, some species move far. *Cordulegaster boltonii,* for example, can be found several metres away from the water and also a few metres up trees. What this means is that any one species has a typical microhabitat where it emerges, and thus ecological studies can focus on these particular areas.

Depending on the type of biotope and the exuviae species, the various types of exuviae are located using different approaches. One can walk along the shoreline of a pond or stream, and search from either the land side or the water side (walking in the water or from a boat) by looking at the vegetation and turning over leaves and branches. Some exuviae may be hidden by dense vegetation, and others are situated under leaves and branches, which means that quantitative searching can be very time consuming. Furthermore, most exuviae are fragile to handle and, when needed for identification, must be removed from the substrate very carefully—by hand or with soft forceps (as in Figure 2.10). For very small damselfly exuviae, a small, moistened brush may be the best choice (always be cautious of puffs of wind that can easily blow them away). Exuviae can usually be found during most of the activity period of the adults, and some may remain for long periods fixed to the substrate, even for many months under bridges. However, they are very prone to destruction in heavy wind and rain.

Exuviae can be stored in any type or size of box, as long as it is well ventilated to avoid going mouldy (if there is concern that this might be the case, a pinch of thymol can be added). Spiders should also be kept out, as clubionid spiders in particular tend to make webs around the exuviae. They can also be pinned like adult insects (or mounted on card and pinned), while diagnostic parts (e.g. the labium) can be placed on a card mount, also on the pin, along with the data label. Where doubt exists as to the identity of exuviae, there is the option of background research by rearing the larvae through to adults and decribing the remaining exuviae.

Value of exuviae in population assessments and conservation planning

Research focusing on the adults can be part of a large-scale, multispecies study, such as conservation planning, as well as focusing on particular aspects of single species. On the other hand, studies on larvae may address issues such as water quality (including environmental variables such as pH, conductivity, and oxygen) or their interactions with other larvae, predators, or prey. However, neither adults nor larvae provide the same level of focus on the success of the species in its habitat as do the exuviae.

A practical example is given here of a species protection and management plan for a threatened dragonfly, the Orange-spotted emerald *Oxygastra curtisii*, based on an assessment of its exuviae (Ott et al. 2007). This species lives in running water, where the larvae may be found mainly among the dense roots of trees (principally *Alnus* and *Salix* sp.) near deposition pools. In Germany, this dragonfly was extinct for several decades, but it was rediscovered in 1999 on the river Our, which remains the only known population in the country. As the species is Red Listed and on the Appendix of the European Commission Habitats Directive, the Federal Agency for Nature Protection (LUWG) requested a detailed study of the species with a view to a protection programme and a management plan. Although a MRR study of the adults and an assessment of the larval population were undertaken, it was the focus on the exuviae that provided the most insight. A 12 km reach of the river was divided into four sections of 3 km, and different researchers assessed each of these sections by walking along the river banks and searching both banks intensively for a few days after their main emergence period in June. The results obtained are shown below.

Comparison of the exuviae collected and the population estimation

Year	No. of exuviae	Population size (males)	Population size (total)
2005	227	210 (± 20)	420
2006	1110	604 (± 26)	1208

It was found to be very important to establish a strong search image for the larvae. Once that had been achieved (in the second year), it was found that the exuviae were a good reflection of the adult population. Only after the detailed study of the exuviae was it possible to conclude from where along the river the adult population was being maintained (as recaptures for females were usually scarce, this is true only for the males). As it turned out, different stretches of the river had different levels of importance, with the exuviae being found only in some parts of the river and under very particular habitat conditions, which often did not correspond with the territories of the males.

The conservation significance of this study is that a few square metres of an optimal habitat (deposition pools with a big alder tree with dense roots) are of very high importance, with as much 5–8% of the whole population emerging from a relatively very small area. Thus, for this species' future survival, it is important that these microhabitats are protected.

The value of using exuviae also has a practical advantage. They can be collected even during cold and cloudy weather, when adults are not active, and at the same time they provide good quantitative data that are much easier to acquire than by the much more demanding process of MRR of the adults.

Jürgen Ott, L.U.P.O. GmbH, Friedhofstrasse 28,
D-67705 Trippstadt, Germany

6.5.7 Transect counts

Transects are simply lines or fixed pathways between defined points or across habitats, along which observers can record information on incidence and abundance of species of interest. Transect counts provide an index of population size, not an absolute measure of abundance, and their value is in confirming species presence rather than estimating population size. Transects also help us detect changes in faunal composition and species abundance in time and space. Their major value is thus in monitoring, in which insects may be recorded by observation (for well-known groups in which identification is unambiguous, as for most of the British butterflies) or by capture (at least of voucher specimens) for later identification (as was done for early surveys of British hoverflies, Syrphidae: Pollard and Yates 1993).

The most familiar application of transects to conservation is the *Pollard walk* (Figure 6.7), pioneered in the UK as the foundation of monitoring butterfly fauna, and leading to the most comprehensive such documentation for any country or insect group (Asher et al. 2001) (http://www.brc.ac.uk; http://www.butterfly-conservation.org). Details of this are given below, but three broad approaches can be incorporated in transect counts: *line transects*, *point transects*, and *belt transects*. One of these may be a preferred option in a given situation. The first is the most commonly employed, and both line and point transects can be undertaken by one person, whereas belt transects require several people.

6.5.7.1 Line transects

Protocols for this method have been developed in some detail, and for a number of insect groups. As in the Pollard walk for butterflies, the recorder walks along a

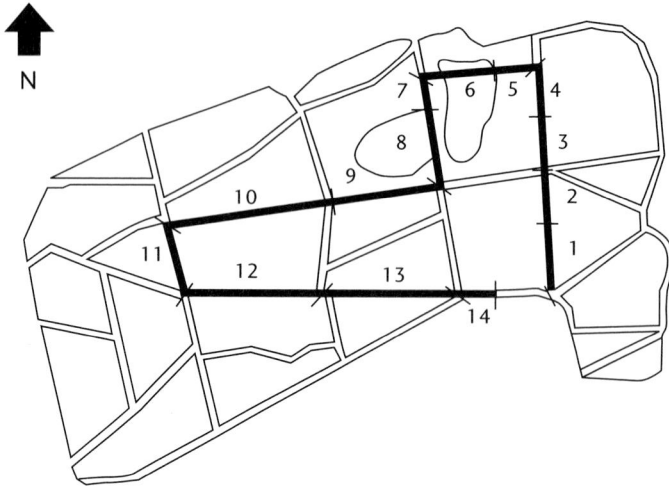

Fig. 6.7 A Pollard walk, which is a transect route used to count butterflies in the UK. This is a route at Monks Wood, Cambridgeshire, 3 km long and divided into 14 'sections', which runs mostly along the major rides (tracks) and through one of two large fields (see text). It takes between 60 and 90 min to record butterflies along this transect (from Pollard and Yates 1993).

predetermined path, which may reflect access (such as along footpaths or wood-land rides) or simply a line such as a compass bearing between map points, and records information en route. The early British trials of this method for insects, at Monks Wood, Cambridgeshire from 1973, used a defined route of about 3 km, which took 60–90 min to record. It was divided into 14 sections, reflecting different vegetation types or route features (such as track intersections) with results from each section recorded separately to facilitate information on habitat associations or other local variation in abundance (Figure 6.8).

The recommended width of the transect within which insects are recorded should normally not exceed 5 m, but the actual width may be relatively unimportant unless strict comparisons of density across sections or sites is contemplated. In the case of Pollard walks, all butterflies seen within the defined boundary and within about 5 m ahead of the observer are recorded. The recorder walks at a steady pace, and various efforts are made to help standardize results from different recorders. Butterflies flying high above the recorder (including canopy species) are not recorded, so are likely to be substantially underestimated, and settled butterflies are not deliberately flushed up to swell the counts. It is necessary also to establish a firm policy for butterflies patrolling up and down any path, hedgerow, or other feature that constitutes part of a transect, to avoid excessive duplicative counting. Some recorders, for example,

Year	2009	Date	9 July	Recorder	E. P.		
1-2		3-5		6-8			
		Site name	MONKSWOOD				
9-11		12-17				18-19	
Start time	11.00	End temp °C	19°	% Sun	90	End wind speed	2
20-23		24-26		27-28		29	

Section	30-32 33-35	36 1	39 2	42 3	45 4	48 5	51 6	54 7	57 8	60 9	63 10	66 11	69 12	72 13	75 14	78-80 15	Total
Brimstone	54																
Common blue	106																
Green-veined white	99		1			1		1	1					2			6
Hedge brown	76					1	7	8	1		2						19
Large skipper	88	1							1								2
Large white	98																
Meadow brown	75	1	1	4		1	8	12	13	14	10	1		2			57
Orange tip	4																
Peacock	84																
Red admiral	122																
Ringlet	8	1	1	1		3	7	17	17		38			2	1		88
Small copper	68																
Small heath	29						3										3
Small skipper	120						4		1								5
Small tortoiseshell	2																
Small white	100						1										1
Wall	94																
White admiral	64		1											1			2
Marbled white	78						1										1
W l hairstreak	113		2														2
Section		1	2	3	4	5	6	7	8	9	10	11	12	13	14	15	
Sunshine		S	S	S	S	S	S	S	C	S	S	S	S	S	S	S	

Notes: Hedge Browns all ♂♂   PLEASE TOTAL EACH SQUARE

Fig. 6.8 Sample of a recording sheet from a Pollard walk, where butterflies are recorded along a transect in the UK (see text). The small printed numbers, including a code for each butterfly species, assist data processing (based on Pollard and Yates 1993).

elect to count only butterflies that pass them from front to back, or the converse. Where there is doubt, Pollard and Yates (1993) recommend counting the butterflies again. A similar approach is used for monitoring other insects, such as bumblebees. The numbers of these insects are measured along transects 2 m wide, with a thin 2 m long plastic tube (sometimes with weighted, dangling pieces of string at either end) strapped to the shoulders to maintain constancy in transect width.

Variation in activity due to time of day and weather conditions must also be considered. *Standard counts* are restricted to around the middle of the day (between 10h15 and 15h45 in the UK). Wind speed during a count is estimated (Beaufort scale) and noted on the data form, with the general recommendation that it is usually inadvisable to record when wind speeds are above force 5 ('small trees in leaf begin to sway'). In turn, standardized temperature requirements are (1) if the shade temperature is >17 °C, counts should be made irrespective of the sunshine, and (2) between 13 and 17 °C counts may be made if at least 60% of the walk is in sunshine. The temperature threshold is reduced to 11 °C at more northerly sites in the UK where such temperatures are the activity norm. Temperature criteria included are comprehensive, because the recorder must have a reasonable chance of encountering suitable conditions in which to work, but also a reasonable proportion of the butterflies must be available to count!

Similarly, Brooks (1993) specified a similar suite of minimum conditions for valid surveys of British dragonflies. For comparison with the above butterfly protocol, these minimum conditions are (1) surveys should start between 11h00 and 13h00, (2) shade air temperature should be above 17 °C, (3) there should be at least 50% sunshine (established by the observer recording at intervals whether a shadow is cast: if at least half the records are positive, the survey is valid), and (4) wind conditions should be light, at most without leaves and branches moving. However, it must be emphasized that these various temperature and sunshine thresholds apply to the UK. They will differ in other parts of the world, with upper temperatures, for example, being limiting for many insects in hot, arid areas. You will have to at least standardize your recordings, and hopefully optimize them, through preliminary sampling.) Results of the Pollard walk are recorded separately for each section of the walk as each is completed (as numbered in Figure 6.7). Handheld digital recorders are now available for performing this task.

Butterflies and dragonflies are perhaps the most conspicuous insects that can be assessed by line transects. However, a variety of others can also be assessed. Grasshoppers, for example, although normally far more cryptic than the above groups, are easily flushed out of the vegetation by a moving observer and can then be identified with close-focus binoculars if need be and counted reasonably accurately. Those that require identification can usually be caught with insect tongs (Chapter 4.4) with little disturbance to your routine or the habitat.

Line transects have also been used for nocturnal insects, using a torch or spotlight to find moths. Either continuous walking (to survey single species: Spalding 1997) or stopping at intervals (as below) (Birkinshaw and Thomas 1999) may be feasible, with the latter perhaps more useful for multi-species surveys, because of the need to capture moths in the dark to identify them. Birkinshaw and Thomas opted to stop every 10 paces, using a spotlight to observe moths within a 5 × 5 m 'box' in front of them at each stop. Practical points favouring this approach over continuous walking include (1) continuous walking at night is very slow if all species are recorded, (2) many moths show evasive or defensive behaviour when caught in the light beam, and (3) it is easy for the recorder to trip and fall in the dark! Other visual signals may form the basis for counts (e.g. flashes of glow-worms) and knowledge of any behavioural idiosyncrasies of the focal species may suggest modifications to improve the methodology of such surveys.

The same principles apply to the other transect methods, but the basic practical procedure differs.

6.5.7.2 Point transects

Rather than continuously recording all individuals seen along a transect line, the observer stops at regular predetermined intervals (such as every 100 m, or at regular GPS points) along the route and, standing still, records all the focal insects seen within a given radius (10 m is often feasible in open vegetation) or within a given time interval (usually within the range of 3–10 min). Close-focus binoculars are very useful for identifying some larger insects at the more distant parts of the radius. A survey transect will usually comprise at least 10 such points, with the entire survey being completed under set conditions and time. As in the case of line transects, any notable habitat details and other features should be noted for each point.

6.5.7.3 Belt transects

These are undertaken relatively rarely, but are useful for assessing the microdistribution of species across larger areas or apparently homogenous open habitats such as grasslands. Belt transects can also help to provide information on relative abundance or density across such areas. The principles and conditions are exactly as for line transects, except for spatial scale. Belt transects are the entomologists' equivalent of 'a line of beaters' as recorders (so can take advantage of groups of volunteers on field day exercises to augment information from line transects). Recorders spread along a starting base line (at, perhaps, 20 m intervals) and walk forward, maintaining the distance between them recording the insects found either continuously or stopping at intervals to do so (perhaps every 5–10 min or every 100 m), together with habitat data and any other information needed.

This method has been used to detect 'hotspots' of sun moths (Castniidae) in grassland in Australia, with the interval counts translating to a grid of numbers that can be used as a GIS layer against vegetation or topography. More experienced and less experienced recorders are spaced along the line, so that advice and quality control is available continuously while recording is in progress. For any of these approaches, it is inevitable that differences will occur between observers, as a result of individual differences in experience, confidence, visual acuity, and patience (loss of concentration), as well as interpretation of verbal or written instructions on procedure. This adds a measure of error to the data. Belt transects are clearly labour intensive, and their main use in practice has been as an adjunct or one-off exercise to augment information from line transects in larger habitat areas.

The major application of transects is generally in assessing changes in phenology and population sizes, through repeated use, such as weekly transects in the same area during the entire summer or more defined flight season of a focal species. Pollard and Yates (1993) suggested that transect monitoring for British butterflies was preferable to MRR methods (Chapter 7) because (1) there is no handling of butterflies, and (2) fewer assumptions are necessary about butterfly behaviour and about the structure of populations. In addition, transect recording can be used for both very common and very rare butterflies (as well as those that are difficult to capture), which are situations where MRR is impracticable.

6.5.8 Quadrat counts

Area-based methods for counting organisms are commonly based on *quadrats*. These are delimited areas of habitat that are sufficiently large to provide reliable information on density, yet are sufficiently small to be logistically feasible. Quadrats must be sufficiently numerous and well dispersed to provide replicated and representative results. These are perhaps of greater value for more sedentary insects than for highly mobile ones, as they also allow for mapping of resources such as plants, including food plants, and of early stages (see Chapter 5.5). A quadrat is a usually quadrangular area, most commonly between 1 × 1 m and 10 × 10 m in size, depending on the density and size of the insects, and the information required. You need first to determine the wider arena for your survey, and then use quadrats delimited by a frame or mapped corner pegs. Individual quadrats should be dispersed either randomly or more regularly (such as on a grid pattern) across the site (see Chapter 3). If they are permanent, to be revisited at intervals, GPS coordinates for labelled corner pegs should be recorded to help you find them in the future.

Box 6.3 Quadrats and caterpillars

The Eltham copper butterfly, *Paralucia pyrodiscus lucida*, a threatened Australian lycaenid, occurs on several very small (1–2 ha) isolated sites near Melbourne, Victoria. It is unusual in having two measures of change between generations for several seasons, i.e. counts of both caterpillars and of adult butterflies. Both are based on quadrat counts, with the major survey colony marked with a 'permanent' 10 × 10 m grid, by metal pegs. Within each quadrat, all individual larval food plants (*Bursaria spinosa*) are mapped, and all those that support caterpillars (in the current or previous seasons) are individually numbered and tagged.

Eltham copper caterpillars are nocturnal, and are shepherded up the plants by tending ants *Notoncus* sp., in whose subterranean nests at the base of the plants they pass the day. Caterpillars are counted by torchlight, with volunteers inspecting each individual *Bursaria* in quadrats (under specified conditions of temperature and wind speed), recording presence/absence on plants, number of caterpillars present, and number of attending ants (by which the cryptic caterpillars are often first detected; the ants are glossy and very conspicuous in torchlight). Over time, information on host plant fidelity, frequency, and duration of caterpillar feeding has been gathered, and such data can be invaluable in population estimates—many caterpillars (marked individually with fluorescent dusts), for example, do not emerge to feed every night (Canzano, personal communication 2006), so that single-night counts may severely underestimate the number present in the area.

Quadrats may be searched for all individual insects, or you can use each quadrat merely for recording whether a species is present or absent (Box 6.3). For small quadrats, much of this can be achieved while you remain outside the quadrat, so that the area within remains relatively undisturbed and untrampled. In the case of larger quadrats, you need to be careful to minimize any within-quadrat damage. Quadrat counts can become laborious if total counts of small insects, such as immature leafhoppers, are needed, but nevertheless provide information that is difficult to obtain in other ways. Quadrat assessments can be augmented using other methods (such as sweeping) over a larger surrounding area, as long as you standardize effort and technique throughout.

6.6 Populations and conservation status

6.6.1 Conservation status

Assessing the *conservation status* of a species implicitly includes considerations of the number, size, and condition of populations to provide a quantification of the level of threat the species faces. Taxonomic distinctiveness is a necessary criterion for delimiting the entity involved. In other words, you need to be very familiar with the species (or, at the lower taxonomic level, *evolutionarily significant unit*, ESU: genetically and usually morphologically distinct species sub-unit) you are dealing with (Box 6.4). In addition to criteria of threat, such as those nominated by the World Conservation Union (IUCN, see below), some protective legislations also recognize the concept of *significant population* (i.e. a large and healthy population of an otherwise very rare species). Warren et al. (2007) provide an interesting discussion on this field.

Population condition reflects the general health or fitness of both the population itself and its members, in both genetic constitution and wider aspects of viability that may well affect vulnerability or population survival in the future. As Clarke (2000) emphasized, considerable background demographic and life-history information on any species is required to assess the *extinction risk* of a population, and these data are usually inadequate for most species of concern. Genetic techniques can reflect both the quality (as diminished genetic variation in relation to other populations), and they can imply levels of taxonomic distinctiveness not apparent from morphological study alone. Nevertheless, these insights help to guide management of the populations for effective conservation (Chapter 8).

Box 6.4 Taxonomic and evolutionary inferences from genetic study of populations: the Golden sun moth

---

Clarke's (2000) study of the Golden sun moth *Synemon plana* (Castniidae) is one of the largest genetic analyses of a threatened insect. It used over 1200 individuals taken collectively from 36 populations across the entire range of the species on native lowland grasslands in south-eastern Australia. This size of sample is exceptional (and may seem excessive) for a threatened species, and was possible only because of confining sampling to males presumed to be near the end of their short life of only 1–4 days, and spreading sampling across the flight season of several weeks. The data helped to indicate which populations may have priority for conservation management, as ESUs (sensu Moritz 1994).

The methods (Clarke and O'Dwyer 2000) involved use of individual moth abdomens stored at −80 °C until processed, and analysis of 16 enzyme systems representing 20 loci. Parameters of genetic diversity measured were: the percentage of polymorphic loci, allele richness (numbers of alleles per locus and polymorphic locus), gene diversity (as expected heterozygosity), observed heterozygosity, and the inbreeding coefficient. Measures of Nei's genetic distance revealed the populations to cluster into major groups corresponding to geographical area. Each major group was sufficiently distinctive to merit management as an independent ESU, with the recommendation that any translocations (Chapter 8) should not be made between groups. The Victorian populations (Group 1 in figure below) were genetically very distinct from all others and possibly represent the greatest levels of allele diversity and genetic variation: they may be sufficiently distinctive to be regarded as a separate subspecies or race.

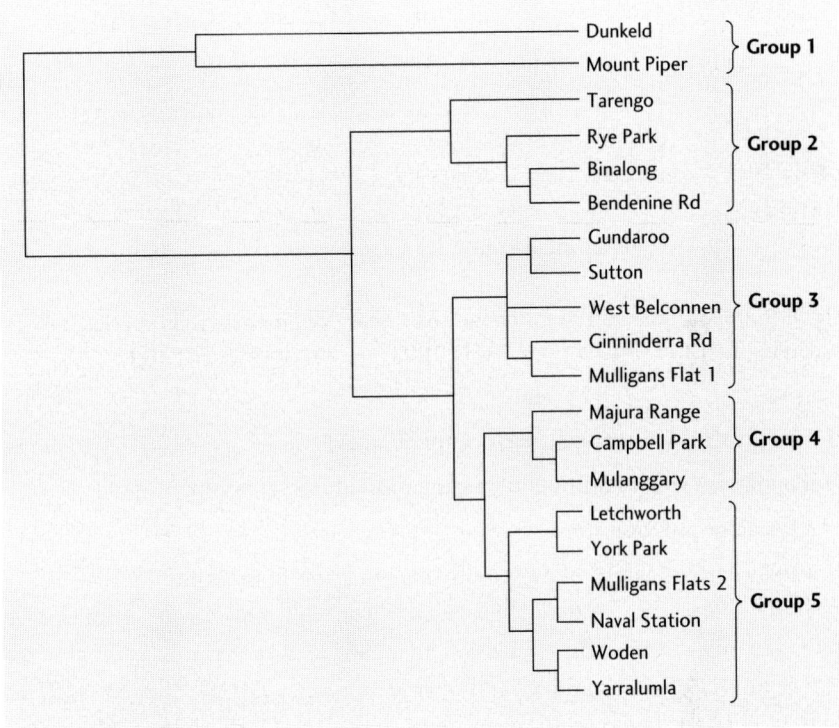

6.6.2 Level of threat

Allocating a *category of threat* (i.e. intensity of threat) to a species, as a measure of its *conservation status*, is a dynamic process with far-reaching implications for allocating priority attention and resources to its management. You need to be very clear that any assessment of conservation status and category of threat are not fixed, and they require continued revision as new information is made available. Generally, assessments need to be redone at least every ten years.

The most recent IUCN Red List *Categories of Threat* (Critically Endangered, Endangered, Vulnerable) (Figure 6.9) are each supported by a range of so-called *Criteria* (A–E, Table 6.2), any one of which may be used to justify the conservation status of a species. The Criteria include:

A Extent of reduction in population size, with thresholds specified for categories of endangerment

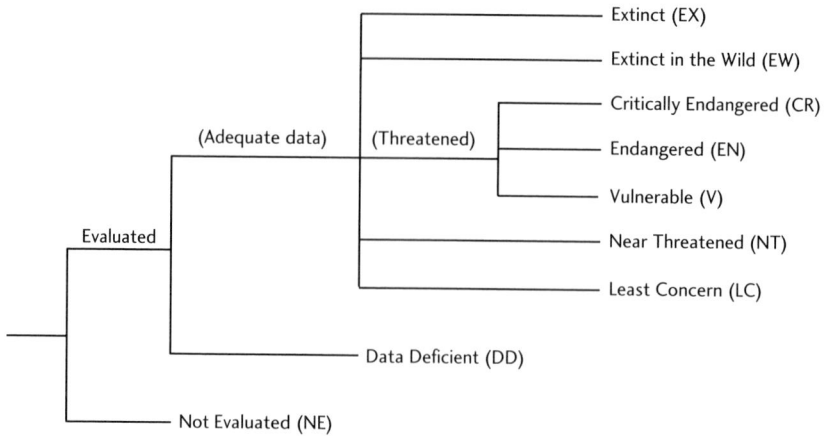

Fig. 6.9 Dendrogram of the Categories of Threat used for assessing species according to IUCN Criteria (from IUCN 2001; see also www.redlist.org).

B Geographical range and areas occupied

C Population size, as number of mature individuals remaining

D Overall population size

E Quantitative analysis of probability of extinction in the wild.

These criteria aim to detail risks facing a species, and involve quantitative estimates of factors such as changes in *extent of occurrence* (*EOO*) (but see Jetz et al. 2008 for a critique) and *area of occupancy* (*AOO*) (Figure 6.10), as well as changes in population numbers and size. Assessed in combination, these changes constitute *extinction risk*. However, such quantitative data is rarely available for invertebrates, and is reflected as a high level of *uncertainty* for most of them. Nevertheless, through the use of reasonable estimates of population size and condition, coupled with estimates of category of threat, insect species are gradually being added to the *Red List of Threatened Species* (see http://www.redlist.org), which, in turn, helps to indicate the overarching scale of need for conservation at the species level (Baillie et al. 2004).

The term *population*, as used in IUCN assessments, differs in its use in general ecology and as used elsewhere in this book. The IUCN definition of population is 'the total numbers of individuals of the taxon', with *population size* then 'the numbers of mature individuals only'. Subordinate *subpopulations* are defined formally as 'geographically or otherwise distinct groups in the population between which there is little demographic or genetic interchange (typically less than one successful migrant individual or gamete per year or less)' (related to

Table 6.2 *Overview of thresholds for the IUCN Red List Categories of Threat (see text) (after Butchart et al. 2005; www.redlist.org).*

Criterion	Critically endangered	Endangered	Vulnerable	Comments
A1: reduction in population size	≥90%	≥70%	≥50%	Over 10 years/3 generations in the past, where causes are reversible, understood and have ceased
A2–4: reduction in population size	≥80%	≥50%	≥30%	Over 10 years/3 generations in past, future or combination
B1: small range (extent of occurrence)	<100 km^2	<5000 km^2	<20 000 km2	Plus two of (a) severe fragmentation / few localities (1, ≤5, ≤10), (b) continuing declines, (c) extreme fluctuation
B2: small range (area of occupancy)	<10 km^2	<500 km^2	<2000 km^2	Plus two, as in B1
C: small and declining populations	<250	<2500	<10 000	Mature individuals. Continuing decline either (1) over specified rates and time periods or (2) with (a) specified population structure or (b) extreme fluctuation
D1: very small populations	<50	<250	<1000	Mature individuals
D2: very small range	N/A	N/A	<20 km^2 or ≤5 locations	Capable of becoming critically endangered or extinct within a very short time
E: quantitative analysis	≥50% in 10 years/3 generations	≥20% in 20 years/ 5 generations	≥10% in 100 years	Estimated extinction-risk. Using quantitative models, e.g. population viability analyses

the metapopulation concept discussed in section 6.4). Besides the definition of population described above, there are other relevant definitions, which are given in Table 6.3. The IUCN definition of subpopulations equates to metapopulation units or, for example, geographical outliers.

There is considerable debate over the suitability of the IUCN criteria for insects, discussed by Samways (2002) and Warren et al. (2007). The challenge hinges on the quantitative aspects of the criteria, which, for insects, often have

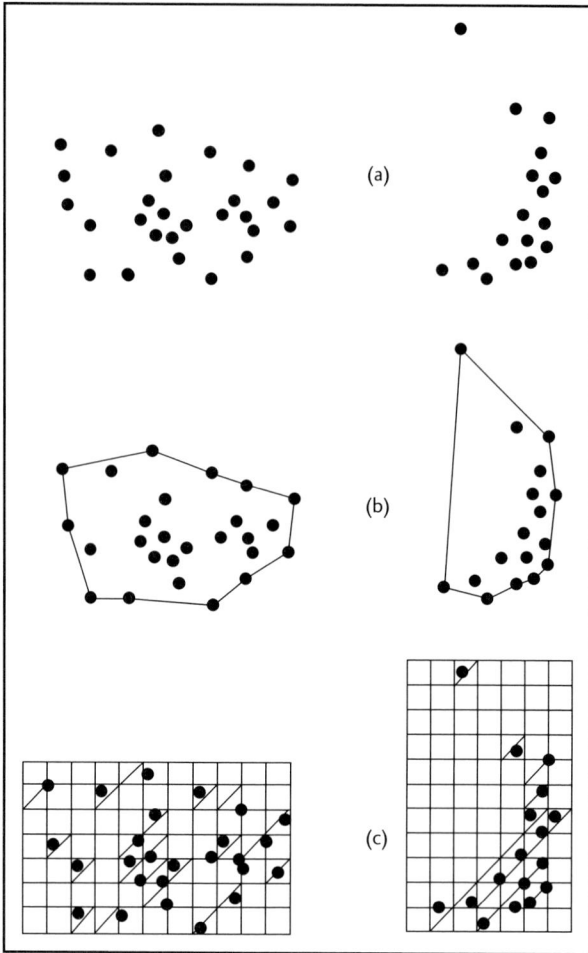

Fig. 6.10 Two examples of the distinction between extent of occurrence and area of occupancy. (a) The spatial distribution of known, inferred, or projected sites of present occurrence. (b) The possible boundary to the extent of occurrence, which is the measured area within this boundary. (c) One measure of the area of occupancy that can be achieved by the sum of the occupied grid (from IUCN 2001; see also www.redlist.org).

high levels of uncertainty attached to them. Criterion A (reduction in population size) is often very difficult to apply, especially as a putative decline may well be masked by the vagaries of the insect's normal high population variation which occurs even in the absence of human impact. In turn, uncertain quantitative data make prediction of future trends in population viability often very

Table 6.3 *Some of the definitions used to interpret and apply the IUCN Red List Categories of Threat (IUCN 1991 2003). Criteria A–E are noted in the text and in Table 6.2. Note that some discussion in IUCN (1991, 2003) not relevant to insects is not included here.*

Term (Criteria)	Definition
1. Population / population size (A, C, D)	The total number of individuals of the taxon, measured as the number of mature individuals
2. Subpopulations (B, C)	Geographically or otherwise distinct groups in the population, between which there is little demographic or genetic interchange (typically one successful migrant individual or gamete per year, or less)
3. Mature individuals (A, B, C, D)	The number of individuals known, estimated or inferred to be capable of reproduction. Qualifications include (a) mature individuals that will never produce new recruits should not be counted; (b) where population size fluctuates, use a lower estimate; (c) in the case of taxa that naturally lose all or a subset of mature individuals at some point in their life cycle, the estimate should be made at a time when mature individuals are available for breeding; (d) re-introduced individuals must have produced viable offspring before they are counted as mature
4. Generation (A, C, E)	Generation length reflects the turnover rate of breeding individuals in the population
5. Reduction (A)	A decline in the number of mature individuals of at least the amount (%) over the time period (years) specified, although a decline need not be continuing. The downward phase of a normal fluctuation will not normally count as a reduction
6. Continuing decline (B, C)	A recent, current or projected future that is liable to continue unless remedial measures are taken
7. Extreme fluctuations (B, C)	Extreme fluctuations occur when population size or distribution area varies, widely, rapidly and frequently, typically with a variation greater than a 10-fold increase or decrease
8. Severely fragmented (B)	Refers to the situation in which increased extinction risk to the taxon results from the fact that most individuals are found in small and relatively isolated subpopulations
9. Extent of occurrence	The area contained within the shortest continuous imaginary boundary which can be drawn to encompass all known, inferred or projected sites of present occurrence of a taxon, excluding cases of vagrancy. Can often be measured by a minimum convex polygon (the smallest polygon in which no internal angle exceeds 180° and that contains all the sites of occurrence)
10. Area of occupancy (A, B, D)	The area within the extent of occurrence occupied by the taxon, excluding cases of vagrancy. In some cases, it is the smallest area essential at any stage to the survival of existing populations of the taxon
11. Location (B, D)	A geographically or ecologically distinct area in which a single threatening event can rapidly affect all individuals of the taxon present
12. Quantitative analysis (E)	Any form of analysis that estimates the extinction probability of a taxon, based on known life history, habitat requirements, threats and any specified management options. PVA is one such technique. Full use should be made of all relevant available data; assumptions (which must be appropriate and defensible), the data used and the uncertainties in the data or quantitative model must be documented

difficult. Nevertheless, in some cases, such as British butterflies, these criteria can be applied relatively effectively.

However, Criterion B involves distribution, and can be far more straightforward to quantify, together with trends in *extent of occurrence* and/or *area of occupancy* being possible to estimate from knowledge of the extent of landscape fragmentation and habitat loss (Fig. 6.10). However, these assessments of distribution may still confound vulnerability when the species is simply just rare (being a state that need not necessarily be associated with threat).

Criteria C and D have many of the same problems mentioned above for Criterion A. Criterion E (quantitative analysis to show probability of being extinct in the wild, EW) has rarely, if ever, been achieved realistically for an insect. This fact highlights the limitations of effective population modelling for this purpose with insects, and the general uncertainty over what might constitute a viable insect population (below).

In view of the above, it is not surprising that the threatened status of insects has most commonly been assessed on the basis of the number of populations and aspects of habitat patch or colony area, rather than on the numbers of individuals present. An estimate of numbers of individuals can be both difficult to obtain and misleading unless tracked over a number of generations, especially when environmental conditions are adverse, which may cause some insect populations to lose *apparency* in some years. For example, some rare and Red Listed dragonfly species can go locally extinct in dry years to return in wet ones, even in a World Heritage Site (Samways 2009). The point is, as Clarke (2000) noted, the basic life history and demographic parameters usual in vertebrate conservation status allocations are difficult to derive for most insects. This is one reason why insects and other invertebrates are under-represented on lists of threatened species.

Various alternative suites of criteria have been advanced in efforts to overcome the perceived shortcomings of the IUCN Criteria for insects, by reducing reliance on quantitative thresholds and emphasizing more the extent of threats, the number of populations, and their distribution and occupied area(s). Notwithstanding the stated IUCN qualification that criteria for threat evaluation should be used 'on the basis of available evidence on taxon numbers, trends and distribution, making due allowance for statistical and other uncertainties', problems can still arise during assessments, particularly of lesser-known faunas. Discussion of European (van Swaay and Warren 1999) and Australian (Sands and New 2002) butterflies illustrate many of the contrasting practical problems and ambiguities that can arise, even for this well-known insect group. Nevertheless, the IUCN criteria and the Red List are generally recognized worldwide as the 'gold standard'. For reasons of global standardization, we strongly recommend that IUCN Red List assessments are undertaken, for both globally and nationally threatened species.

For many developed countries there are national initiatives already in place (see Warren et al. 2007 for the situation in the UK) and you, as an individual, may only be able to participate in an ongoing framework. However, there is much to do in other parts of the world, where in fact biodiversity is under greatest threat. You really need to be resident in one of these countries, or at least work in a team with citizens of the country, because Red Listing requires considerable knowledge of the focal species as well as knowledge of the local socio-political system if the Red Listing is to be translated into meaningful conservation action.

As regards the Criteria, we recommend that for insects you focus mainly on Criteria A, B, and D. It is particularly important when assigning a category of threat, that you justify fully why you are assigning the species to a particular category. Furthermore, it is not simply a case of assigning an A, B, or D, but also a subcategory according to the ruling of the IUCN. These are spelt out in the *IUCN Red List Categories and Criteria*, which is available on the web (http://www.redlist.org); an example is given in Table 6.4.

When you undertake such an *Assessment* you are known as an Assessor. However, you cannot just submit your findings to the IUCN Red List office. The assessment must be verified, or, in IUCN-speak, *evaluated* by an impartial expert (an *Evaluator* or *Reviewer*) on the focal taxa or geographical area, and preferably both. This is done by contacting the Focal Point (i.e. focal person) of the Terrestrial and Freshwater Invertebrate Red List Authority (TIRLA), who will advise on who could do the evaluation. Taxon evaluators have priority over geographical ones, so where there is already an Reviewer for a well-known taxon, such as dragonflies, the dragonfly Reviewer would be assigned the task. But in the case of more obscure taxa, the geographical Reviewer would assist. So, for example, where an Indian earwig requires Red Listing, the Reviewer would be the South Asian Focal Point.

Another issue that was touched on above, but which needs further emphasis, is that the Red List is both dynamic and precautionary. Species can be moved up and down the Red List. The move down to a category of lower threat can sometimes simply come about through better knowledge of the species, which means that it turns out not to be as threatened as previously thought. Alternatively, the conservation status of the species may improve through conservation action. This has been the case, for example, for some South African dragonflies where the key threat was riparian invasive alien trees, and removal of these trees had an immediate positive effect on the species of concern.

Table 6.4 *An example of an abbreviated assessment of two damselfly species, the Ceres streamjack (see Figure 5.1) and Basking malachite (see Figure 1.4) for entering on the IUCN Red List (based on Samways 2006b).*

Class	Insecta
Order	Odonata
Sub-order	Zygoptera
1) Family	Platycnemididae
Taxon name	*Metacnemis angusta* Sélys, 1863
Common name	Ceres streamjack
Status	VU D2
Distribution	South Africa
Rationale	Formerly known from only two female specimens, last one collected in 1920. This species was thought possibly to be extinct, until a population was discovered on the Du Toit's river, near Villiersdorp, Western Cape [Literature Reference]. This population occurs where invasive alien trees (*Acacia mearnsii*) have been removed along the river course [Literature Reference]
Range and population	Known only from one site in the Western Cape
Habitat	Weedy, bush-fringed pools in river braids
Threats	Invasive alien plants are known to be a primary threat. Drainage of pools may also be a threat
Conservation measures	Further searches should be made. Removal of invasive alien trees is a key conservation measure
2) Family	Synlestidae
Taxon name	*Chlorolestes apricans* Wilmot, 1975
Common name	Basking malachite
Status	EN B2ab(i,ii,iii,iv)
Distribution	South Africa
Rationale	Qualifies because of small number, and size of populations, several of which have been lost between 1974 and 2001 through habitat loss and landscape transformation, and area of occupancy of less than 500 km
Range and population	In the early 1970s this species was known from 10 sites [Literature Reference], whereas by 2001, it was known from only 2, showing a decline in extent and quality of habitat. Today it is known only from the Kubusi river at Stutterheim and Thorn River, Eastern Cape. Possibly <1000 adults exist
Habitat	Clear, shallow, rocky streams with riffles and glides, and with an abundance of tall grass, herbs, and indigenous bushes (oviposition site) overhanging the water
Threats	Highly threatened by the synergistic effects of cattle trampling of banks, shading of streams with invasive alien *Acacia mearnsii*, and possibly detergent and alien rainbow trout. This species is not known from any formally protected area
Conservation measures	Searches for further localities are urgently required. Removal of *A. mearnsii* should continue. Liaison with local farmers is essential so that cattle only enter the streams at certain points, fencing off other areas of the stream

The issue of acting in a precautionary way (i.e. applying the *precautionary principle*) is of particular interest to insect conservationists. There are many species and landscape surrogates (e.g. large forest patches) that can play a major role in insect conservation without us being absolutely sure as to what extent. Good conservation involves some foresight, which means we must act before the threatening process continues any further, especially as such a threatening process often has momentum and slowing it down requires planning ahead of time.

Following through this precautionary approach to the species level requires some important considerations. The Red List is indeed a list of *threatened species*, including those that *might* be threatened but for which there is enough information to assign them to a particular category, i.e. they are Data Deficient. However, you can also submit to the Red List office, through an Reviewer, assessments of species that are not necessarily threatened, i.e. of Least Concern (LC). The information is added to the IUCN data base, known as the Species Information Service (SIS). However, when you assign LC to a species, this assumes that you have already taken the precautionary principle into consideration. While the species may be safe right now, there are, in fact, increasing threats on the horizon, in which case you would allocate the category of Vulnerable (V).

Much conservation management activity takes place using political rather than biogeographical boundaries. Although the Red List is essentially about global conservation, it also provides guidelines for regional or national Red Listing, which are on http://www.redlist.org (Figure 6.11). Of course, there are other levels of Red Listing, e.g. State, Province, County, but the process in principle is very similar at all spatial levels. Any *globally* Red Listed species will automatically qualify a species for *regional* Red Listing, although this may not necessarily be done in practice. The aim of regional Red Listing is to provide national legislators with an opportunity to protect a species that may be threatened within a national boundary, even though it may be common elsewhere. Often these nationally significant conservation subjects are at the margin of their geographical range, which generally increases the challenges to conservation, as environmental conditions may not be optimal as at the centre of the geographic range. In some cases, the populations may be isolated, and even separate evolutionarily significant units (ESUs), compared with elsewhere in the range. Nevertheless, most of the issues with regional Red Listing are the same as global Red Listing, although with clear recognition of the relationship between the national population and that over the national border. Two examples of regional Red Listing are given in Table 6.5.

Red Listing automatically draws attention to a species and gives it charisma that it may not have had prior to the assessment. This often leads to a positive

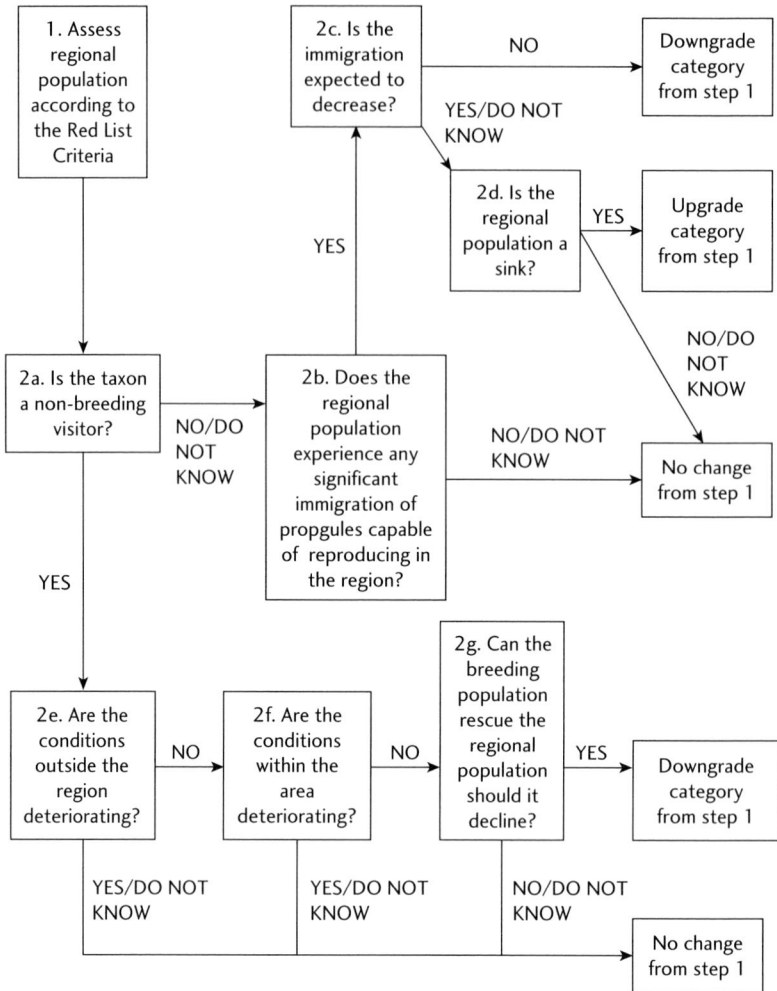

Fig. 6.11 Conceptual scheme of the procedure for assigning an IUCN Red List Category of Threat at a regional or national level. In step 1, all data used should be from the regional population, and not from the global population. The exception to this is evaluating a projected reduction or continued decline of a non-breeding population. In such cases, conditions outside the region must be taken into account in step 1. Likewise, breeding populations may be affected by events in, for example, sink habitats in the case of migratory insects, that must be considered in step 1 (from IUCN 2003; see also www.redlist.org)

feedback loop where more attention is given to future refinements of its Red List status, through further assessments, and often to the desired goal of species-level conservation action. Sometimes this increased effort also leads to the discovery

Table 6.5 *An example of an abbreviated assessment of two damselfly species, the Makabusi sprite and the Hairy duskhawker for South African National (Regional) Red Listing of the species. These assessments may still be entered on the IUCN Species Information System, even though the two species are not globally threatened (based on Samways 2006b).*

Class	Insecta
Order	Odonata
1) Suborder	Zygoptera
Family	Pseudagrionidae
Taxon name	*Pseudagrion makabusiense* Pinhey, 1950
Common name	Makabusi sprite
National status	VU B1ab(i,ii,iii,iv)
Global status	LC
Percentage of global population	c.5%
Distribution	D R Congo, Mozambique, South Africa, Zambia, Zimbabwe
Rationale	In South Africa, it was recorded at Hangklip, Soutpansberg in May 1979 [Literature Reference]. This locality is now severely disturbed with plantation trees and overgrown with invasive alien trees. Despite intensive searches at this site, it has not been rediscovered there. However, it has been rediscovered at several river localities in the Limpopo Province, and where it is susceptible to the adverse impact of invasive alien trees
Range and population	Currently this species is only known to occur in the rivers of the arid north-west of the Limpopo Province
Habitat	Grassy-fringed, slow rivers with partial canopy
Threats	These are distinctive and immediate, and include the poor forestry practice of planting into the riparian zone, and allowing proliferation of invasive alien bushes and trees
Conservation measures	Searches should continue. No specific conservation measures are in place, not currently planned, although the national Working for Water Programme, which is removing alien trees along watercourses, will greatly benefit this species
2) Suborder	Anisoptera
Family	Aeshnidae
Taxon name	*Gynacantha villosa* Grünberg, 1902
Common name	Hairy duskhawker
National status	VU B2ab(ii,iii,iv)
Global status	LC
Percentage of global population	c.1%
Distribution	South Africa, north to Tanzania and Senegal
Rationale	Although widespread and common elsewhere in Africa, this species is only known from two records in South Africa [Literature Reference]. Although much of its habitat is in reserves, these are currently subject to land claims and are under development for eco-tourism. The localities where it was recorded in the past have also been subject to severe drought in recent years, that has led to drying out of its breeding areas. However, with South Africa being at the southern end of its geographical range, it is likely to be marginal, and may be prone to periodic colonization and local extinction, following natural wet/dry cycles
Range and population	In South Africa, known only from swamp forest in coastal, northern KwaZulu-Natal
Habitat	Sluggish streams in swamp forest
Threats	Loss of habitat through tree removal, and construction of infrastructures for tourism
Conservation measures	No specific measures are in place, and none are currently envisaged. It is essential to continue to monitor overall conservation of coastal swamp forest

of new, threatened species (Dijkstra et al. 2007), and also often to better knowledge of closely related species in the same physical area.

6.7 Population distinctiveness

Table 6.6 expands on the idea of ESUs to indicate especially significant populations that might be accorded priority for conservation, and how these might be differentiated for independent management. Such units may represent geographical or ecological outliers that lack opportunity to exchange genes with other populations. They are usually particular or unusual phenotypes (subspecies) or distinct ecotypes. Considerations of *exchangeability* are perhaps particularly important in seeking to define or delimit such entities. Crandall et al. (2000) derived a scheme whereby populations can be ranked or categorized according to whether ecological (immigration and emigration) and/or genetic exchangeability can be demonstrated within recent and/or historical time. The conditions are (1) whether adaptive differentiation has occurred, (2) whether gene flow occurs, and (3) whether this is historical or recent. Their scheme (Table 6.6) allocates populations a '+' or '−' in each of 4 cells, to give 16 categories of divergence between populations. The signs denote rejection (+) or acceptance (−) of the null hypothesis of generic and ecological exchangeability. The greater the number of '+' signs, the greater is the differentiation.

One insect example is that of Legge et al. (1996), who investigated Cryan's buckmoths *Hemileuca* species (Saturniidae), a group of moths allied to *H. maia*, but not formally described. These moths occur on a small number of bogs and fens in eastern North America, and have attracted attention as ESUs possibly meriting management. Legge and his colleagues were unable to differentiate the moths by molecular genetic studies (involving mtDNA and allozymes), so that gene flow between populations was adequate and there was no reason to reject genetic exchangeability. However, the buckmoth differed from all other *Hemileuca* tested, in that caterpillars could develop on *Menyanthes trifoliata*. This meant that recent, but not historical, ecological exchangeability could be rejected, to give the code corresponding to category 6 of Table 6.6. This ecological differentiation was regarded as having adaptive significance, so that separate management units (or ESUs) were deemed to occur.

Frankham et al. (2002) commented that this approach helps to bring some order to the largely more subjective array of designated ESUs, by focusing on objective evolutionary criteria to elect units that appear to merit special notice. A second insect example discussed by those authors is the Puritan tiger beetle

Table 6.6 *Defining management units (evolutionarily significant units, ESUs) within species based on genetic and ecological (immigration and emigration) exchangeability. Categories of population distinctiveness (see text) are based on rejection (+) or acceptance (−) of the null hypothesis of genetic and ecological exchangeability for recent and historical time frames. The numbers down the left side reflect decreasing evidence for significant population differentiation (from Crandall et al. 2000). Notation in the form of 'recent genetic/ recent ecological' (top line), 'historical genetic/ historical ecological' (bottom line).*

Category: Relative strength of evidence	Evidence of adaptive distinctiveness	Recommended management action
1 (strong)	+/+ +/+	Are separate species
2	+/+ +/+ −/+ +/−	Treat as separate species
3	−/+ +/+	Treat as distinct populations (recent admixture, loss of genetic distinctiveness)
4	+/− +/+	Treat as distinct population: natural convergence (treat as single population), or anthropogenic convergence (treat as distinct population)
5	(a) +/+ (b) −/+ (c) −/− −/− −/+ +/+	(a, b) recent ecological distinction so treat as distinct populations; (c) allow gene flow consistent with current population structure
6	−/+ −/−	Allow gene flow consistent with current population structure; treat as distinct population
7	+/− +/−	Same as category 6
8 (weak)	+/− −/− −/− +/− −/+ −/− −/− −/+ +/− −/+ +/− −/−	Treat as single population; if exchangeability is due to anthropogenic effects, restore to historical conditions; if natural, allow gene flow

Cicindela puritana, classified to category 2 (Table 6.6). The two populations examined were not genetically exchangeable (mtDNA: low gene flow, significant differentiation) nor ecologically exchangeable, based on habitat features. These populations, from Connecticut River and Chesapeake Bay, were thus strongly distinct and were recommended for separate management regimes (Vogler et al. 1993).

Although the concept of ESUs may have practical value in drawing attention to the often marked differences between populations of nominally the same species, an opposing view is that the debate over their value is 'intellectually empty' (Ehrlich et al. 2004), and distracts from proceeding with effective conservation (see debate in DeWeerdt 2002). However, there are, among insects, some very significant ESUs (such as subspecies of lycaenid butterflies) that require specific and concerted attention. Ehrlich et al. (2004) also overlook the fact that often what were thought to be simply separate ESUs in fact turn out to be different, but cryptic species. We recommend that as an insect ecologist and conservationist, you remain very aware of the concept of ESUs and their conservation.

6.8 Population condition

Young and Clarke (2000) emphasized that traditionally there have been two rather distinct fields of endeavour in population studies. The ecological viewpoint sees demographic processes as central to determining population size. This has been by far the more common approach in conservation studies. However, the genetic viewpoint can be equally important in helping to emphasize loss of *genetic diversity* or *population fitness*. This, in turn, recalls Caughley's *small population paradigm* (section 6.1), i.e. considerable genetic change and stress may go hand in hand with low numbers of reproducing individuals, and may be far more deleterious in a small population than in a large one, and manifest as decreased individual fitness.

Lowered genetic diversity reduces a population's potential to adapt to environmental change. This is of course essentially a long-term, evolutionary perspective (although with the rapid generation times of several insects adaptation may occur over comparatively short time periods). However, any short-term reduction in fitness due to inbreeding is usually of more immediate concern. Genetic concerns also promote use of a concept rather different from that used by many ecologists—that of the *effective population* (N_e, the number of individuals contributing to the next generation) rather than simply the total number of individuals present (the *census population*). All the consequences of small populations, and many of the concerns underpinning the broader development of conservation genetics, are couched in terms of the effective population. The concept is thus related to that of *minimum viable population*, the minimum effective population needed to maintain genetic variability.

Frankham et al. (2002) suggested that for a threatened species to retain its evolutionary potential, a N_e of 500–5000 individuals is required, figures that are usually far greater than those of most populations that arouse conservation concern. The two most relevant issues to consider are (1) does the species (population) have enough genetic diversity to evolve in response to environmental change, and (2) is the population sufficiently large to avoid loss of reproductive fitness by avoiding inbreeding depression. A third goal of genetic management is to avoid the continued accumulation of new deleterious mutations.

Management policies for populations therefore seek to maximize genetic diversity, while at the same time attempting to reduce *small population effects* of inbreeding depression and *genetic drift*. Genetic diversity is likely to be related directly to variability in fitness of individuals within a population, and the ability of a population to adapt to environmental change or disturbance. Brakefield (1991) noted several generally accepted management policies likely to help conserve genetic diversity in populations:

- Maintain population size, minimizing incidence and duration of any bottlenecks.
- Maximize the proportion of breeding adults.
- Preserve relevant habitat and environmental heterogeneity.
- Minimize the nature and extent of any anthropogenic environmental change.
- Retain natural patterns of gene flow.

We strongly recommend that careful consideration of each and every one of these five points in your research planning. You should optimize ways of measuring them, and determine whether they are ample enough to conserve populations of the focal species.

6.9 Population modelling

The predictive modelling of animal populations, to assess their size, viability, and chances of extinction, has become a popular exercise, with a number of powerful software packages being devised for this purpose. Sound application of any such model demands high-quality data, coupled with good understanding of a species' biology and the factors that affect its abundance and reproductive dynamics. Section 6.6 on conservation status noted the almost universal lack of such quantitative information for insects, in the context of being unable to reliably quantify extinction risk. Thus, application of one important program, RAMAS-Red (R), devised for this very purpose, commonly leads to ambiguity, as the outcomes may polarize between the categories Critically Endangered and Data Deficient (Clarke and Spier 2003). This emphasizes that the best information available is still insufficient to do more than suggest the need to apply the *precautionary principle* for many species. Such an approach is intuitively obvious, even from less sophisticated observations, and is usually accompanied by a recommendation that you should 'find out more' to remove the ambiguity.

Frankham et al. (2002) made the perceptive comment that the most important contributions of risk assessments using *population viability analysis* (PVA) do not necessarily come from the quantitative assessments of extinction risks themselves but, rather, from the process of conducting the exercise. Thus, it is necessary to (1) summarize all available information about the life history and biology of the target species, (2) identify threats to the species, (3) assess their likely importance, and (4) identify potential recovery strategies and evaluate their relative impacts. Many PVAs have been developed by teams, commonly through workshops involving all (or most) experts on a given species, so that all available points of view are canvassed. One problem with such exercises for insects is simply that often there are few people with practical experience of any one species in the field, so that adequate informed peer-review by independent parties may be difficult to achieve. The process helps to identify significant gaps in knowledge and understanding, and may lead to increased support to redress this. PVAs for insects may have very limited practical applications at present, and there is much scepticism over their value, not least because of the levels of stochasticity and uncertainty that must inevitably be superimposed on poor foundation data.

The widely used software packages for PVA incorporate demographic and environmental stochasticity, and fall into two main groups:

- **Individual-based**, tracking the life of each individual in the population (e.g.VORTEX, GAPPS)
- **Matrix-based**, tracking only the number of individuals (e.g. RAMAS). RAMAS Metapop and VORTEX can incorporate fragmented populations.

Brook et al. (2000) provide a useful critique of the performance of several PVA models.

6.10 Summary

The population is often the unit of particular interest in ecological and conservation studies of insect species. Of particular significance for insect conservation are declining populations and small populations, representing pressure on them and associated extinction risk. Regrettably, much species-focused conservation is *crisis management*, emphasizing that more strategic thought and planning is required in the future.

Assessing changes in insect population levels is not easy, because so many insect populations are dynamic and highly variable over time. Furthermore, the assessable forms of the insect species, usually the adults, are only *apparent* at certain times of the year, and for rare and threatened species the numbers may be very low and difficult to assess. Nevertheless, procedures are available for estimating insect population sizes and levels, as well as their viability, which are also linked to aspects of behaviour and habitat conditions of the focal species. When this is done for several species country-wide, as for British butterflies, a very powerful national assessment tool becomes available.

The concept of *metapopulation*, which relates to natural extinction/recolonization processes across the landscape is particularly applicable to insects. It is significantly related to habitat quality and *connectivity* of quality habitat patches.

Several methods are available for estimating insect population size. However, it is rarely possible to obtain accurate estimates, as populations of insects naturally fluctuate considerably in both time and space. Furthermore, sampling must be done when environmental conditions for assessment are good. Choosing which method you should use depends on your exact goal and carefully estimating the resources you have available for attaining it at an actionable level of accuracy.

Threatened species present some special issues as they and their habitats should not be harmed, and their population existence, let alone its level, may be very hard to determine. Nevertheless, comparative methods are available.

Assessment of conservation status is an important exercise, and with it comes an estimate of category of threat to populations of a species. The 'gold standard' for assessments of threatened species is the IUCN *Categories of Threat* (based on five *Criteria*) that leads to the placement of species on the *Red List*. Of the five criteria, only three are generally practicable to insect conservation: extent of reduction in population size, geographic range, and overall IUCN 'population size'. Recommendations are given here for conducting Red List assessments both for global and for national Red Listing. Such assessment activities can also be applied to species subunits, known as *evolutionarily significant units* (ESUs). In turn, populations can be assessed on their condition (i.e. viability), and in some cases, with good data, their population dynamics can be modelled.

7

The population and the landscape

7.1 Introduction: the relevance of landscapes

Organisms, the places in which they live, and the resources they need, occur across *landscapes* (or waterscapes in the case of aquatic species). The term 'dispersion' is used to describe the distribution of individuals within and across their habitat or the landscape, and organisms are commonly unevenly or patchily distributed (aggregated). The major concerns for the well-being of many insects come about from changes to the structure (especially of plant types), species composition (*ecological integrity*), and quality (*ecosystem health*) of these landscapes. Extensive human-induced changes have led to loss and fragmentation of habitats, and hence reduced quality or loss of critical resources.

Processes such as removal of indigenous vegetation, the incursion of alien species, or of structural simplification of habitats such as agricultural modification or urban development, are instrumental in affecting insect abundance and distribution and the species composition across the landscape. These changes may decrease the *extent of occurrence* and *area of occupancy* of insect species through fragmentation of previously more extensive and continuous populations (Fahrig 2003) (Figure 7.1). In addition, quality of habitat can be influenced by the products of human activity, such as fertilizers, pesticides, and pollutants.

Many insects that are a major focus of conservation action now occur only in small fragments of their former distributions, and with loss in number of populations, may decrease in population size, and have reduced potential for continued survival. Two parameters are particularly important for conservation: (1) the amount (e.g. area) of habitat present and the quantity and quality of the resources it provides, and (2) the extent to which *habitat fragments* (*patches*) become functionally isolated in the landscape, in relation to the ability of organisms to move between them.

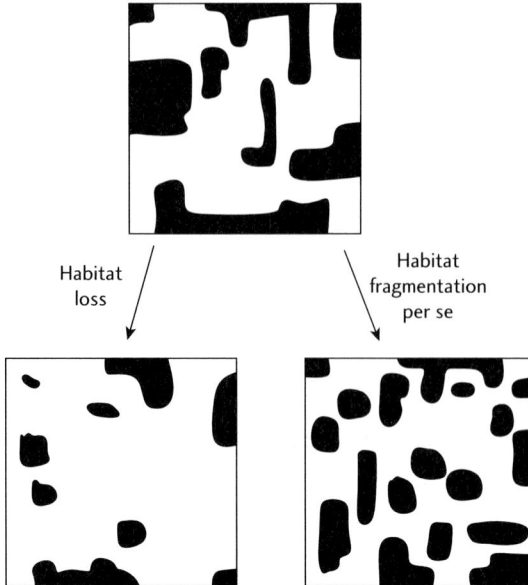

Fig. 7.1 Smaller patches (black) in the matrix (white) come about through habitat loss and habitat (landscape) fragmentation per se. This means that reduced patch size can come about from the combined effect of loss of habitat and of increased habitat fragmentation. Habitat loss also increases patch isolation (from Fahrig 2003).

Both the size of the patch and its distance from other populations (i.e. other patches) are important for colonization success by dispersing insects. Colonization success therefore also depends on the *dispersal* ability of the species. The size of habitat patches also influences the *carrying capacity*, as well as the extent of the impact of *disturbances* from outside the patch (i.e. the *matrix*). Much conservation at the landscape level involves reducing the detrimental effects of *fragmentation* through attempts to improve landscape structure, design, and quality. The twin strategies dominating this approach are improving *connectivity* between habitat patches (i.e. functionally reducing their isolation) and protecting them from further despoliation. Reducing the inhospitality of the matrix is a potentially effective measure for improving landscape connectivity. Simulating natural disturbances, such as fire or megaherbivore impact, is another option for maintaining habitat quality. All these approaches arise from the realization that much insect conservation has to be considered at the landscape, rather than species, level if it is to be truly effective. Furthermore, insects may be useful for monitoring the success of landscape restoration efforts for biodiversity in general (*bioindicators*; Chapter 10). Intact landscapes are, however, also important for the conservation of individual species, in which the structure of the population and its ability and tendency to disperse may be important attributes to clarify.

7.2 Effects of landscape structure

Habitat fragments, which by definition are isolated physically, may not necessarily be isolated functionally, and landscapes can, to some extent, be designed to facilitate movement of insects between such patches (see Chapter 6), by providing *stepping-stone habitats*. However, at some stage in the trajectory from the original extensive habitats, through fragmentation, to the final stages of attrition, such fragments become too isolated, too small, or too degraded for certain insects to survive (Forman 1995) (Figure 7.2).

There is no universal rule as to when such irretrievable loss thresholds are reached (see Figure 7.3). Indices relating habitat extent, connectivity, and quality are often elusive, and can be considered only on an individual species basis. For example, a large, dispersive birdwing butterfly may need hundreds of hectares of tropical forest to sustain a viable population, whereas a small sedentary lycaenid butterfly may need only a hectare or less. However, precise knowledge of these dimensions for most insects is not available. This emphasizes the significance of the *precautionary principle* for land management for insect conservation. This principle acknowledges that (1) undue landscape fragmentation or habitat loss is undesirable, (2) increased connectivity between patches of habitat is desirable, and (3) imposition of barriers (broadly, impediments to free movement) is undesirable. These three components all contribute to maintaining insect diversity, given that there are so many species and complex interactions that are virtually unknowable within the time frame available for conservation. This means that insect diversity must necessarily be maintained by adopting a *precautionary approach* (Samways 2005).

The rationale behind the precautionary principle is illustrated by studies of insects in agricultural landscapes. Aspects of cultural management and the biological control of crop pests draw upon ecological principles to reduce the adverse effects of land-use intensification on natural enemies (New 2005). Such approaches implicitly involve assessment of insect dispersal and of the factors that facilitate or impede this in the landscape, as well as assessment of the incidence and persistence of focal insect species in the habitat array (Tscharntke et al. 2007).

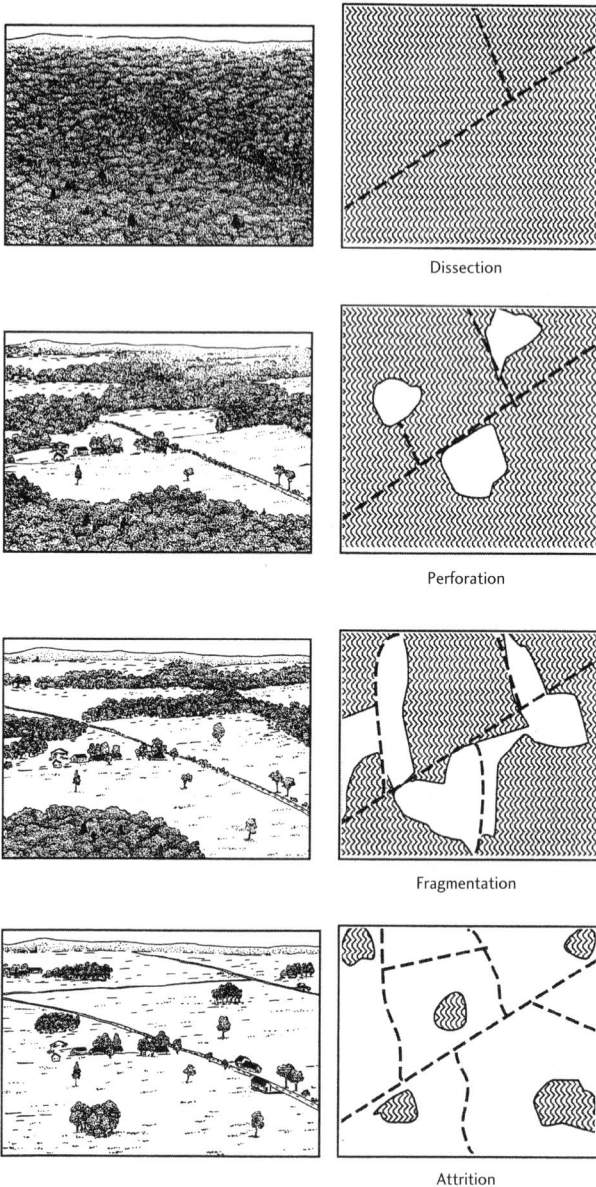

Dissection

Perforation

Fragmentation

Attrition

Fig. 7.2 The building of a road across the natural landscape dissects the landscape, and with agricultural development the landscape becomes gradually perforated. With more and more agricultural conversion, the disturbed areas become larger, and merge together, so isolating the remaining natural remnants, indicating fragmentation. Then with even more agricultural (or urban) pressure, the remnant patches become smaller, a process known as attrition (from Hunter 1996, following the terminology of R. T. T. Forman, with kind permission of Blackwell Publishing).

(a)

(b)

(c)

Fig. 7.3 Fragmentation of the landscape in southern France has led to reduced separation of two sibling species of bush cricket, *Platycleis affinis* (above) and *P. intermedia* (below), resulting in their coming into acoustic range of each other, due to lack of a gradual ecotone separating them. The result is that the song of *P. affinis* (a) influences the song of *P. intermedia* (b), with temporary modification of the song of *P. intermedia* towards that of *P. affinis* (c).

Establishing the suitability of fragmented or remnant habitat patches for insects depends on surveys to determine the density and population sizes, and the presence of other species on which these species depend. The critical resources needed for the focal insect to thrive must also be quantified. All the various survey methods noted in Chapter 6 may apply, and practical questions include:

1) Is a habitat patch of sufficient size or quality to support a typical assemblage or viable population of a nominated species?
2) Can the patch be colonized naturally from existing populations in other habitat patches to provide genetic interchange?
3) If not, is it necessary to inoculate it with founder stock (Chapter 8)?
4) Will it need management as an independent site?
5) Can changes to the surrounding landscape area improve its effective size or carrying capacity (such as by better buffering and reducing edge effects), its quality and accessibility?

Knowledge of population structure and species-specific dispersal ability is clearly needed for any particular insect that becomes the target of such an endeavour.

The effects of fragmentation and habitat loss are thus essentially twofold. *First*, a small habitat patch, by definition, will only be able to sustain a few individuals. Inevitably, there is a high risk that such populations will not be robust enough in the face of chance adverse environmental impacts and, in addition, have less genetic resilience (see below). When these effects are translated across all populations of different species confined to small patches, inevitably, given time, some of those populations will go extinct. Thus a small patch will have relatively low species richness. This process of gradual loss of populations leading to local extinction of species is known as *ecological relaxation*.

From the basic tenets of island biogeography theory (MacArthur and Wilson 1967), species richness is related to island size, with the analogy of habitat patches being 'islands' in a 'sea' of inhospitable terrain. This model predicts that halving the habitat area may lead to loss of 10% of resident species. Additionally, decreasing the size of a habitat patch increases its vulnerability to edge effects, such as invasions by alien plants and animals, with likely further loss of habitat quality and dangers from competition and other interactions. In essence, the amount of unaffected interior habitat may be reduced substantially. Even though a patch may appear natural to the casual observer, its suitability for indigenous species may be diminished (Klein et al. 2002) (Figure 7.4).

Second, isolation reduces genetic exchange between populations, leading to deleterious genetic processes, such as inbreeding, that influence the fitness of the resident population. Sato et al. (2008) give a good example of this. They showed

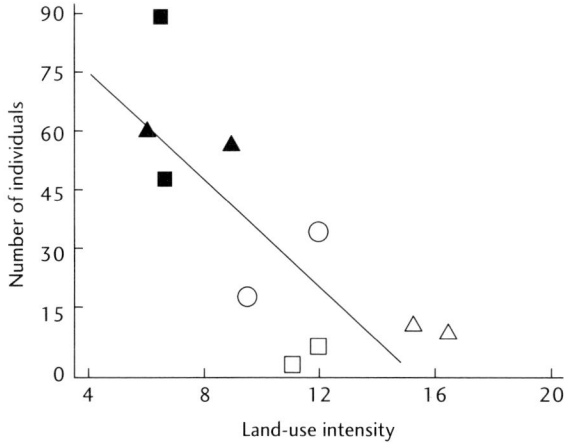

Fig. 7.4 Land-use intensity in relation to numbers of individuals of social bees, indicating a decline in overall abundance with increasing intensity of land use. The different symbols represent different land-use types from low-intensity, near natural forest through intensively managed agroforestry (from Klein et al. 2002).

that populations of three sympatric damselfly species that were isolated in urban patches had only half the genetic differentiation of those in more connected rural populations. Conservation involves assessing and reducing these genetic bottlenecks through reducing population isolation within the landscape. It also involves establishing the functions of particular landscape features, particularly (1) their influences on dispersal, by being barriers or corridors, respectively impeding or facilitating movements between habitat patches, and (2) whether they remain suitable habitats for insects even though changed from their original form and composition, as in the case of agro-ecosystems.

Landscape hospitality to insects must be considered at spatial scales far different than for most vertebrates. Features such as hedgerows only 1–2 m wide can furnish interior habitat. However, this only applies when there is suitable vegetation and other conditions. In turn, such features may be harmful to certain other species by constituting barriers. Alternatively, they may be beneficial by being *movement corridors* for dispersing insects. Even features such as single-lane roads or mown strips (such as below power lines) may interrupt dispersal and cause fragmentation of a previously continuous range (Mader 1984).

The microclimate within a monoculture of an agricultural crop or intensively managed pasture may also be much changed from that of natural vegetation on the same site. Even transformed landscapes that may appear benign to humans can present insects with formidable problems and threats (Figure 7.5). The

Fig. 7.5 An example of fragmentation of a natural landscape, with extensive natural habitats only present today at the higher elevations, with the lowlands extensively and intensively modified, especially for agriculture. Knowing the value of such natural fragments and their connectivity (the extent to which insects can move between them) is an important challenge in insect conservation.

ability of insects to move within the landscape is essential for the maintenance of *ecosystem services*, for example the movement of pollinators. The consequences of isolation within an alienated landscape include *cascade effects*, where loss of one component of the food chain affects other components of the community, increasing the risk of further local extinctions.

The extent of isolation of a putative habitat patch can be assessed only in relation to the dispersal capability of species seeking that patch. The study of insect dispersal capability is thus an integral part of insect conservation at the landscape level, and sometimes essential for formulating appropriate conservation management strategies.

Changes in geographical range of insects are commonplace, but understanding these has received more attention in recent years because of the influences of climate change, whereby ranges may expand or contract along geographical or climatic gradients. This syndrome differs in implications from understanding more local insect movements within a species' current range, which may have more immediate application for habitat management. Indeed, viability of populations or metapopulations is highly dependent on local movements, which enable colonization of good-quality habitat patches. However, when such movements are prevented extinction may result, illustrating how landscape features can determine survival or loss of a local population.

The most thorough documentation of such phenomena comes from the British Butterfly Monitoring Scheme. The data for this scheme are derived from transect surveys (Chapter 6) and illustrate the appearance and loss of species from long-term transect studies. These findings have also subsequently been validated by more intensive searches. The following premises were used to recognize colonization or extinction in the survey:

- A breeding population was deemed to be 'present' when a butterfly species was recorded over four successive flight seasons/periods, or 'absent' when there were no records for four successive flight periods. Other combinations of records were considered inconclusive.
- Extinction at a site was assumed when 'presence' was followed by 'absence', as determined above.
- Similarly, colonization was assumed when 'absence' was followed by 'presence'.

Using these criteria for British butterflies, Pollard and Yates (1993) identified 123 colonizations and 79 extinctions over 15 years, mostly relating to within-range changes of species. These results support the concept of metapopulations, rather than signalling any major changes in geographic range. In contrast, geographic range changes are seen at the edges of a species' range, and usually occur along latitudinal, elevation, or other ecological axes. Since these early studies, there have been further substantial changes in the geographical ranges of some of these butterfly species as a result of accelerated climate change (e.g. Hill et al. 2002) combined with local extinctions from habitat loss (Warren et al. 2001).

7.3 Resources and landscape geometry: the insect's habitat

There is much loose terminology surrounding the concept of habitat, which is so often thought of as some sort of fixed spatial entity. Such a spatial perspective that describes the physical domain of the insect is really its biotope. Another slant, and of great significance for conservation, is an appreciation of the habitat rather as a set of resources required by an insect. Dennis et al. (2007) provide a detailed discussion of the concept, which we strongly recommend that you consult. They point out that a convenient way of categorizing such resources for insects is under each stage of the life cycle. This is forcefully brought home when one considers that, functionally, a caterpillar is a very different animal from the butterfly. This concept is well known to entomologists as developmental polymorphism. Dennis et al. (2007) point out that an adult butterfly, for example, would require a certain set of resources for egg laying, mate location, resting, roosting, feeding, and predator escape. Other stages can be treated using the same approach, and their sources mapped. The habitat then becomes the intersection and union of these resources (Figure 7.6). The resources required by each stage may be visualized as belonging to two groups:

- **Consumables**, such as host plant parts and nectar
- **Utilities**, including (a) conditions for existence and persistence, such as sites for thermoregulation, mate location; and (b) suitable conditions for development and activities, such as suitable microclimates and enemy-free space (Figure 7.7).

Determining all these features and establishing which ones are critical, and even important, under all conditions, is not an easy task. There may be some unknown vital factor which comes into play under certain conditions. For example, while certain microclimatic conditions may be ideal for proliferation of a certain life stage, such conditions often have a trade-off. Warm, moist conditions may, for example, stimulate an increase in the population, which is offset by epizootics as a result of high population density, and such trade-offs might be one of the reasons why some species are rare. Nevertheless, the conceptual framework provided by Dennis et al. (2007) has great merit when developing a species conservation research programme.

7.4 Insect movements and MRR methods

Much of our understanding of the influences of particular landscape features has come simply from watching insects and their reactions to these features. For example, the phenomenon of *hilltopping* (Shields 1967) by some butterflies and other insects has important ramifications in conservation management, in that such topographically defined assembly sites may be a critical resource for species otherwise distributed widely and at low densities across a wide landscape (Figure 7.8). Understanding of this behavioural phenomenon, primarily a mate-finding exercise, has come from watching the insects, and sometimes marking them to trace their movements.

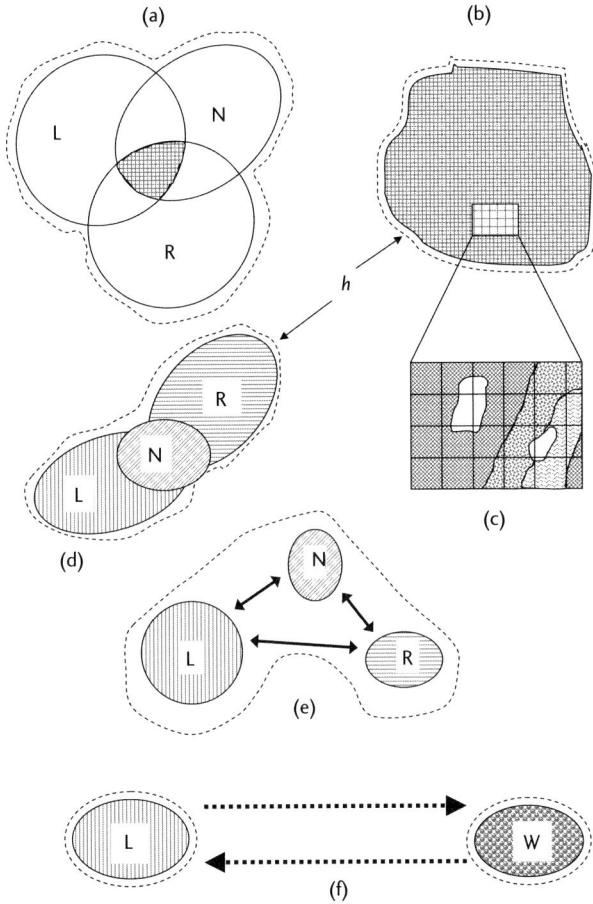

Fig. 7.6 A habitat model based on resource distributions and individual movements. For simplicity, resources are shown schematically as sets or envelopes; the elements of sets are arbitrary units of ground space, based on fine-scale responses of individual butterflies, e.g. 1 m units, as illustrated by the grid in (c). A maximum of three resources is illustrated in each diagram: N, nectar resource; L, larval resource; R, roost sites; W, over wintering sites; h, habitat boundary. Resources are combined by daily search flights to give habitats by: (a) intersection and union; (b) equivalence and equality (e.g. a host plant which is also a nectar source and roost substrate for a butterfly); (d) contiguous union; (e) disjointed union linked by back and forth daily flights; and (f) disjoined non-union linked by seasonal migration. Shading distinguishes resources (c) illustrates a small part of (b) in which light shading is host plant, medium shading density is nectar, heavy shading is jointly nectar and host plant, and white is neither, but can be used for other activities, such as pupation, adult resting, etc. The habitat core is illustrated by cross-hatching (from Dennis et al. 2007).

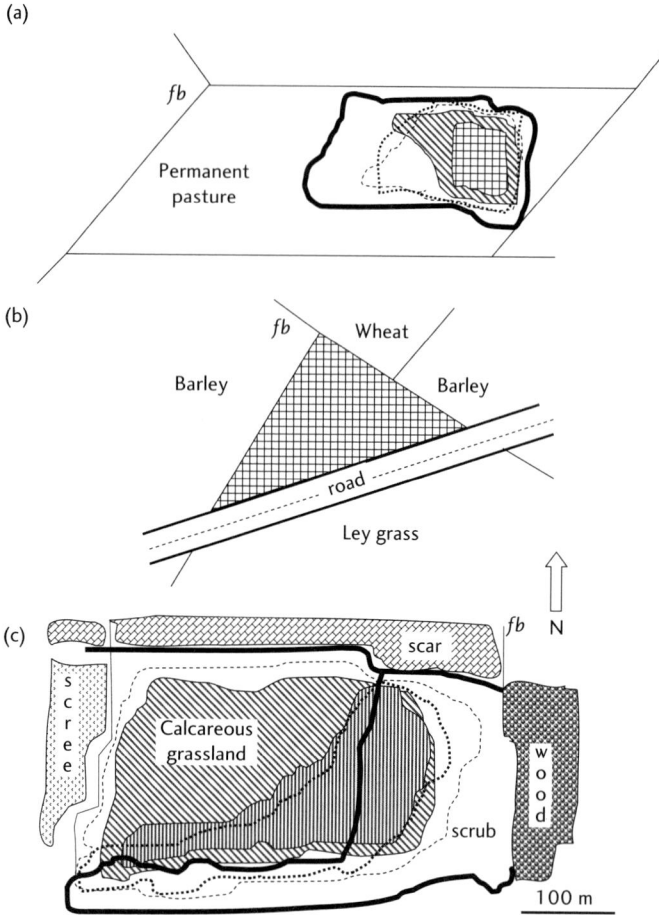

Fig. 7.7 A practical interpretation of the habitat model presented in Figure 7.6. Here the relationships between resource distributions, forming habitats for butterflies, vegetation units and field boundaries are shown. (a) Resources coinciding with but occupying less area than both the vegetation unit (valley mire) and field boundary; e.g. the Meadow brown butterfly *Maniola jurtina*. (c) Resources overlapping vegetation units but within old field boundaries; e.g. Silver-studded blue *Plebejus argus*. Lines and shading: thin lines, field boundaries; thick continuous lines, vegetation boundaries; diagonal shading, host plant; horizontal line shading, host plant suitable for egg laying and larval development in period of study; speckled line, area of nectar sources; dotted line, area of mate location; cross-hatching, coincidence of suitable host plants, nectar sources, and conditions for mate location (from Dennis et al. 2007).

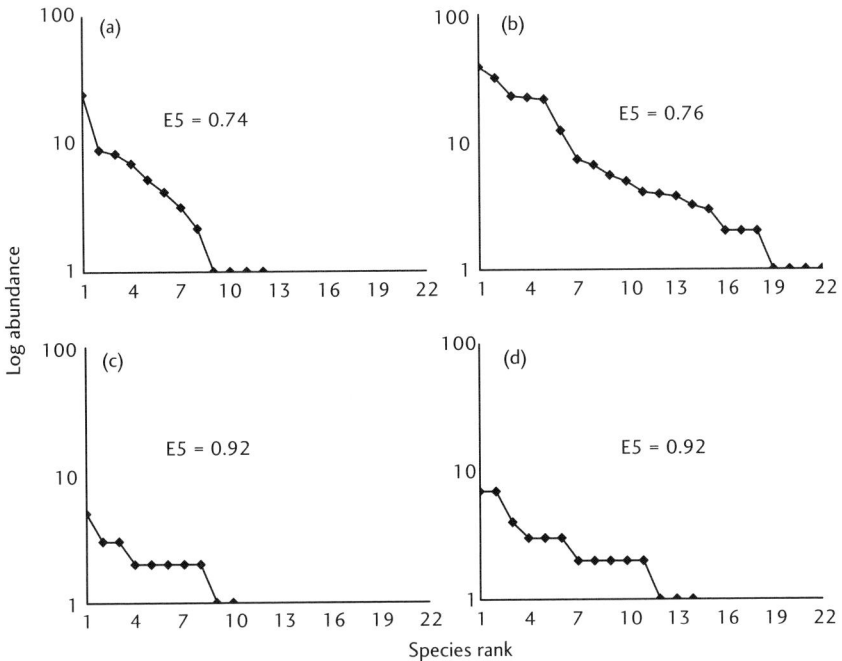

Fig. 7.8 Rank-abundance curves for butterfly species hilltopping on hills with different vegetation cover. Hills covered with (a) alien *Eucalyptus* trees, (b) natural grassland, (c) alien *Acacia* trees, and (d) natural forest. The alien trees are planted on grassland and reduce both the species richness and the evenness (E5) of the butterfly assemblage. Note that it is always important to compare similar vegetation structures with each other to determine if it is the alien plants per se, or merely a change in habitat structure that is causing the change, i.e. trees of any sort, whether indigenous or alien, reduce hilltopping behaviour (from Lawrence and Samways 2002).

The extent of dispersal (or sedentariness) of an insect population will determine the levels of individual movements in and out of the habitat or site, and subsequently its population structure (Chapter 6). Evaluating the frequency and distance of these movements, particularly the ability to disperse between habitat patches, is a key aspect of insect conservation planning. (The scale of such movements across the landscape is different from the mass, long-distance migration undertaken by some insect species.) A commonly used approach to help estimate movement is with the use of mark–release–recapture (MRR) methods, which have a number of pertinent applications in this context as well as in estimating population size. Before embarking on a project on insect movement, we strongly recommend that you consult the excellent review by Osborne et al. (2002).

Although not necessarily part of a MRR protocol (as it does not prescribe recapture), we draw attention to the rapidly developing field of radio telemetry which is a marking technique that tracks the movements of individual insects (Box 7.1). The principle is simple enough, although the technology is sophisticated. A minute radio transmitter is attached to the insect and it is then tracked using conventional radio telemetry techniques on foot (Hardersen 2007), or from vehicles or light aircraft (Wikelski et al. 2006) (Figure 7.9). Should you wish to embark on this field of research, we strongly recommend that in addition to consulting the literature you also consult experts in the field.

Box 7.1 Radio telemetry in insects: unlocking biological secrets using microtechnology

Radio telemetry (RT) is a technology that allows the remote measurement of one or more variables of interest. Radio telemetry generally consists of a transmitter placed on the animal (see figure below) and a hand-held receiver unit with an antenna used by the researcher to monitor the animal's position, or to record a particular variable in real time (e.g. body temperature) (see table below). More elaborate telemetry technologies include automated receiver units and direct satellite uplink to the transmitter or receiver.

One of the major constraints of RT technology is the mass of the transmitter relative to the body mass of the species investigated. The general rule of thumb is that the transmitter should be 10% (or less) of the animal's body mass, and that it should not alter the animal's natural behaviour or physiology, although this depends to some degree on the questions asked during the study. Therefore, an essential component of RT studies is the demonstration that the individuals studied are unaffected by the additional load of the transmitter (see rigorous methods described in Lorch et al. 2005, Wikelski et al. 2006, Negro et al. 2008).

Two main types of telemetry devices are generally recognized, namely passive and active devices, termed for the absence or presence of an onboard power supply. Passive telemetry devices, such as radio-frequency identification (RFID) or passive integrated transponder (PIT) tags, transmit unique frequency signature data. They can be smaller and lighter because they have no battery, and are generally used for identifying a particular animal. However, for most active devices every model has a fixed upper limit to data capacity and therefore these devices tend to trade off performance against data storage capacity (i.e. the device can record many parameters for a small number of total samples or a larger number of samples for a single parameter) and both of the latter are traded off against the lifespan of the RT device. Critically, the use of a passive device requires obtaining the data from the transponder, thus some recapture of, or at least gaining close proximity to, tagged individuals becomes a priority. Active devices, on the other hand, allow the remote collection of data which is dependent on the battery size of the transmitter and the radio-frequency environment of the study site. Active

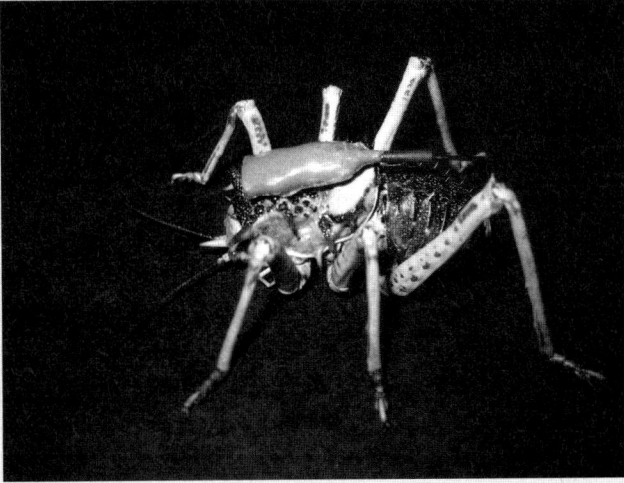

Insect with RT device for recording body temperature (Holohil Systems, Canada) attached to a king cricket (Orthoptera: Stenopelmatidae) (photo J.S. Terblanche).

devices also tend to be more expensive because of the production costs of small batteries. It is probably worth mentioning, however, that there are other telemetry devices, which may be categorized in between purely active or passive, that have various commercial applications but with huge potential for biological research (see O'Neal et al. 2004, Naef-Daenzer et al. 2005).

Research areas that have benefited most from RT are studies using large animals (mammals, fish, reptiles, and birds) and include movement patterns (such as migration routes), home range sizes, activity times, and foraging behaviour (reviewed in Cooke et al. 2004). For example, major breakthroughs in physiological ecology, such as field metabolic rates and diving behaviour of marine mammals and birds, have been aided by RT technology (Cooke et al. 2004). However, notable achievements in insect biology research have also been produced using RT. For example, the evolutionary benefit of synchronized group behaviour in insects, such as periodic mass outbreaks (or swarming), is generally poorly understood. Recent work by Sword et al. (2005), which relied heavily on RT to recapture the mormon cricket *Anabrus simplex* (Tettigoniidae) after marking a sample of individuals, revealed that synchronized outbreaks and migratory bands used by this species can reduce individual risks by diluting predation pressure. In addition, much recent work on insect migration has owed its success, at least in part, to the development of radio microtransmitters (Wikelski et al. 2006). Such advanced technology has also contributed to novel insights into the free-flight behaviour of the hawkmoth *Agrius convolvuli* (Wang et al. 2008).

However, to date RT has generally played a relatively small role in insect biology research. This has probably been a consequence of technological and/or financial constraints, in terms of the size limitations in production of microprocessors and accompanying power sources (i.e. batteries). Currently, the smallest RT device with an onboard power supply is ~0.2 g (Naef-Deanzer et al. 2005). Nevertheless, it seems that with the increasing present-day commercial drive, and the decreasing production and financial limitations which were once considerable, potential applications are opening up in numerous research areas (e.g. Kutsch 2002, Ando et al. 2002). The use of harmonic radar may, however, allow this size barrier to be broken. Indeed, lighter passive devices (tag weight range 0.0004–0.08 g) that are able to yield at minimum spatial location data, have been used in entomology and may prove fruitful for certain applications (O'Neal et al. 2004). If RT data can be obtained by remote satellite relay, rather than by the costly and time-consuming field readings generally in use, the applications to biology are limitless and will prove invaluable in small-animal conservation (Wikelski et al. 2007). Undoubtedly, RT will be one of the major research tools unlocking the secrets of insect biology in the future.

John S. Terblanche, Department of Conservation Ecology and Entomology, Stellenbosch University, South Africa

Some examples of insect research using RT. The species' body mass, transmitter mass, and manufacturer of the RT device are shown

Order	Species	Species mass (g)	Transmitter mass (g)	Transmitter manufacturer	Research activity	Reference
Coleoptera	*Carabus olympiae*		0.3	Biotrack Ltd, UK	Dispersal patterns	Negro et al. (2008)
Coleoptera	*Carabus coriaceus*	1.195–2.083	0.6–0.7	Holohil Systems, Canada	Dispersal patterns	Riecken and Raths (1996)
Coleoptera	*Osmoderma eremita*	1.7	0.48–0.52	Holohil Systems, Canada		Hedin and Ranius (2002)
Coleoptera	*Scapanes australis*	6.43-8.95	0.47–0.49	Holohil Systems, Canada	Flight patterns	Beaudoin-Ollivier et al. (2003)
Odonata	*Anax junius*	1.2	0.3	Sparrow Systems, USA	Migration	Wikelski et al. (2006)
Lepidoptera	*Agrius convolvuli*	0.94	0.23		Free-flight control	Wang et al. (2008)
Orthoptera	*Schistocerca gregaria*	3.00	0.3		Free-flight muscle potential	Kutsch (2002)
Orthoptera	*Anabrus simplex*		0.45	Biotrack Ltd, UK	Dispersal patterns	Sword et al. (2005)

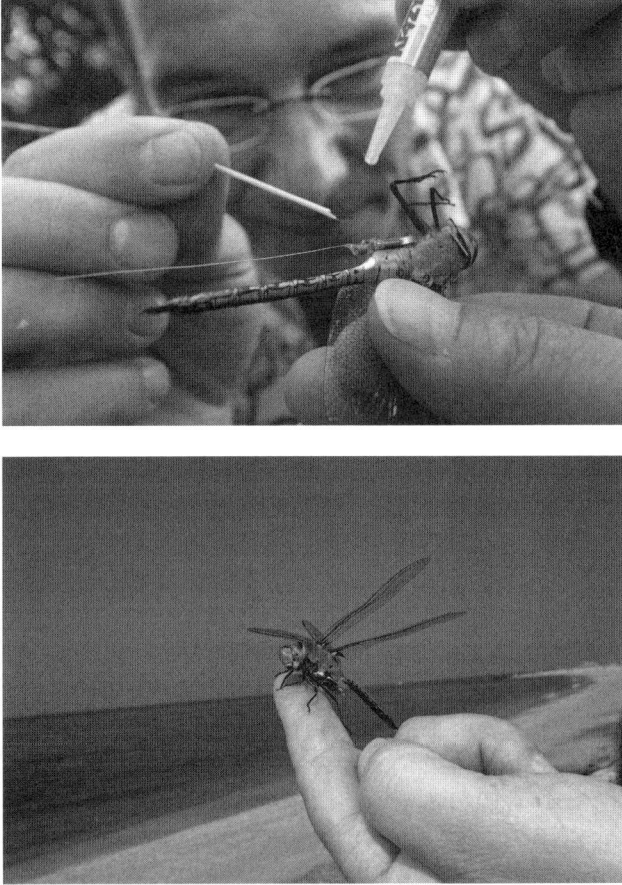

Fig. 7.9 Attachment of a 300 mg radio transmitter to the thorax of a Green darner dragonfly *Anax junius* for determining its migration pattern along the east coast of the USA (Courtesy Christian Ziegler).

The principle of MRR is fundamentally simple. Individuals are captured, marked in some way (see Hagler and Jackson 2001) so that they can be recognized in the future, released at their point of capture, and allowed to disperse into their parent population. On a later occasion, insects are recaptured from the site, and the proportion of marked to unmarked individuals used to calculate population size. Newly captured individuals may then be marked, differently from the first marked cohort, so that with repetition of the exercise over a number of occasions, information on longevity, sex ratio, individual distance movements (if individual marks are used), and other population metrics can be accumulated. Any prior knowledge of population structure of the focal species may influence the precise procedure of a study.

A considerable variety of indices have been devised to calculate population size and turnover of individuals in an insect population. We recommend that you consult Southwood and Henderson (2000) and, in particular, Begon (1979) and Krebs (1989) to see the various techniques and analyses available for MRR. For most insect populations, Jolly–Seber method is the most appropriate, as it is for open populations where there are births, deaths, immigration, and emigration, i.e. it relates to the metapopulation model, which is characteristic of most insect populations. Samways and Lu (2007) give a practical worked example of MRR of such a population, in this case of a highly threatened butterfly and its common, sympatric sibling (Figure 7.10).

Capture and marking are processes demanding great care, because delicate or small insects can easily be damaged. Nevertheless, some very small insects, such as *Drosophila melanogaster* flies, can be marked with fluorescent micronized dust (Kristensen et al. 2008). MRR methods also make several assumptions, which it is important as far as possible to validate at the outset. These are:

- That the insects are not harmed by handling, and that their subsequent behaviour and lifestyle will not be compromised, by, for example, marks affecting flight capability or rendering them more susceptible or conspicuous to predators
- That the marks are sufficiently durable, probably permanent, to allow the study to proceed
- That the marked animals do indeed disperse freely into the population after release
- That the principle of *equal catchability* (see Krebs (1989) for a detailed discussion) applies, namely that the probability of catching a marked individual is the same as that of catching any other member of the population
- That the population is sampled at discrete time intervals and the time taken to sample is relatively small.

7.4.1 MRR process

The stages involved in MRR, and the decisions needed, are outlined here, and assume that the purpose of the study has been clearly defined, together with the arena over which it is to take place. Some preliminary study is usually needed: as Southwood and Henderson (2000) put it, 'At the outset of study neither the size of the population nor the correct assumptions to make about capture

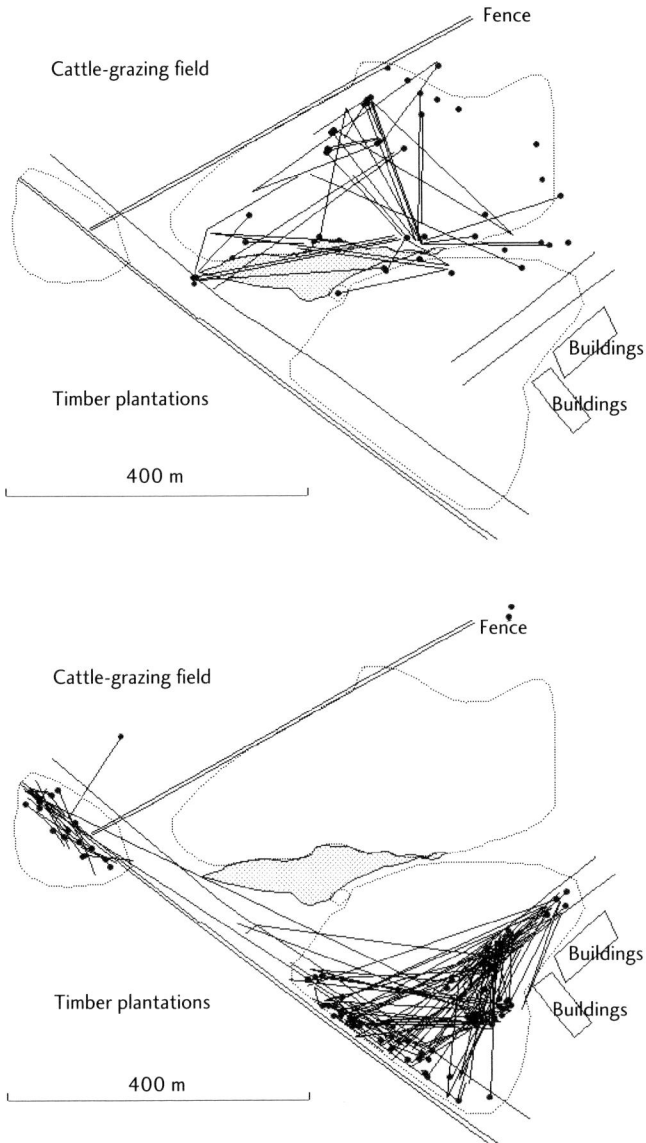

Fig. 7.10 A detailed mark–release–recapture study of two sympatric blue butterflies, one of which is highly threatened (the Karkloof blue *Orachrysops ariadne*; above) and its common sibling (the Grizzled blue *O. subravus*; below) shows how differently the two species respond to landscape pattern. Although the Karkloof blue is the stronger flier, it is severely limited in distribution by a very rare prostrate form of its food plant (from Samways and Lu 2007).

probabilities are known'. With a wide variety of models now available, deciding which one(s) to use may itself be a complex task. MRR methods are labour-intensive so that the purpose and accuracy is an important consideration.

Decisions to be made before embarking on a MRR study:

- **Capture:** How? Is it necessary to hold individuals in cages before marking, e.g. so that they can be processed in batches? Capture at single point, series of points, or more extensively across site? All available individuals captured, or only some—if the latter, how many? One or both sexes, when the species is dimorphic?
- **Marking:** Batch marks (all individuals on occasion with same mark) or marking for individual recognition? Is marking technique decided (together with any need for repeat marking on later sample occasions), and tested for safety to insects involved? Durability and legibility of marks assured? Is a post-marking holding period required—if so, are field facilities organized and adequate?
- **Release:** Conditions (weather, time of day) suitable? Are released insects unharmed and capable of behaving normally? Site of capture or more widely? Single release of whole batch or more dispersed releases?

- **Recapture:** Will one or more recaptures be needed in the future, or are marks detectable otherwise, e.g. through binoculars? Is provision for any re-marking assured? Interval between observations established?

7.4.2 Capture

Simple netting, using a normal butterfly net (terrestrial) or dip net (aquatic) is the most usual approach, with the initial sample usually based on the greatest number that can be caught over a particular time interval or at a given site, either by walking around or from fixed points. More rarely, samples may be more systematic, such as 'x individuals at each of y points' at intervals across a site. Sometimes, other means such as pitfall traps for carabid beetles or bait traps (either terrestrial or aquatic) are employed, or insects are retrieved from more general samples taken by beating or sweeping, or aquatic methods such as Surber samples. Such methods can have a considerable impact on the focal population, and although their use may be appropriate in ecological studies, they are not recommended for known threatened species. Nocturnal insects, such as many moths, can be retrieved individually from light traps, light towers, or lighted sheets, and held in individual containers until marking, usually the following day.

Some workers have incorporated some measure of specimen condition at the time of capture, inferred as an estimate of relative age. For butterflies, Arnold (1983) followed a rating scheme devised by Watt et al. (1977) as:

- **Very Fresh (VF):** recently emerged, wings still shiny and soft
- **Fresh (F):** wings and cuticle dry and hard with no visible damage
- **Slightly Worn (SW):** noticeable wear of scales from wings and body
- **Worn (W):** wings showing fraying or tearing in the cuticle
- **Very Worn (VW):** wings with extensive scale wear and cuticle damage.

Lepidoptera lend themselves well to such rankings, although many other insects do not. However, many dragonflies undergo characteristic wing wear and wing colour changes as they age, making it possible to allocate age categories for these insects as well (references in Corbet 1999).

7.4.3 Marking

Although marking should normally be undertaken in the field within a very short time of capture, it is sometimes necessary to hold or cage individuals for some time before marking. For example, it may be more convenient to capture a number of individuals and mark them in a batch, rather than process each individual as it is captured. Not least, this minimizes handling and obviates the possibility of same-day recapture of individuals. However, it may also mean that the insects may not be released at the exact point of capture. For very small populations, low numbers of insects, or highly dispersed individuals this may not be necessary, and the insects are marked as they are caught, and released. Sometimes however, individuals need to be retained after marking and before release (below).

Discrete marks are usually needed, taking care that they do not 'blot' or become too large—particularly if several individual marks are needed to constitute some form of numerical code (below). However, a few larger marks are preferable to more and smaller marks, because single small marks may be lost on damaged individuals (Ehrlich and Davidson 1960). Paints and the like can be applied with an entomological pin or very fine brush, and inks with fine-tipped indelible pens, most conveniently felt-tipped pens. For batch marking, relatively imprecise marks may be adequate, as the aim may be simply to determine whether the individual was one of a group captured on a previous occasion. In such a case, a simple contrasting colour may be all that is needed. Thus, some workers have used spray guns to mark insects such as locusts (Davey 1956) in the field, without capturing them, as an aid to tracking movement of swarms.

If there is any risk of damage from the marking procedure, or if you feel that you are insufficiently experienced, preliminary work or practice on common species of lesser conservation interest is strongly recommended. Simply, just try it! As with observer acuity training for transect counts (Chapter 6.5.7), you will find such investment pays dividends in improving the methodology employed in a study.

The two broad approaches to marking insects in MRR, each of which can be achieved by a variety of methods, are batch marking and individual marking. The first can be applied in studies such as determining movements between sites (e.g. to clarify population structure), and simple measures of population size, and involves all insects taken on a given occasion or from a particular site receiving the same mark (e.g. of the same colour), so that each can be recognized subsequently as a member of that particular capture set. The disadvantage is that the marked insects cannot be individually recognized. However, Griffiths et al. (2005) developed a method for rapidly mass marking individual insects using a laser marking technique which holds great promise. The second approach caters for this, but is generally more time consuming and laborious, so that the extra information needed or accrued must be balanced against this. It is perhaps more common, as the finer details of individual biology, longevity, and movement become incorporated increasingly into conservation assessments.

For batch marking, the requirements are for a readily visible, identifiable and unambiguous mark which does not harm the insect, and which can be applied easily and rapidly, preferably without need for any subsequent holding of the insect. This is usually achieved by a single colour of paint or ink applied to the insect's wing(s); in Lepidoptera it is sometimes necessary for some wing scales to be brushed off. Where possible, colour materials should be selected to avoid this extra step. One commonly overlooked side effect of this approach for species of direct conservation interest subject to collector pressure (some butterflies, in particular) is that marking may render the insects less cryptic and more susceptible to collectors. Conversely, as Gall (1984) suggested for the Uncompahgre fritillary *Boloria acrocnema*, marking may render the insect less desirable to commercial collectors because the specimen is then blemished.

Individual marking has two basic approaches, numbers and 'codes', with the common need for each individual to have unique marking so that it is not confused with any other insect. Either method should allow for marking up to many individuals, perhaps up to several thousand. Direct marking of insects is much more common than attaching labels to them, although numbered adhesive paper tags have been used in some classic studies of butterfly migration, such as that of the Monarch butterfly *Danaus plexippus* in North America. Small photographically produced plastic tags (1–2 mm diameter) have been used on flies, and numbered 'bee tags' may be available for larger insects.

Numbers or other marks may be written individually in sequence on insect wings using fine felt-tipped pens, for which no underlying wing support is needed. In the past, a variety of acrylic paints and inks have been used with varying degrees of success and permanence. Some of these (such as some model aeroplane paints) were prone to flaking off. Fluorescent pigments have been used for caterpillars, and use of different colour combinations can then enable individual recognition, even if these marks are not highly localized. As in any similar exercise, notebook recording/data logging (e.g. with CyberTracker) of all marks allocated is needed, and should be made contemporaneously.

'Codes' are based on series of marks, of characteristic (diagnostic) colour and position, with the number of colours sometimes helping to reduce the number of individual marks needed for higher number combinations. Two such well-tried examples are shown in Figure 7.11. One, for grasshoppers, allows up to 999 individuals to be identified with continuous marks and, with addition of another marking site, this number could expand by a further 10 000 insects. In this case, and others, one practical point to consider is that application of several marks with one colour may be much easier than changing colours, particularly if more than two colours are to be used on the same individual The second example

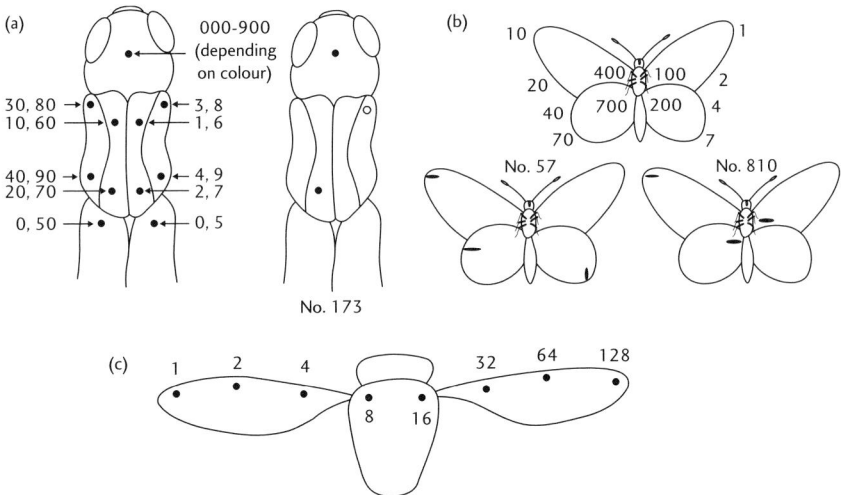

Fig. 7.11 Systems for marking insects individually using number codes applied as dots (e.g. with felt-tip pen) to the surface of (a) a grasshopper, (b) a butterfly, and (c) a fly (redrawn from Southwood and Henderson 2000).

given is for butterflies, also allowing up to 1000 individuals to be distinguished, here by marking with a single colour. Because such systems depend on a variable number of marks, any loss of any mark may cause breakdown.

We strongly recommend that you consult the excellent review by Hagler and Jackson (2001) as a prelude to marking the insects in your study. However, we must emphasize that any marking methodology used for threatened species must be very non-intrusive, and unlike the situation with pest species, where ample preliminary tests can be performed, there is little room for any error that may reduce population viability. For this reason, you may well find that traditional marking techniques, such as simple marking with a felt-tipped pen, work best in practice.

7.4.4 Release

In general the release of marked insects should occur after the shortest possible holding period, with the twin considerations of insect health and suitable release conditions assured to maximize their chances of survival. Immediate release is often possible, although the insects should not be released immediately when still under the effects of any anaesthesia or when still cold after cooling (which may have been necessary to reduce the shock of the capture and marking process), or when marks are not fully dry. Likewise, nocturnal insects should not be released until the following night, even if processed in the early morning, otherwise they may be severely disorientated and susceptible to predators or to overheating through not finding adequate refuges. Any holding of marked insects in the field must be in cages sheltered from direct sunlight or precipitation.

MRR releases are usually direct into the habitat and as close as possible to the capture site(s). Whether point or more dispersed release is done, it should reflect the earlier capture pattern. In releasing unfamiliar insects, perhaps with some novel marking scheme, it may be useful to undertake a trial release before the main programme begins, to observe the behaviour of a few liberated individuals and confirm that the procedure is normal.

7.4.5 Recapture

The basic recapture procedure is the same as it is for 'capture' in section 7.4.2 above, with due attention to whether physical recapture is actually necessary or whether some form of sighting without capture might be a viable alternative. It is important that the presence of marks does not bias the operator, and the principle of equal catchability (Krebs 1989) of all individuals underpins the process.

7.4.6 Outcomes from MRR studies

MRR approaches allow you to evaluate population size and how it changes over time (Box 7.2). It also enables you to quantify a number of aspects of insect movement at the individual and population levels. Perhaps the largest-scale example of the latter has been quantification of the long-distance movement patterns of the Monarch butterfly in North America. Urquhart (1960) pioneered the technique of attaching paper labels to monarchs in Canada to trace their movements. Other potential applications of MRR include:

- Detecting differential movement between the sexes and at different ages
- The relative numbers of individuals moving within a habitat patch and amongst patches, as in movement between populations in the metapopulation
- Effects of landscape features on individual movement
- Measuring longevity
- Extent of migration
- Any other context in which individual recognition may be valuable, such as mating frequency, flower or other food visitations, and wider behavioural studies, such as territoriality or habitat fidelity. These investigations require observations of individuals (with or without marking) (Figure 7.12).

Box 7.2 Dispersal of the rare Southern damselfly *Coenagrion mercuriale* **and its relevance to the conservation management of the species in the UK**

The Southern damselfly is restricted in the UK to two biotopes: small lowland heathland streams, and ditches in old water-meadow systems on chalk streams. Within these, it is restricted largely to patches of early successional habitat with relatively warm microclimates. Designing rotational management to conserve these habitats (such as by rotational mowing or cutting of bankside vegetation, and hand-clearing channel vegetation) requires knowledge of dispersal of the adult insects.

Purse et al. (2003) used MRR surveys to determine the main features of dispersal in relation to interpatch distance, habitat type, and scrub barriers, as aspects of landscape connectivity influenced by heathland management. Adult *C. mercuriale* were marked individually by a number on the fore wing and a dot of paint on the thorax, at each of a number of habitat subsections at two major localities for the insect (Preseli, Pembrokeshire: subsections average 47 m long; New Forest: subsections average 25 m). Recaptures, based on total marked individuals of 1988 (Preseli) and 2947 (New Forest) enabled calculations of net unidirectional movement of individuals over their life, and distance between recapture n and recapture n + 1, and its relation to interval time. In both localities, most individuals moved within the main patch, with 20% and 47% not moving at all. Most recaptured individuals moved over short distances, with only a few moving up to about 1 km (Figure 7.12), and individuals were capable of moving up to about 0.5 km in 1–2 days—that is, in a relatively short part of their adult life span of around 7–8 days.

In summary, the damselfly exhibited low emigration rates (1.3–11.4%) from patches and low colonisation distances, with movement between patches more likely over shorter distances (50–300 m), and with scrub barriers reducing dispersal. This appears to be quite typical for many insect species. Management efforts should be directed to enable *C. mercuriale* to recolonize sites within 1–3 km of other populations. Specific conservation measures include removal of scrub boundaries between sites of existing populations to promote stepping-stone dispersal movements.

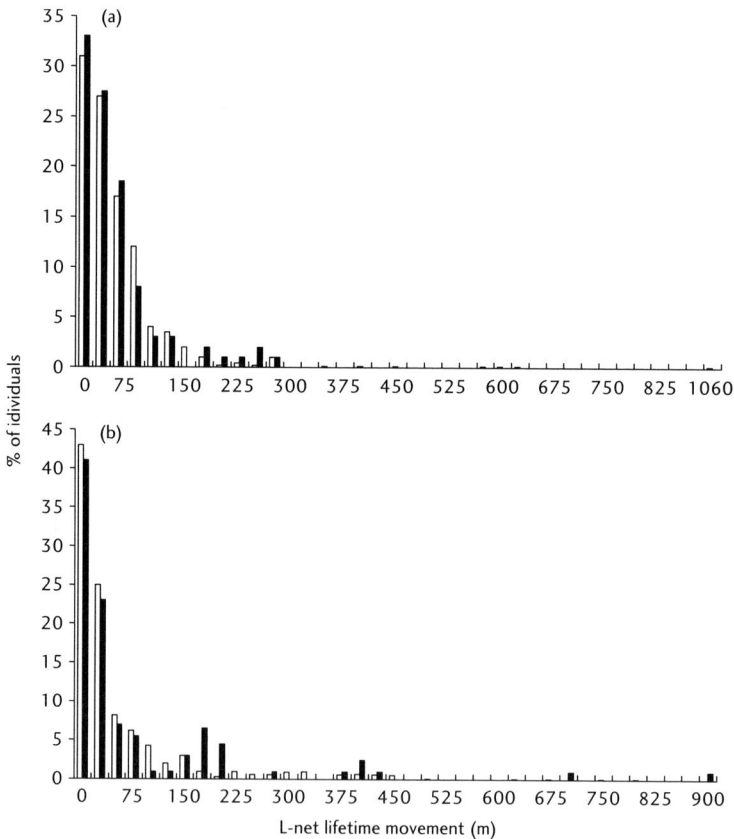

Fig. 7.12 Percentage distribution of net lifetime movements (L) for males (open bars) and females (closed bars) of the Southern damselfly *Coenagrion mercuriale* at two sites (a and b) in the UK. At both sites, most recaptured individuals dispersed over only short distances although a few moved up to 1 km. (From Purse et al. 2003).

7.5 Landscape geometry and insect populations

Ecological engineering and ecological landscaping are increasingly important agricultural practices which involve modifying the landscape to improve the activity of beneficial organisms (Gurr et al. 2004). These landscape manipulations to improve beneficial insect abundance have also had wider relevance in insect conservation. Habitat creation and resource enhancement, such as by improving the extent and quality of field margins as refuge habitats (conservation headlands; Dover 1991), improving access by predators to crop pests (e.g. beetle banks; Thomas et al. 1992), and reducing the isolation of cropping areas in the landscape are all components of this (New 2005). Field margin management has many functional implications (Fry 1994), and observations of butterflies such as the European Scarce copper butterfly *Heodes virgaureae* reveal clearly their close association with edge habitats and apparent reluctance to cross open spaces (Fry et al. 1992, Figure 7.13).

It is often useful to determine these flight paths, as they illustrate how the focal species use the landscape. Their use varies enormously from one species to the next, even within the same higher taxon. A first step in quantification of this is to develop *continuity diagrams*, where the flight path of each species is compared across the same landscape (Figure 7.14). Quite simply, the width of the illustrated continuity diagram of the flight path of any one species represents the number of individuals following that path.

Some insects use field margins as flight corridors, or sites along which to establish territories and patrol for mates. The structure of field margins in the landscape has considerable influence on insect movements For example, the barrier effect of tall vegetation was demonstrated by Fry et al. (1992), who used an artificial hedge of green tarpaulin stretched between wooden stakes, and compared the numbers of butterflies crossing or turning away from this at heights of 1, 2 or 3 m, and controls without it. Even a 1 m barrier substantially reduced passage of butterflies in that study. Presence of such barriers may thereby help to functionally define a habitat patch in a given study. For example, Thomas and Jones (1993) recognized habitat patches for the British Silver-spotted skipper butterfly *Hesperia comma* as separate from one another if they were divided by any continuous barrier of woodland or scrub, or by at least 25 m of unsuitable grassland. For the metapopulation of this species, groups of habitat patches were further defined as colonies if <50 m apart when separated by scrub or woodland or <150 m apart when separated by unsuitable grassland. As with some other studies (see Holloway et al. 2003), the probability that a patch would be colonized by the skipper decreased with distance to the nearest other populated patch.

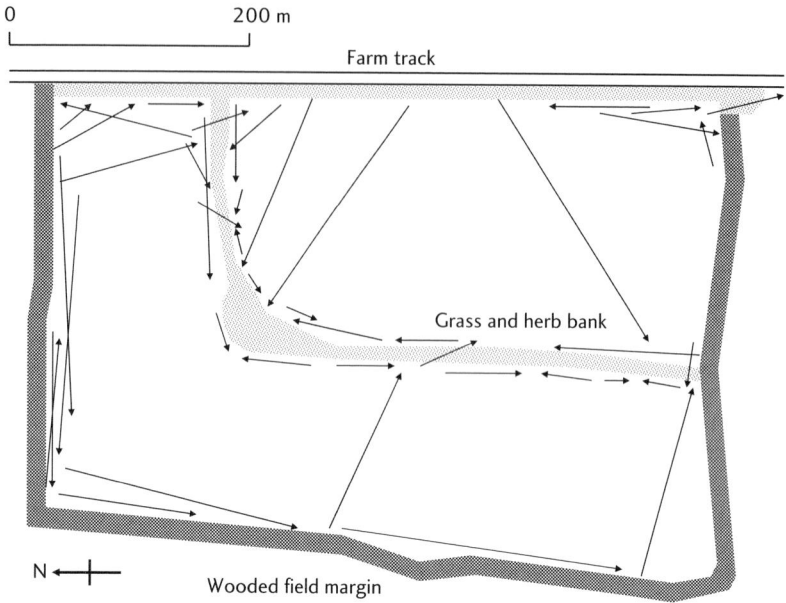

Fig. 7.13 Movements of the Scarce copper butterfly *Heodes virgaureae* in arable fields. Each arrow shows the track (15 min observation) for an individual insect. The butterfly is associated closely with edge habitats, and very few individuals crossed boundaries such as the bank in the centre of the field or the wooded/track perimeter. (From Fry et al. 1992).

A very useful conceptual approach when considering how landscape features affect insects is that developed in the field of corridor ecology (Hilty et al. 2006). Corridors, or any landscape feature, may be viewed as having one or more of six (but not all) features. We strongly recommend that you consider this conceptual six-feature approach when deciding on how to conserve a species across the landscape in general and not necessarily just in the case of corridors:

- **Conduit**, i.e. facilitates movement
- **Barrier**, i.e. prevents or deflects movement
- **Habitat**, i.e. can be a place to live and breed
- **Filter**, i.e. some individuals can pass through, and others not (e.g. one of the two sexes) (this is also a multispecies comparative feature where one species can pass through but not another)
- **Source**, i.e. there is a net gain of individuals, with some perhaps migrating out of the area

(a)

(b)

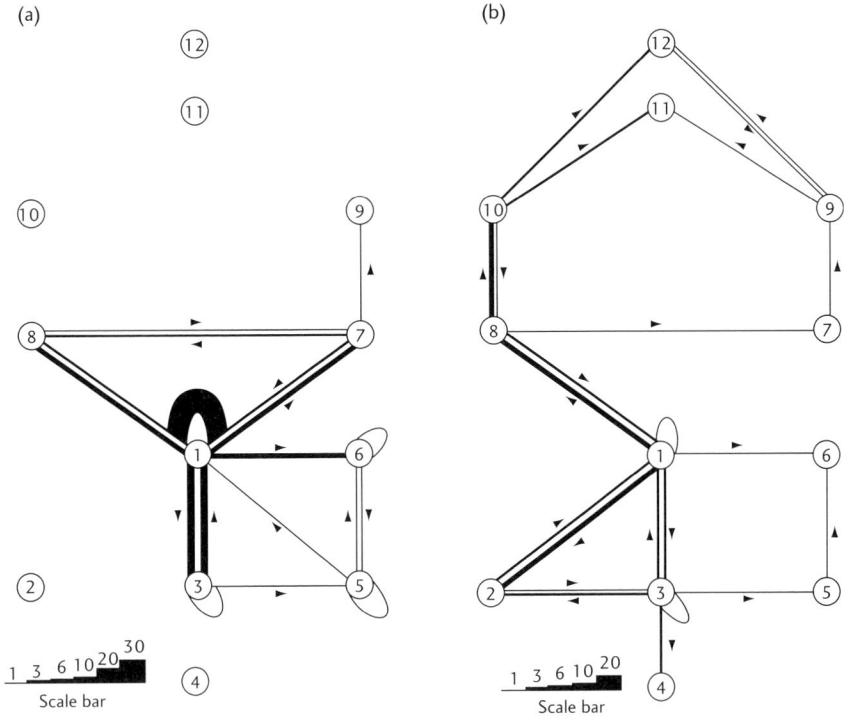

Fig. 7.14 Continuity diagrams representing the flight response of two butterfly species, the White-barred acraea *Hyalites encedon* (left) and the Mocker swallowtail *Papilio dardanus* (right) as they cross the same landscape. Widths of the lines indicate numbers of individuals following that path, while arrows represent direction. The numbers in the circles represent particular landscape elements: 1, water edge; 2, plane trees; 3, cut grass; 4, grass and flower beds; 5, uncut grass; 6, hollow; 7 and 8, reservoir; 9 and 10, forest edge; 11, forest interior; 12, above forest (from Wood and Samways 1991).

- **Sink**, i.e. there is a net loss of individuals in the focal area as a result of entry but insufficient breeding to maintain the population at the same level (Hess and Fischer 2001) (Figure 7.15).

The concept of the landscape being a filter for various life stages, and for different species, is an important one in insect conservation. It is at the interface between landscape and species conservation, i.e. the *coarse filter* and the *fine filter* respectively. The focal species and management for its conservation is a trade-off between making conditions suitable for that species, but possibly making the habitat patch unsuitable for other species in doing so. Thus it is essential that

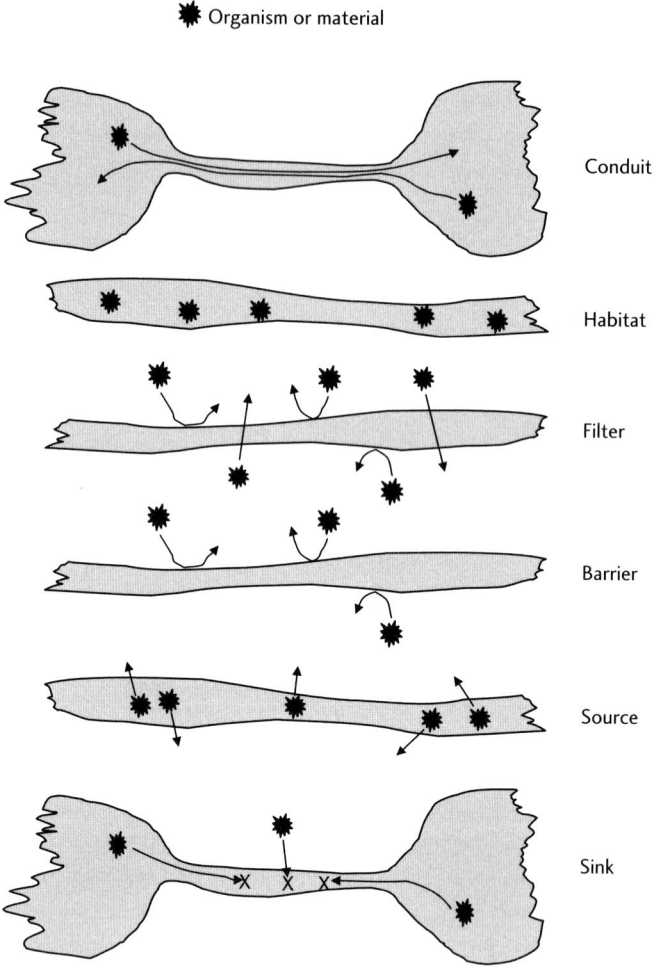

Fig. 7.15 Diagrammatic representation of corridor function. *Conduit*: organisms are able to pass all the way along the corridor. *Habitat*: the conditions in the corridor are good enough for the species to live there relatively permanently. *Filter*: some individuals and some species are able to cross the corridor and others cannot; the same would apply to the length of the corridor. *Barrier*: the individuals or species of concern are not able to cross or move along the corridor. *Source*: the population is able to undergo, through breeding, a net increase in the corridor, with some individuals even emigrating. *Sink*: the population undergoes a net decrease, with natality and immigration insufficient to maintain the local population (from Hess and Fischer 2001).

you establish your conservation goal with absolute clarity. The real dilemma is when there are several rare and threatened species living in the same local area that have slightly different habitat requirements. Do you make conditions really suitable for just one, or for several? This leads again into the concept of *dynamic conservation management*, where any action must be cognisant of changing conditions at any one location.

Population distribution is dependent on the prevailing environmental conditions at the habitat. Davies et al. (2006) showed that the habitat choice and utilization by the Silver-spotted skipper butterfly Hesperia comma in the UK is temperature dependent. Females actively adjust microhabitat selection in response to temperature variations, with host plants being chosen for oviposition depending on ambient temperature. Habitat requirements are likely to change with global climate change, implying that management must consider these changing conditions when managing habitats for species. Indeed, selecting fixed habitat sites may become outdated as the years go by, with various environmental factors limiting local butterfly abundance (Figure 7.16). This emphasizes the importance of corridors which allow populations to gradually shift according to changing ambient conditions over many years, in addition to providing for more immediate ecological requirements. Hopkins et al. (2007) give an excellent overview of this adaptive strategy.

Consideration of landscapes as *variegated*, that is, as *differential filters*, allowing some individuals or some species to pass but not others, is central to insect

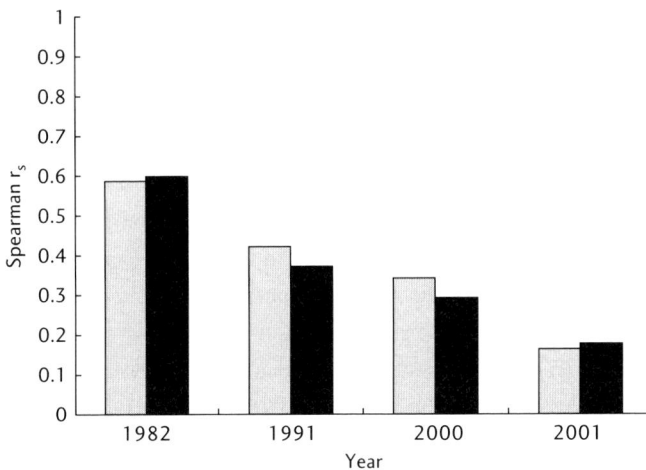

Fig. 7.16 Spearman's rank correlation coefficients between the adult population density of the Silver-spotted skipper butterfly *Hesperia comma* and habitat suitability using two different models (grey and black bars) for two different types of habitat patches in the UK over a 20 year period. These results suggest that factors other than simply availability of suitable microhabitats, but quality of microhabitat, for egg-laying are now also a limiting factor (from Davies et al. 2006).

conservation (Ingham and Samways 1996), because it determines the level of conservation that can be achieved. High connectivity builds in precautions to overcome adverse stochastic events and genetic restrictions imposed when populations become isolated. Predators, parasitoids, and pathogens can also follow these lines of population continuity, but that is offset by the populations having more vigour than those which are isolated.

Against this background, it is very useful to consider some of the tenets of landscape ecology when developing a foundation for a landscape approach to insect conservation that considers the many populations of many species across those landscapes. We recommend that you first consult the classic paper by Wiens et al. (1993), where they consider various landscape descriptors which affect population size or species richness, evenness (see Figure 7.8), dispersion, predictability, and dispersal. Good introductory texts to the subject include Forman (1995), Farina (2000), Gergel and Turner (2002), and Lindenmayer and Fischer (2006).

These descriptors are:

- **Connectivity**, as discussed above
- **Patch size distribution**, also as discussed earlier in this chapter, with the special aim to have large patch size, high patch quality and reduced patch isolation
- **Perimeter/area ratio**, which is related to shape and dictates amount of interior relative to amount of edge
- **Boundary form**, e.g. how hard is the boundary which will be a differential filter
- **Patch orientation**: this can be important as a patch on one side of the hill can have very different conservation value from a patch on the other because, for example, it has very different climatic conditions; see Figure 5.6
- **Context**: this refers to the type and quality of the neighbouring area, or surrounding matrix in the case of a patch
- **Contrast**: this refers to the level of difference of two adjacent land-scape elements, or in the case of a patch, between it and its matrix) (Figure 7.17).

7.6 Dispersal and dispersion (aggregation)

As noted above, landscape geometry has major influences on the *dispersal* (i.e. movement) of insects. In addition, it affects the distribution (*dispersion*) of individuals (and species; Chapter 9) across those landscapes. Individuals often move

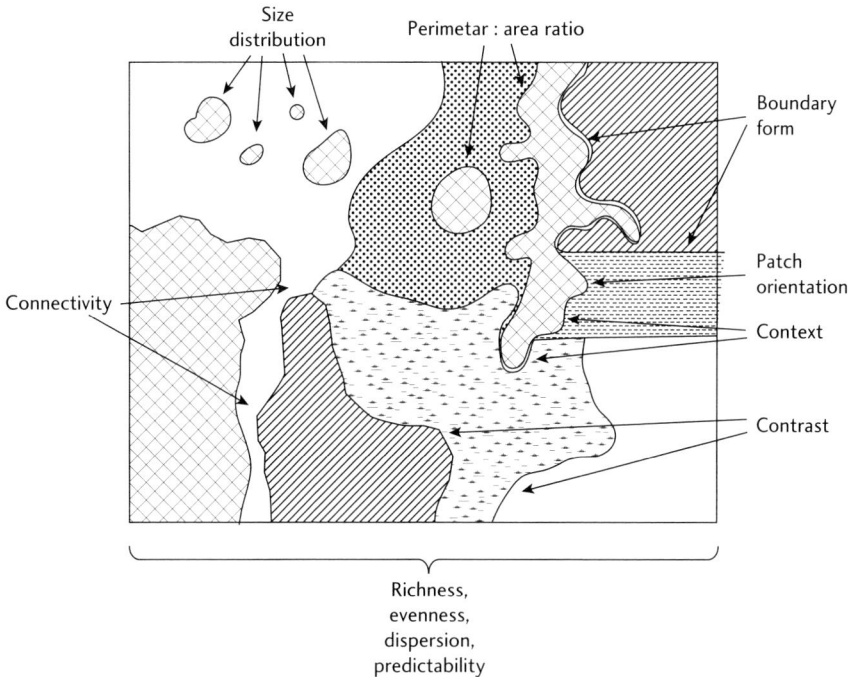

Fig. 7.17 It is conceptually useful for the practice of insect conservation to appreciate the descriptors of spatial structures that can occur in the land mosaic (see text) (from Wiens et al. 1993).

to find optimal resources, where they may then settle. However, the dispersion pattern of individuals in a population is also determined by insect life history and behaviour (e.g. oviposition behaviour, larval congregation, territoriality, sociality, predator avoidance), as well as by resource quality and quantity. The rate and distances of individual movements, e.g. within and across plants, across habitat edges, or across landscapes, will determine patterns of dispersion. This raises the important point that the dispersion pattern of a population is highly scale-dependent, i.e. when quantified at one scale (say within a plant) individuals may be evenly distributed, but when quantified across plants in the landscape individuals may have a clumped (or aggregated) distribution (Figure 7.18).

Understanding patterns of dispersion across the landscape is fundamental to insect conservation, because where the individuals congregate is usually a sign of

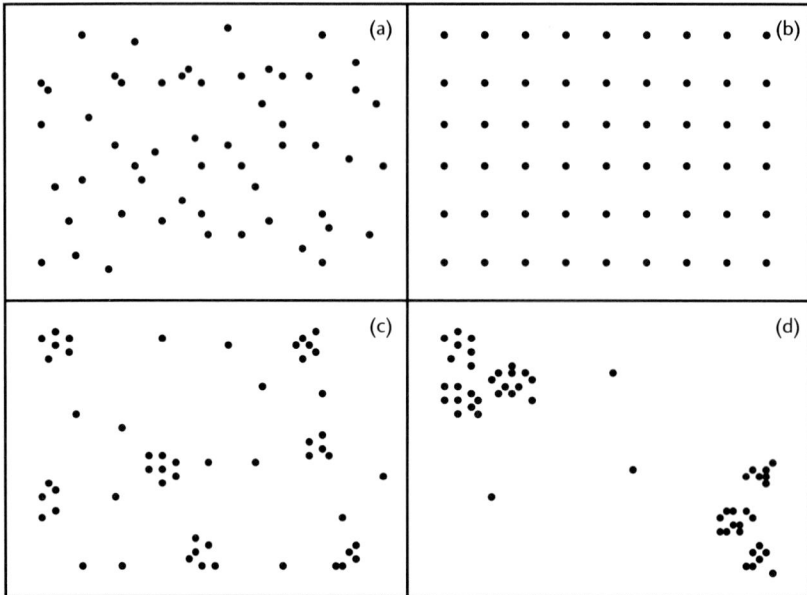

Fig. 7.18 Examples of patterns of dispersion of individuals across the landscape. Top left: random; top right: regular (even); bottom left clumped (aggregated); bottom right: aggregations of clumps. Generally only the bottom two are seen in insect populations, although territoriality (e.g. in crickets) can result in a pattern close to regular.

optimal environmental conditions (see Chapter 5) and the conditions towards which you should manage the wider landscape to increase the size of the optimal patch. This is seen, for example, in the UK where bracken is removed to *improve* grassland for certain target butterfly species. When conditions are not optimal, individuals often move quickly through the landscape to find those optimal conditions. For example, various South African butterflies in *disturbed* grassland corridors move up to 13 times faster through poor-quality than through high-quality corridors. In the high-quality linkages, they dawdle, searching for food resources and oviposition sites (Pryke and Samways 2001). This emphasizes that movement is intimately related to the dispersion pattern of settled individuals across the landscape. Thus, elucidating dispersal and dispersion are themes of which you need to be continually aware.

Quantifying dispersion patterns (often more generally referred to as *aggregation* or *spatial heterogeneity*) has a long and venerable history in entomology, as described in an excellent early review by Taylor (1984). More recent work makes the distinction between spatially implicit, spatially semi-explicit, and spatially explicit measures of aggregation, which as their names imply incorporate different types and amounts of spatial information (i.e. none, averaged across space, and point-specific respectively) (Veldtman and McGeoch 2004). *Spatially implicit measures* include, for example, the exponent *b* in Taylor's power law, and the recommended measure Morisita's I_M (Hurlbert 1990). These measures include no explicit spatial information (no *x* and *y* locality coordinates are used in the calculation). By contrast, *spatially semi-explicit measures* (e.g. spatial auto-correlation) encompass locality data and include, for example, Moran's *I* as used in the explanation and example of spatial autocorrelation in Chapter 3. Finally, and Perry (1995, 1998) has developed a *spatially explicit index*, namely spatial analysis by distance indices (SADIE), based on the distance that would be required to move individuals to achieve a random distribution. This approach has been successfully used to understand the distribution of insect pests in fields and, for example, the distribution of silk moth pupae across trees in a landscape. The advantage of this approach is that it allows one to identify precisely where in a landscape aggregations, or clusters, of individuals are occurring. We refer you to the literature cited in this section for further details.

It is commonly necessary to monitor the incidence of an insect species and to relate the presence of breeding populations to landscape (broadly, habitat) features. Any of the surveillance methods noted in Chapter 6 may be employed, but some additional applications are given here. These are all relevant to popula-tion estimates as well as to more widespread detection of incidence and relative abundance across sites. The emphasis in Chapter 6 was on population size, while here it is more on population distribution, both being important parameters of insect conservation. Management aims to maintain connectivity among these population groups, which is one of the main reasons why *corridors* and *ecological networks* are employed (Chapter 11.3). The philosophy for using ecological net-works, and a major example how they may be implemented in practice, is given by Samways (2007a).

7.6.1 Activity patterns

Quantifying dispersion patterns means making sure that you are detecting all the individuals in a population. Indeed, appreciating the complexity of insect life cycles and activity patterns has a significant influence on our ability to detect and monitor pattern and sizes of populations. As discussed in Chapter 6, many insects pass much of the year in stages that are either inactive (eggs, pupae) or are difficult to identify (most immature stages of terrestrial taxa), and the usu-ally more conspicuous and more easily identified adult stage has been the major focus of investigations.

Many insects have very restricted periods of activity, often reflecting circadian rhythms, in addition to being influenced strongly by weather conditions. Thus, the tropical Evening brown butterfly *Melanitis leda*, as its name implies, is active mainly around dusk (Parsons 1999). It is also active at dawn, unlike the lepidopterists who might observe it! The Australian golden sun moth *Synemon plana*, conversely, flies only in the middle of the day, around 11h00–14h00, so sampling must be undertaken within that short interval to gain valid information on presence or abundance. Many hilltopping insects are active for only a few hours each day, so that particular species of Lycaenidae, for example, are sometimes characteristically found only in late morning, or early afternoon, or later afternoon.

It follows that knowledge of any such defined activity pattern must be incorporated into your assessment or monitoring programme. Failure to recognize such activity patterns can lead wasted effort and misleading conclusions. This is particularly important when you are undertaking cross-site investigations, or doing Red List assessments (see Chapter 6). In such studies, comparisons must be conducted at the same time of day as well as under the same environmental conditions (e.g. less than 10% cloud cover). Preliminary sampling is strongly recommended, with repeated observations at a single site over a day (or other predicted sampling period), to determine the optimal time to conduct the full study.

7.6.2 Recording activity and abundance

Information on presence of individuals can be accrued over time by repeated observations. This will give an estimate of the population size, but is only valid when comparing sites based on the same sample effort. Timed records, taken at stated intervals and with defined sampling effort provides information for comparing sites, as long as equal sampling effort is applied to each site (such as '10 minutes every 2 hours from 08h00 to 16h00 along 20 m of river bank' or '30 minutes between 10h00 and 12h00 every Tuesday morning from April to September inclusive, with adjustment to Wednesday or Thursday if weather conditions unsuitable over 1000 m² of grassland'). The second, longer-term example here will provide information on changes in population levels over the year, i.e. the phenology of the species. Temporal observations, as discussed in Chapter 6.5.2, can provide essential information on apparency over time, which must, in turn, be supplemented with an awareness and an adjustment for differences in apparency between taxa (Dennis et al. 2006). Nevertheless, a series of assessments over time can be added together when making comparisons between sites as long as there has been equal sampling effort per site. You may, for example, add the 'Thursday samples' for each week for April and May, and again for June and July, etc. You may wish to compare sites for each of these sampling periods, or over the whole sampling period. The replication comes about from repeating these observations at different sampling units (SUs) spaced to avoid pseudoreplication (see Chapter 3), and not from repeating measurements over time for which the motivation is different.

7.7 Summary

Insects are distributed unevenly, often patchily, across landscapes. Their well-being depends on the structure, composition, and quality of landscapes. Yet anthropogenic impact is reducing the quality of these areas through landscape fragmentation, invasion by alien species, and input of noxious substances from human activity. Conservation thus focuses on maintaining as much habitat as possible and reducing its isolation. As populations of a whole host of species occur across any one landscape, it is essential to consider ways of maintaining as much insect diversity as possible by employing the *precautionary principle* of keeping as much high-quality habitat as possible as a buffer against further change. The principle also accommodates dispersal between habitat patches, as well as maintaining areas of residence. If movement is prevented, and as fragments decline in quality, there is gradual loss of populations, leading eventually to loss of species (from adverse environmental effects and genetic impoverishment), through a process known as *ecological relaxation*. Rapid changes in climate exacerbate this loss.

How the challenge of conserving populations is addressed relates very much to encouraging the dispersal of individuals across the landscape in the face of hurdles imposed by human activity. Mark–release–recapture and radio telemetry are methods being used to investigate this, as discussed in some detail here.

Various aspects of landscape geometry need to be appreciated for insect conservation. Landscape geometry has a strong influence on insect dispersal, and thus on options for how insects might be conserved. This appreciation relates as much to applied entomology, where certain modifications to the landscape are used to encourage the activity of indigenous and other natural enemies, as it does to conservation entomology. However, in conservation, maintaining good-quality habitat is not a fixed management process, as species may need to move to survive climate change. One response is to use corridors that encourage movement. However, any single corridor will not necessarily benefit all species equally. An understanding of continually changing conditions, and making plans to deal with them, is *dynamic conservation management.* This approach builds on an understanding of the effects of the various facets of the landscape, and insect responses to them, in addition to the connectivity provided by corridors.

Where individuals settle across the landscape is known as their pattern of dispersion, and is commonly quantified using a range of possible measures of aggregation or spatial heterogeneity. Dispersion is intimately associated with insect dispersal dynamics, as well as aspects of behaviour and resource quality and availability, or 'habitat', across the landscape. Quantifying dispersion also involves understanding how best to record insect activity patterns and forms an important basis for conserving them.

Ex-situ conservation: captive rearing and reintroduction programmes

8.1 Introduction: conservation objectives

Ex situ conservation refers to conservation undertaken outside the natural habitat. It is usually applied to the process of taking organisms into captivity to enhance their security, through breeding from this foundation stock, and, in due course, by releasing them or their offspring, perhaps after many generations, into the wild. This process becomes necessary (1) when populations in the wild have declined to such low levels that they may not be self-sustaining, (2) where threats to populations and/or their habitats are so severe that extinction is deemed likely, and/or (3) where captive individuals and their offspring can be protected from natural enemies or other factors causing high mortality, so that numbers can be built up either to augment source populations or to found new populations by translocation or other controlled release. It is thus a highly interventionist aspect of practical conservation. As such, it cannot be considered lightly, although in some instances it may be the only practical option in an attempt to save populations or species that would otherwise be almost certainly doomed.

If a critical habitat for a species is likely to be destroyed through urban or agricultural development, one conservation option is to salvage as many individuals as possible. Where it is not feasible to transfer these directly to another site (with due care to avoid mixing disparate genotypes when moving them to an existing population), it will be necessary to sustain them in confined, but favourable, conditions until release elsewhere is possible. This may be after breeding in captivity for several generations, so that any release is of descendants. Where release is intended into a *disturbed* habitat, this can only be done after substantial habitat restoration and where all threats have been removed. In extreme cases where a wild population is likely to be *extirpated* because of some inevitable human disturbance (e.g. the building of a dam or establishment of a residential area), the rescue process may involve substantial duty-of-care and highly competent husbandry, notwithstanding that the population would have been lost anyway if it had been left. Under less extreme cases, captive insect colonies may be founded from relatively secure *donor populations* for research leading to more informed management, or to build up numbers for translocation or protection. As Pearce-Kelly et al. (2007) noted, many insects can be bred up in numbers for field releases in relatively short periods. Furthermore, the basic biological knowledge accrued during these ex-situ breeding programmes can also be useful in other facets of the species' conservation.

Almost invariably, such programmes involve identified species of specific conservation concern. It is thus inevitable that details of their biology may be imperfectly known, so that these exercises may be intrinsically of high risk. Even with reasonable foundation knowledge, possibly augmented by information on related species (although great care is required when extrapolating from one species to another), fostering an insect population in captivity over several generations or longer without substantial mortality or loss of individual quality is inevitably experimental. Therefore, it is essential to record all details of conditions, food and so on, for the benefit of future workers. The vast body of applied entomology literature (e.g. the biological control literature), and hobbyist accounts (the last particularly for well-known groups, such as Lepidoptera and phasmatids for which a wealth of such compendia exists) can be consulted for relevant background on any insect species for which captive rearing is being considered. Many zoos and similar institutions maintain live invertebrate exhibits, some of them contributing importantly to insect conservation programmes. Consultation with any local zoo or commercial butterfly house may therefore be very useful when formulating a project. Some of the most significant ex-situ conservation programmes for insects have been initiated through zoos. In short, to reduce the risks to the species involved any available relevant knowledge must be brought to bear in ex-situ conservation programmes. Maintaining the health and fitness of the individuals is vital, so that genetic deterioration (Chapter 6) or *behavioural* and *ecological capability* are not compromised. Lewis and Thomas (2001) suggested that parallel measures to reduce or slow adaptations to captivity should be introduced alongside measures to reduce deleterious genetic effects. This translates into creating as near-natural conditions as possible in captivity, especially as adaptation to captivity can take place in just a few insect generations.

Strange as it may seem, this may not always mean that environmental conditions in the cages must always be at the same 'perfect' level, otherwise *captivity ennui* can set in. For example, if a regular light/dark cycle, 80% humidity, and an abundance of food is supplied to *Platycleis* bush crickets, as the weeks pass they stop singing and then start to die off rapidly. However, when the light/dark regime, humidity, and temperature are varied a little, the insects seem to fare much better.

An immediate consideration is that costs of establishing a captive breeding or husbandry programme for an insect may be considerably higher than expected. It may, for example, be necessary to plant and rear specific host plants for herbivorous insects, or prey for predators. Strict sanitation to control diseases such as latent viruses must also be introduced.

Ex-situ conservation is a familiar strategy for vertebrates, for which inter-institution coordinated studbooks are usual in helping to optimize genetic quality control. This approach is still unusual for insects. However, insects may have considerable positive attributes for captive breeding programmes, not least that their small size and commonly high fecundity ensures that many can be kept in far less costly conditions than larger vertebrates. Their numbers may be built up over relatively short periods, again a marked contrast with most vertebrate programmes. Sherley (1998b) states that 'Captive breeding of an endangered invertebrate needs to be recognized as a formal area of study in its own right that must be perfected before it can be integrated into trans-location programmes'. However, selection of optimal candidates for ex-situ programmes may be needed, as not all worthy taxa can be accommodated, or is it even possible for many of them. Dragonflies, for example, do not generally lend themselves to captive breeding over many generations, as they appear to need ample open-air space in which to fly, thus requiring enormous airy cages. Even the Cassava hornworm moth *Erynnis ello*, which is a pest in South America, is notoriously difficult to breed unless the adults are given a large arena in which to fly.

A scheme suggested for collection planning (for invertebrates and lower vertebrates: Visser et al. 2005) recognizes three major categories of programmes:

1) Conservation breeding programmes:
- **Ark.** Species now globally extinct in the wild and that would become completely extinct without ex-situ management.
- **Rescue.** Species that are in imminent danger of local or global extinction and are managed in captivity as a recommended conservation action.
- **Supplementation.** Species for which ex-situ breeding for release may benefit the wild population as part of a recommended conservation action.

Each of these includes provision, or planning, for field release where appropriate, or a plan to develop a field component.

2) Research:

- **Conservation research.** A species undergoing specific applied research that contributes directly to the conservation of that species or a related species and/or their habitats in the wild.
- **General research.** A species recommended for clearly defined pure or applied research that increases knowledge of natural history, population biology, taxonomy, husbandry or disease/health management.

3) Education:

- **Conservation education.** A species (or group of species) recommended for a clearly defined educational purpose of inspiring visitors, raising awareness, or increasing knowledge of conservation issues or projects associated with that species or its habitats. Conservation education species can be used to promote changes in public attitude, or to generate financial or other support for field conservation projects.
- **General education.** A species (or group of species) recommended for clearly defined educational purposes based on novel or otherwise remarkable characteristics, such as appearance, natural history or behaviour.

It is important to recognize that not all attempts may succeed. One example, based on detailed knowledge of the species concerned, is given in Box 8.1. However, this should not be a deterrent, as some other species have been reintroduced with great success arising out of detailed knowledge of the species of concern. The Large blue butterfly *Maculinea arion* reintroduction programme in the UK, although involving release of butterfly individuals from a foreign subpopulation, has been an immense success as a result of detailed knowledge of the insect's biology (Thomas 1999).

Box 8.1 Extinction of the Essex emerald moth *Thetidea smaragdina maritima* in captivity in the UK

This geometrid moth had always been regarded as a 'notable find' by collectors in England, but became presumably extinct in the wild in 1991 and in captivity in 1996. The history of this unusually well-documented case (Waring 2005) provides considerable insights into ex-situ conservation of a threatened insect. It is also a salutary reminder that the outcome may not always be favourable, despite application of scientific 'best practice' to species whose basic biology had been known for a century and more.

T. s. maritima was one of the first five moth species to be formally listed as 'Endangered' in the UK (under the Wildlife and Countryside Act 1981). This did little to protect the insect's habitat, as it emphasized prohibition on collecting and trade by imposition of a penalty of £20 000 per specimen! The moth had earlier (March 1979) been given statutory protection under the Conservation of Wild Creatures and Wild Plants Act 1975, again with penalties for collection.

Annual monitoring of the one known colony from 1978 gave highly variable numbers (1–90 in different years) up to 1984 (51 larvae), after which the colony became extinct. However, a second colony was discovered in 1987, and stock from this was used by Waring to found a captive breeding colony, and for subsequent attempts to re-establish the moth in the wild. This second wild colony became extinct in 1991.

Methods for breeding the Essex emerald had been pioneered by moth collectors in the nineteenth century, and included use of a second larval food plant, *Artemisia abrotanum*, more easily cultivated in non-coastal environments than the usual *A. maritima*. The 11 larvae found in 1987 formed the basis for the captive stock. Waring (2005) noted that these were then considered to be the last remaining wild larvae, and commented on the sense of trepidation and responsibility attending their capture. The captive colony increased to 135 larvae in 1988 and to more than 600 in 1989. The colony was then divided to be maintained in four different localities to reduce risk of total loss, and subsequently drawn on for establishment trials. Concerns arose over the fitness of captive individuals, possibly from inbreeding within the small founder population. Fecundity was much lower than expected, and many eggs failed to hatch. In the final year (1996) only 30 eggs were laid; none hatched and the moth became extinct in captivity and in the UK.

In this case (1) the biology of the moth was reasonably well understood; (2) known field colonies throughout a century of documentation were all small, usually no more than, at most, a few hundred individuals; (3) captive larvae were kept on growing food plants; (4) active steps were taken to reduce risk of disease, and to shelter the caterpillars from rough weather; and (5) attempts were made to maintain the phenology of the captive stock as close as possible to that of wild populations.

The ultimate aim of any ex-situ breeding programme is to assure the security of a species in the wild through release of individuals. This requires consideration of the target area to where the insects are to be transferred (Watts et al. 2008).

8.2 Farming or ranching insects for conservation

These two terms are often differentiated, particularly for butterfly rearing programmes, and both have place in conservation programmes for a wide array of insects. *Farming* refers to captive, caged populations in which the insects are reared in conditions varying from artificial to as natural as can be achieved. *Ranching* refers to practices involving resource enhancement in indigenous habitats, sometimes with augmented protection by limited or temporary caging, and so building up numbers in essentially wild populations.

Perhaps the best known, and pioneering, exercise on this theme has been for commercially desirable birdwing butterflies (Papilionidae) in New Guinea (Parsons 1992, New 1997b), in which the vine food plants of caterpillars were planted in densities far greater than normal occurrences. Female butterflies (*Ornithoptera*) are attracted to these, with augmented presence also of cultivated nectar plants, to oviposit, and the resulting caterpillars can be *sleeved* or caged to protect them from natural enemies, and to ensure high-quality reared specimens for commercial sale. This commercial need was the driver for this extensive programme, lessening both uncontrolled take of wild specimens and the destruction of forest habitats by providing alternative incomes for local people. The farming manual for the butterflies (Parsons 1978) suggests that up to 500 individual *Aristolochia tagala* vines can be maintained in a garden area of only around 0.2 ha, with butterflies effectively attracted to these by companion nectar plants. About half the caterpillars (or reared butterflies) are left or later released to sustain field populations, so that the programme is based on sustainable harvest from augmented wild populations.

This example has had important local economic and conservation benefits: (1) through centrally coordinated marketing, it has provided livelihoods for many people through demonstrated reward from a forest product whose sustainability depends on habitat conservation, (2) in some places, by providing entry to a cash-based economy, reducing needs based on shifting agriculture, and (3) populations of rare butterflies can be enhanced both through habitat enrichment and reduced loss of habitat, as well as by reducing collecting pressure by provision of specimens through organized, monitorable commerce. These principles led to suggestions (e.g. by New and Collins 1991, New 1997b) that such operations could become a powerful conservation tool for other rare butterflies, with Parsons (1992) noting parallel operations in China. The approach essentially treats butterflies (and, in the future, probably other commercially sought insects) as a wild crop to be harvested rationally and sustainably.

The midway conservation strategy between farming and ranching is exemplified by a long-term programme, from the 1920s, to introduce the Dutch form of the Large copper butterfly *Lycaena dispar batavus* to the UK to replace the extinct indigenous subspecies, *L. d. dispar*. At Woodwalton Fen, conditions for the butterfly were modified by growing and planting larval food plants (great water dock *Rumex hydrolapathum*), and clearing weeds and other vegetation from around existing docks. Despite some initial successes from releasing caterpillars, conditions after the initial trials were generally not adequate for outdoor survival. Fully grown caterpillars were subsequently brought indoors for the winter to complete development under protected conditions. Butterflies were also encouraged to mate in captivity before release in spring. This exercise therefore became intensive in that protection was given every year in order to maintain the introduced population (Duffey 1977, Pullin et al. 1995).

8.3 Process

The phases involved, and the major decisions that must be considered, when carrying out ex-situ conservation of an insect are given below. The entire process is overridden by consideration of the wisdom and need of the exercise, and whether it is indeed the best approach available. Thomas (1995) suggested that in the more common context of reintroducing insects to habitat fragments, such a move should generally take third place to strategies with wider and, perhaps, more long-lasting effects, namely (1) maintaining existing habitat connectivity and (2) restoring habitat connectivity by management that provides new habitat, including *stepping-stone habitats* (Chapter 10). Whichever option is chosen, it is likely that there will be little or no prior experience with handling the particular insect concerned, so that thorough documentation should be routine as an investment in future exercises.

Practical and ethical problems of translocating or otherwise introducing insects are many and varied, and the objectives for any particular exercise must be spelled out very clearly. In the UK, JCCBI (1986) took an important initiative in publishing a code of conduct for this form of endeavour, and that code is important reading for anyone contemplating such an exercise, in setting the foundations for later developments.

A flow scheme for insect translocation strategies, based on extensive experience with New Zealand Orthoptera, although of much wider relevance, is summarized in Figure 8.1 (Meads 1995), with the prerequisite knowledge and skills needed at each stage given in Table 8.1.

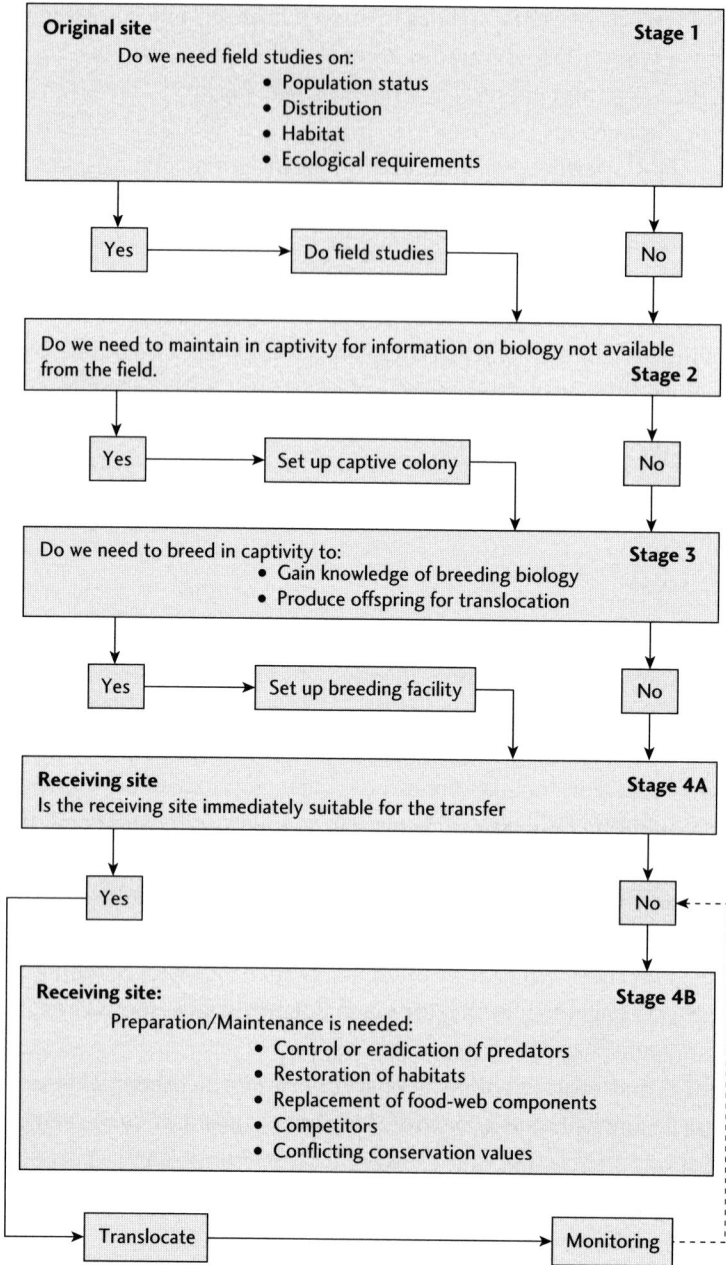

Fig. 8.1 Captive breeding and reintroduction as components of a six-stage process (from Meades 1995).

Table 8.1 *The knowledge and skills needed at various stages (see Fig. 8.1) of an insect translocation programme (after Meads 1995).*

	Knowledge needed	Skills needed
Source population		
	Taxonomic position	Field survey methods
	Distribution	Population estimation
	Population dynamics	Conservation status
	Conservation status	Other environmental studies
	Life cycle	
	Feeding biology	
	Breeding biology	
	Predator/prey relationships	
	Habitat requirements	
	Environmental conditions	
Captive maintenance		
	Life history	Culture methods (food, shelter, substrate)
	Feeding behaviour	Behavioural observation
Captive breeding		
	Breeding habitat	Culture methods
	Breeding environment	Behavioural observation
Receiving locality		
	Status of predators	Species recovery
	Status of competitors	Restoration
	Food availability	Predator control (and monitoring)
	Refuge availability	
	Ecological and environmental parameters	
	Sustainability of release environment	

- **Capture:** How? How many? All or some? Particular life stages or ages? Any particular sex ratio?
- **Transport and holding:** How? How far and for how long? Are controlled environmental conditions needed? Separate cages or all together? Reception facilities guaranteed, and suitable?
- **Maintenance:** Are adequate conditions of caging, climate control, and feeding available for the entire period envisaged? Indoor or outdoor conditions? Are natural foods available, or is some reliance on artificial or semi-artificial diets likely—if so are these proven? Are any quarantine guarantees

or precautions needed? Is enough technical help (e.g. or feeding and maintaining sanitation) available?

- **Breeding:** Does the project involve simply holding short-term or longer-term breeding programmes? If the latter, is there a defined project duration, or is it open-ended? Is there provision for dividing reared stock across more than one site? Are there any controls over genetic deterioration? Is there a target number of individuals for the programme?

- **Release:** Where? When? Single or more sites or releases? Is the receptor site (or sites) adequately prepared and secure, with any earlier threats removed? Is purpose to establish new populations or to augment existing ones? Will captive stock be kept after initial release(s)? *Hard* or *soft release*? Any minimum number of individuals, and at what growth stage?

- **Monitoring:** Adequate provision for *post-release monitoring* to determine success or failure? Over what period? What level of monitoring is needed?

Each of these may require original investigation to ensure that enough funding and other support is available to support the enterprise to a conclusion.

Individual cases range from short-term holding of insects for release within a short time, to long-term captive breeding over many generations. For captive breeding, additional considerations of logistics and quality maintenance may arise. Founder numbers for breeding stocks of threatened insects are usually small, and it is thus largely inevitable that genetic problems (such as inbreeding) may arise. The IUCN (1995) recommended that captive populations for ex-situ conservation should be founded well before the wild populations drop below 1000 individuals. Perhaps particularly for insects (but also for most vertebrates for which such programmes have been initiated) far fewer founders than this ideal have been available, so that captive populations are likely to lose genetic diversity and become inbred far more rapidly than desirable. It is likely that genetic considerations, in addition to simply attaining numerical targets of individuals, will increasingly become core components of insect captive rearing programmes. However, the primary aim is invariably to increase the captive stock as rapidly as possible, by breeding from all available individuals, so that genetic management is usually considered only once a reasonable number of individuals have been bred.

Honan (2008) found that morphological deformities appeared, nymphal survival rate decreased, and egg size became smaller in captive stock of the Lord Howe Island stick insect *Dryococelus australis* within four generations in captivity. In response, males from different cultures in different insectaries were introduced, resulting in remarkably improved survival. Honan then separated individuals of one of the cultures into 18 genetic lines. Adult females were kept in groups of 10, to which 2 adult males from other genetic lines were added and rotated at regular intervals. Despite this intervention, there was no major amelioration of inbreeding depression, and the only solution may be to introduce new genetic material from the remaining wild population.

A captive breeding programme can have six major steps:

1) Decline of the wild population to a level where conservation concern is aroused/expressed
2) Founding a captive population
3) Growing that population as rapidly as possible
4) Maintaining it over an indeterminate period, possibly over many generations
5) Selecting individuals for reintroduction
6) Managing the reintroduced population in the wild.

Although it is rarely considered specifically in programmes of this nature, it is important to mention that many people are allergic to insects or their products. The effects range from having strong reactions to stings, bites, or urticating hairs to incurring respiratory problems through inhaling airborne particles in insectaries. Of particular note, is that continual exposure to insects and the particles and substances that they give off often leads to an allergic reaction over time. We therefore recommend that goggles, gloves, and breathing masks are worn at all times when you are confined with the captive insects.

8.3.1 Capture

In cases of salvage for a threatened species, the aim is usually to rescue (capture) as many individuals as possible—ideally, all of those that would otherwise be lost. In most other contexts, the situation is different, with it usually being important to take only as many individuals as will not affect the viability of the donor population. The number of insects may, of course, be important in the constitution of the *founder population* to be established elsewhere, and a general working rule, in the absence of a realistically defined value for a minimum viable population, is to err on the side of 'too many' rather than 'too few'. This is to help counter the unknown chances of genetic deterioration and/or other demographic declines likely in a small population. At the lower extreme, a single gravid female may be enough to found a captive breeding colony and, for very rare species targeted for such an exercise, perhaps only very few individuals may be available, so that genetic deterioration is a serious consideration. Relatively high fecundity and rapid development are advantageous in captive breeding, but the parent stock may need to be nurtured very carefully, and captive conditions should ideally approach natural ones wherever possible.

Great care is needed to avoid harming the insects during capture and subsequent handling, and biological knowledge may be very helpful in reducing stress. As a simple example, when dragonflies are being transferred from one pond to another, it must be borne in mind that the larvae are voracious predators, and should be kept singly, or if in small groups, with plenty of waterweed or other shelter to avoid cannibalism. Furthermore, adult dragonflies are strong fliers and can easily damage their wings if kept in cages that allow them to move. They are better confined loosely in individual, airy envelopes, with a small wad of paper tissue to avoid crushing them. Many other insects are far more delicate than Odonata, and physical handling should be kept to a minimum. If the option is practicable, it is often worth capturing a few individuals above the minimum number required, to compensate any handling or transport mortality.

8.3.2 Transport

The major considerations relate to the duration and mode of transport, and protecting the insects from physical and biological harm, such as losses from desiccation or heat stress, and (if necessary) provision of suitable food for longer journeys.

Road transport is the most common mode, and usually for no more than a few hours, so that some of the above conditions are of relatively little importance. Overheating is perhaps the most commonly encountered problem, and is countered by putting the insects (in containers) in an ice chest or similar container, such as a polystyrene drinks cooler. Insects should never be placed in direct sunlight during transport. Express air courier shipment that takes only 1–2 days can usually be sustained by most insects, as long as conditions are suitable for them. However, longer times, especially >3 days, tend to result in reduced survival and fecundity of the insects (Valentin and Shipp 2005). Biological control workers recommend a maximum of 48 hours travel time. For air travel, additional arrangements may be needed:

- Insects must be carried in the cabin, rather than in unpressurized and cold holds.
- They must not be X-rayed, so the cages must allow for close inspection by airline officials without allowing the insects to escape, perhaps by providing an outer, secondary container that can be opened easily to view the insects through a (secure) transparent plastic inner box; if in doubt, consult relevant officials beforehand to determine specific needs.
- They must not be exposed to any cabin-spraying with pesticides.

- Any permits required for cross-border transport or quarantine must be arranged well in advance, to avoid delays in transit.

Reception facilities (whether aquarium facilities or large terrestrial enclosures or cages) must be planned and ready before the insects arrive. It may be necessary to assure supplies of particular food plants or prey or host organisms and, on occasion, this may involve a separate husbandry exercise, with high costs.

8.3.3 Maintenance

Should captive stock be collected from several sites, it is wise to maintain each of the stocks in separate cages, at least for a time, to avoid mixing of possibly significant genetic variants and to help screen for diseases or parasites that can easily spread through a captive population.

The ideal of maintenance, one that flows also into captive breeding, is to provide the insects with conditions that physically and biologically resemble their natural environment, but without any threats to survival. In practice, this translates to minimizing stress on the insects, and seeking to allow them to pursue a normal way of life, under similar climatic conditions, with natural foods, and at similar densities to those of a natural population. Overcrowding is a frequent occurrence in captivity and may result in the appearance of normally latent viruses (as experienced by many lepidopterists rearing caterpillars in small or crowded containers), and cage sanitation is always a prime concern in maintaining good health. Other, unexpected pathogens may also appear. An ex-situ population of the Frégate Island giant tenebrionid *Polposipus herculeanus* for example, developed a fungal infection (Pearce-Kelly et al. 2007), which emphasizes the importance of keeping more than one captive population, preferably at different locations. It also underscores that the original donor population is also susceptible, and even more so where there is genetic inbreeding and possibly less resistance than would be the case with larger populations.

When insects are kept outdoors, some form of shelter from extreme weather may be needed for the culture. When kept indoors, some form of climatic control may be required so that, for example, insects are not kept at temperatures far higher than those they would normally experience in nature. Thus, establishment of the Lord Howe Island stick insect at the Melbourne Zoo necessitated lowering the temperature of the holding room from its usual tropical regime of 30 °C to resemble more closely that of Lord Howe Island, rarely more than 25 °C (Honan 2008). Sometimes, however, the converse applies, with dragonfly larvae often doing much better in insectaries that are open to the air, as long as the temperature is in normal range experienced by the insects.

Provision of natural foods is far preferable to reliance on any substitute diet, but may not always be possible in the longer term. It may not even be possible to determine the natural food requirements. For example, herbivores may be collected from stands of vegetation containing many plant species, and predators from environments with several to many possible prey species. In such uncertainty, a choice of possible foodstuffs can be offered, preferably from among those present at the collection site(s). Additional clues on possible suitable foods can be gained for some insects from compendia such as those of caterpillar food plants published by lepidopterists and, more rarely, for other groups, so that possible foods can be listed and sought before the focal insects are captured. Even many apparently *monophagous* insect herbivores may survive, at least for a short while, on other plants, and one of the central problems of longer-term insect maintenance is assurance that the plants offered and eaten are indeed the most suitable ones for the insects. Where this is not the case, fecundity, fertility, longevity, and population and individual growth rates may decline. If the established, optimal plant (simplistically, that on which the species normally feeds in the wild) is not available, some close relative is sometimes a viable alternative. Should that optimal plant (or other food) not be found close to the maintenance site, it may be necessary to propagate or grow a stock to ensure its continued availability.

A variety of artificial or semi-artificial diets have been devised for insects (below). These may be very valuable to aid in temporary maintenance but are viewed mostly as a reserve strategy for maintenance, although they may make the difference between success and failure during periods of temporary shortage of natural food.

8.3.4 Breeding

Captive breeding of insects for conservation purposes usually involves only relatively small founder populations and small-scale exercises, with the primary aim of building up numbers as rapidly and effectively as possible. Such exercises are viewed commonly as open-ended or indefinite. However, in order to minimize poorly understood genetic effects, we recommend that only a few generations (as few as two have been suggested in particular cases) should be bred without some replenishment from wild or other populations to counter possible inbreeding effects. Many of the conditions for captive breeding coincide with those needed for 'maintenance', but with the added risk of keeping the insects for considerably longer periods, at progressively increasing numbers and, possibly, higher densities.

The vast and complex literature on insect rearing encompasses many scales of operation, from the small-scale productions of greatest relevance here to factory-level operations of commercial insectaries producing biological control agents at rates of up to millions a week. Every level of operation has to contend with similar demands: provision of adequate housing and environmental control; provision of food, with supplies assured; and quality control of the insects reared. For larger operations in particular, operator health is a major consideration, as mentioned above.

The many contrasts between enclosed and natural environments (Table 8. 2), each of which may affect the insects in some way, merit consideration in planning a long-term rearing environment. These constraints apply predominantly

Table 8.2 *Differences in factors affecting insects reared in laboratory and natural environments (after Bartlett 1985).*

Factor	Laboratory	Natural
Temperature	Stable or periodic	Diel and seasonal fluctuation
Light	Artificial, constant or photoperiodic	Natural, sidereal photoperiodic
Humidity	Stable or linked with temperature control	Diel fluctuation
Food	Provided artificially	Actively sought
Shelter	Not provided or inadequate	Actively sought when needed
Predator-parasite	No pressure: absent	Constant or variable pressure: present
Competition	Normally absent	May be present
Microflora/microfauna	Concentrated, may lack necessary taxa	Diffuse, variety present. Pathogenic pressure
Human presence	Intense, continuous exposure	Sporadic exposure
Microenvironment	Average	Optimum sought
Mate searching	Easy	May take effort/energy
Mate acceptance	Promiscuous (forced)	Natural
Oviposition	Artificial, restricted	Natural, free-choice
Dispersal	Restricted	Natural
Wind	Absent or low velocity	Variable, generally present
Chemicals	Pheromones concentrated, others absent, new stimuli	Pheromones diffuse, some others concentrated
Free water	Usually one source	Many sources
Mechanical vibration	Present, may be intense and continuous	Generally absent or sporadic

to laboratory or indoor insectary conditions, and using outdoor enclosures or field cages as rearing units obviously reduces many of the possibly harmful influences. In many programmes, the presumed suitable or optimal environmental conditions will have been defined. This may be done by trial and error, during an earlier maintenance phase.

Each species may have different optima, and recipes (such as those given by Singh and Moore 1985 for representatives of many insect orders) provide background information on rearing environments, containers and cages, diets, control of contamination, etc. However, as Waage et al. (1985) stated for entomophagous insects, only a tiny fraction of species have been subjects of rearing attempts, and it is impossible to give a universal cookbook formula for insects in general. Nevertheless, some artificial diets have been devised specifically with conservation rearing in mind. For example, Morton (1981) described some for butterflies to facilitate production of large numbers of insects in captivity. Indeed, he suggested the development of a 'Captive Breeding Institute' whereby the various disparate conservation breeding exercises for butterflies of one geographical area could be brought together for greater economy and efficiency.

Quality control in laboratory colonies of insects becomes increasingly important with increased time or numbers of generations in captivity. It may manifest in changes in any aspect of insect vitality or morphology (see Box 8.2), and such changes (including genetic deterioration) can be assessed in two main ways:

- Observing trends in laboratory populations over a sequence of generations.
- Comparing the performance or structure of laboratory and field populations.

For assessing the quality of reared insects, Moore et al. (1985) recognized four kinds of rearing system, mainly at scales larger than those likely to be of primary concern here. These are (1) rearing insects for stock supply, as sources for colony propagation; (2) insects produced for research, mainly as experimental material; (3) insects mass-produced for field release, such as biological control agents; and (4) insects reared as usual or factitious hosts for these. For all, control of environment, avoidance of contamination, and appropriate diet is routine and vital, while the various production systems may be continuous or seasonal in operation. Variations in vigour (with a wide range of possible measurements) are recorded and measured against standards based on wild material wherever possible. For example, Moore et al. (1985) noted that for stock supply, fecundity, pupal weight (size), and survival are important parameters. In such larger-scale operations, performance testing may fall into the categories of field testing, behavioural, and clinical measures but, while these have only minimal relevance in small-scale operations, they indicate the wide range of monitoring options possible.

Box 8.2 Adaptation to captivity in a butterfly

Lewis and Thomas (2001) used the Large white butterfly *Pieris brassicae* in the UK to examine the possibility that populations may undergo adaptations to the captive environment, and that these might render the insects less well adapted to survive in the wild—a theme of considerable significance in ex-situ conservation programmes involving insects. The authors compared six traits related to reproduction and dispersal between a wild population and a culture of *P. brassicae* that had been continuously bred for at least 25 years without any augmentation from wild stocks. They considered the culture to represent around 100–150 generations.

Lewis and Thomas (2001) compared two reproductive traits (the prereproductive period and the reproductive allocation) and four morphological features related to flight: total mass, thorax mass, wing area, and aspect ratio (wingspan^2/wing area). They predicted that:

- Captive conditions were expected to favour evolution of a shorter prereproductive period and increased reproductive allocation, so that females laid earlier and produced more eggs
- Total mass should be greater in captive cultures, but interpreting this is difficult (heavier insects might reflect longer development, and could be thought of as an adaptation to longer dispersal)
- Thorax mass trend was similarly difficult to interpret
- A large wing area related to body mass is related to sustained flight.

This means that in small cages selection might rather favour investment in body mass than in large wings, and captive individuals should have smaller wings than wild individuals, together with a reduced aspect ratio.

The prereproductive period (of ~3 days before oviposition) did not differ significantly between the two cultures, and females developed similar numbers of eggs at similar rates. However, wild females laid far fewer eggs, consistent with the proposition that reproductive behaviour might have become adapted to the captive environment. Of the morphological features, captive butterflies were heavier than wild ones, consistent with selection for increased fecundity, but there was no correlation between numbers of mature eggs from ovariole counts and total mass. Both wing area and aspect ratio differed in the trends predicted; the captive stock having relatively small, broad wings suited to higher manoeuvrability in small cages. This case study illustrates that captive reared individuals will differ from wild populations in some fitness-related traits, and that over several generations captive populations may actually evolve to life in a cage.

Box 8.3 Propagation handbooks: the Karner blue butterfly *Lycaeides melissa samuelis*

A few notable species of threatened insects amenable to captive breeding have become the subjects of 'propagation handbooks', which bring together all available practical information on how to handle and maintain that species. The handbook for the Karner blue butterfly (published by the Toledo Zoo, see www.fws.gov/endangered/aza/Kbbweb/kbbprop.pdf) notes that *L. m. samuelis* has declined by 99% or more of its range during the last century, and emphasizes its importance as a flagship species for the globally endangered 'oak savanna ecosystem' of North America. It deals with propagation of the sole host plant (*Lupinus perennis*), how to catch and transport butterflies, housing conditions, rearing techniques, record keeping (including stud book maintenance with data sheets tracing the fate of the offspring of each female), the control of predators in captivity, release protocols, and assessment and maintenance of site condition. Similar treatments have been compiled for several threatened Orthoptera, and each contains abundant practical advice relevant to many related taxa, and others.

Rearing conditions may include, for example, unnatural climatic regimes selected to avoid induction of diapause, the use of semi-artificial or artificial diets with unknown physiological effects, and the use of prey reared on relatively unnatural plant hosts. These factors and others, influence the insect's lifestyle in the interests of enhancing rearing success as the primary objective. As King et al. (1985) pointed out, the lessons from large-scale insect production systems have traditionally been production-orientated rather than behaviour-orientated, but the latter may be of at least equal importance in predicting the vigour of the insect stock. In short, many insect species are biologically labile, so that stock that has been reared in captivity over many generations may differ from the wild stock in a spectrum of attributes, many of which have unknown adaptive values. For conservation, it is important to avoid such changes.

8.3.5 Release

A variety of contexts for release of insects have been distinguished, as summarized in Table 8.3, and the context may influence the needs and process details for a successful outcome. Universal needs include having adequate numbers of healthy individuals for release, and the suitability and security of the receptor site or sites.

A further issue is where a release should be made when the aim is (as commonly) to restore or establish a new population. There is broad consensus that such releases should be made only within the historical or natural range of the taxon involved (Pyle 1976, IUCN 1995). Many incidences, many of them poorly documented, involve hobbyists releasing surplus reared stock 'close to home' rather than in the more distant places where they were originally collected, and

Table 8.3 *Terminology for insect introductions and related operations (JCCBI 1986).*

Re-establishment	Deliberate release and encouragement of a species in an area where it formerly occurred but is now extinct
Introduction	Attempt to establish a species in an area where it is not known to occur or to have occurred
Reintroduction	Attempt to establish a species in an area to which it has been introduced but where the introduction has been unsuccessful
Reinforcement	Attempt to increase population size by releasing additional individuals into the population
Translocation	Transfer of individuals from an endangered site to a protected or neutral one
Establishment	Neutral term to denote any attempt made artificially and intentionally to increase numbers of any insect species by transfer of individuals

such releases potentially confuse survey results and may lead to indiscriminate mixing of local and distinctive gene pools.

In general, releases are best undertaken as close as possible to (1) historical sites for the same species and/or (2) as close as possible to the population from which the captive stock was derived. If releases are to a site from which the taxon has been lost, that site should have been adequately restored and safeguarded against future threats. Although not formulated formally in comparable detail for application elsewhere, or for other insect groups, the points listed for consideration in butterfly restoration in the UK are clearly of much wider value in summarizing many of the general issues involved in the release process (Table 8.4). Although that table makes reference to particular UK organizations for centralizing plans and advice, equivalents occur in many other places.

Two main procedures for release are available. A *hard release* simply exposes the insects to the new environment and allows them to disperse, or may more actively deposit the insects into their new site. On the other hand, a *soft release* entails releasing the insects into a field cage or enclosure at the release site, and perhaps maintaining them there for some time to assure climatic compatibility, then allowing them to disperse after removing the cage or opening up the cage to the outside (Figure 8.2). The two methods may have rather different consequences. From a hard release, the insects may simply disperse into the surrounding area without the investigator knowing their fate and even whether they can or do survive. If only small numbers are involved (not unusual for

Table 8.4 *Points to consider when developing a species restoration strategy for butterflies in the UK (Butterfly Conservation 1995).*

- The species should have declined seriously (or be threatened with extinction) at a national or regional level.

- Remaining natural populations should be effectively conserved, and the restoration plan should be an integral part of a species action plan.

- The habitat requirements of the species and the reasons for its decline should be broadly known and the cause of extinction on the receptor site (where re-introduction is contemplated) should have been removed. There should be a long-term management plan that will maintain suitable habitat, and the site should be large enough to support a viable population in the medium to long term.

- Extinction should have been confirmed at the receptor site (at least 5 years recorded absence), the mobility of the target species should be assessed, and natural re-establishment should be shown to be unlikely over the next 10–20 years.

- Opportunities to restore networks of populations or metapopulations are preferable to single site re-introductions (unless the latter is a necessary prelude to the former).

- Enough numbers of individuals should be used in the re-introduction to ensure a reasonable chance of establishing a genetically diverse population.

- As far as possible the donor stock should be the closest relatives of the original population, and genetic studies should be carried out where doubt exists.

- The receptor site should be within the recorded historical range of the species.

- Removal of livestock should not harm the donor population (donor populations may have to be monitored during the re-introduction programme).

- The re-introduction should not adversely affect other species on the site.

- If captive-bred livestock is used, it should be healthy and genetically diverse (e.g. not normally captive bred for more than two generations).

- Re-introduced populations should be monitored for at least five years, and contingency plans should be made in case the re-introduction fails, the donor population is adversely affected, or other species are adversely affected.

- Approval should be obtained from the Conservation Committee of Butterfly Conservation and all other relevant conservation bodies and organizations (including statutory bodies in the case of scheduled species, Sites of Special Scientific Interest (SSSIs), etc).

- Approval must be obtained from the owners of both receptor and donor sites.

- The entire process should be fully documented and standard record forms completed for Butterfly Conservation and Invertebrate Link (JCCBI) (www.royensoc.co.uk/InvLink/Index.html).

threatened species programmes) *over-dispersion* (spreading out) may reduce the chances of mating. A soft release allows observations on the apparent suitability of the receiving environment, through observing insects' behaviour and acceptability of the immediate surroundings. Keeping the potential releasees together

Centre post 100 × 50 mm
2.4 m high
Aluminium flashing
Wire netting (19 mm)
Top rail
100 × 50 mm
Aluminium flashing
Post 50 × 50 mm
Gravel on top of shade cloth
Removable wire corner pegs
Shade cloth (70% nylon)
Access: Netting flap lifts when
Steel post
corner pegs are removed

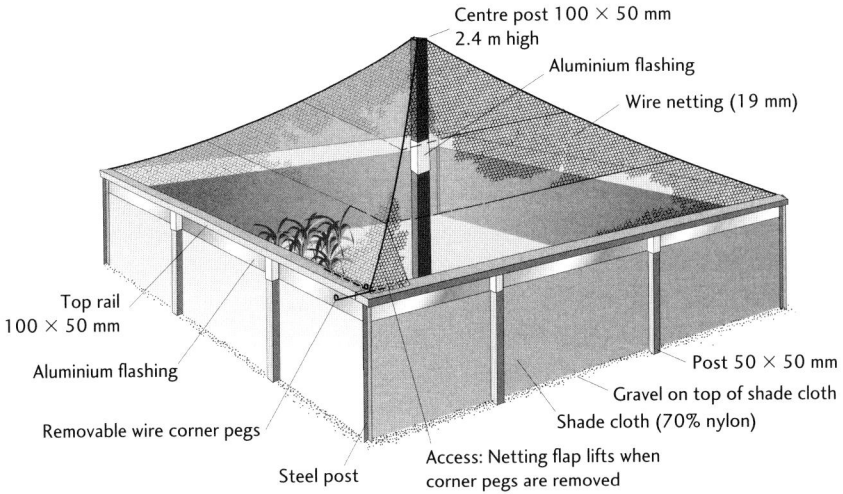

Fig. 8.2 A cage constructed on Red Mercury Island, New Zealand, for retaining the highly threatened Tusked weta *Motuweta isolata*, for soft release of captive bred individuals. More details are given in Stringer and Chappell (2008). (Courtesy Chris Edkins).

Box 8.4 Planning a translocation for a New Zealand weta

The Mahoenui giant weta *Deinacrida* sp. is a member of a notable flagship insect group (the Orthoptera) endemic to New Zealand, where their decline has been attributed to three factors: introduction of mammalian predators, habitat disturbance by people, and modification of habitat by introduced browsers. A group recovery plan was devised by Sherley (1998a). The Mahoenui weta has declined markedly, probably due largely to the third of the above factors, but with predators also implicated. It was the first weta for which translocation was undertaken, from stock reared in captivity.

Two wild populations were known, both very small, and most background research was undertaken on the larger population near Mahoenui (in the western North Island of New Zealand). Three mainland and one island site were selected to receive translocations, and these were selected for ecological suitability, security of land tenure, accessibility for vertebrate predator control, and attitudes of local people. Weta were collected for translocation in groups of 4–7 nymphal individuals and carried in well-aerated containers with fresh gorse *Ulex europeus* for shelter. They were released within 24 hours, or held in an enclosure nearby for later release. Immediate releases were of groups of 12–18 nymphs (with sex ratio unity, and of similar age) to the same bush, and adults found together were released together onto their 'own' bush. Delayed releases followed holding in an enclosure 6 × 10 × 3 m (high) over coppiced gorse bushes, clipped to maintain the weta's preferred dense foliage; the sides were dug into the ground to prevent access by predatory rodents. Up to four translocations were made at each site, but self-sustaining populations had not been confirmed after 7 years from the initial transfers in 1989.

Sherley (1998b) noted various problems that arose. Initial capture of wild weta was often under suboptimal conditions, as a crisis management exercise while the habitat was being destroyed for forestry conversion or fire breaks, so that treatment of the insects was sometimes less than ideal. Monitoring of translocated individuals was difficult because of lack of knowledge of their dispersal. The extra handling imposed by measuring each insect before release was probably stressful, as indicated by the insects' defensive behaviour. Captive-reared individuals were probably less stressed than wild insects, but consideration of acclimation effects then becomes necessary. Sherley emphasized the importance of advance planning and that the success of translocating a weta involves coordinating and timing of six phases:

1 Capture of wild animals to start a founder captive population

2 Timing the production of captive offspring for translocation to the wild at a season when survival is most likely, and corresponding with natural phenology

3 Building an enclosure and managing plants in it at a suitable site well before transferees are due

4 Controlling predators at the translocation site, and undertaking any other habitat enhancement needed

5 Monitoring success at each stage of the programme

6 Planning for repeated introductions.

for a time may help to ensure that matings have occurred and that seasonal development is in tune with the environment. At least, it is a useful adjunct in monitoring and predicting the likely fate of the insects after full release and, hopefully, establishment.

8.4 Practical outcomes

Post-release monitoring should always be undertaken (Watts et al. 2008). As with the other widespread context of releasing insects into a new environment, i.e. classical biological control, many early releases were never monitored effectively, and the projects were deemed complete at initial release, irrespective of the outcome. It is important to understand as fully as possible why any conservation-motivated insect release fails or succeeds and, when feasible, whether a newly established population flourishes and expands its range. Surveillance methods noted in Chapter 6 may be used, and monitoring should be done at suitable intervals and times (such as the flight period of butterflies) for several years. Stringer and Chappell (2008) successfully tracked released individuals of the Critically Endangered Middle Island tusked weta *Motuweta isolata* using radio telemetry.

Founder population size is also a relevant consideration in introduction programmes. Many reintroductions of butterflies in the UK have involved very few individuals, either from reared stock or translocated directly (Oates and Warren 1990). However, there have been very few investigations into the desirable size of such populations, and the reasons for success or failure of the exercises are usually not wholly clear. Brakefield and Saccheri (1994) undertook a range of laboratory rearing trials with the tropical satyrine butterfly *Bicyclus anynana*, designed to indicate genetic effects and implications. They suggested that introductions involving at least 10 gravid females from an outcrossed captive stock or a large natural population will (1) be more likely to succeed than when very few founders are used, but (2) will also be as likely to persist as those involving many more founders. However, they cautioned also that laboratory-based guidelines should be substantiated by proper field-conducted experiments. The genetic problems attendant on small populations in general (Chapter 6) are all relevant in reintroduction programmes and captive rearing.

Genetic considerations are clearly important also to minimize uncritical mixing of different genotypes. This is especially so when seeking to augment existing populations, or to establish new populations, when these may intermingle (e.g. as metapopulation units) with others in the wild. When the aim of a reintroduction is to establish or re-establish a working metapopulation, landscape features are also a central consideration (Chapter 7), because the need for such an exercise may commonly reflect a past history of habitat loss through fragmentation, to the extent that inter-colony dispersal has become impossible or too infrequent. Witzenberger and Hochkirch (2008), in a study of translocation of the Common European field cricket *Gryllus campestris* in Germany, clearly illustrated the importance of translocation of a high number (>200) of individuals, in their case nymphs, from different subpopulations to provide enough genetic diversity in the founder population for long-term survival. However, it is also important to consider the size of the population from which the translocated individuals were taken. Schmitt et al. (2005) showed that the translocation of only 50 individuals of the Mountain ringlet butterfly *Erebia epiphron* from a population of >100 000 individuals to a new location was enough to transfer most of the genetic variety of the original large population.

Table 8.5 *Polygon categories of suitability for the reintroduction or recolonization of the Heath fritillary butterfly* Mellicta athalia *in Kent, UK (Holloway et al. 2003).*

Category	Polygon criteria for selection
1. Occupied polygons	Containing extant *M. athalia* colonies (all category 1 polygons also exhibited the following criteria: broadleaved or mixed woodland, managed for conservation, coppiced, minimum areas 6–58 ha)
2. Recolonization	Broadleaved or mixed woodland; coppiced; <150 m from at least one category 1 polygon; not separated from category 1 polygon by barrier; minimum area 5 ha unless part of larger complex
3. Recolonization	As for category 2 polygon except: <150m from at least one category 2 polygon; not separated from category 2 polygon by barrier
4. Reintroduction	Broadleaved or mixed woodland; managed for conservation; coppiced; minimum area 5 ha unless part of larger complex
5. Colonization from reintroduction	Broadleaved or mixed woodland; coppiced; < 150 m format least one category 4 polygon; not separated from category 4 polygon by barrier; minimum area ha
6. Colonization from reintroduction	As for category 5 polygon except: < 150 m format least one category 5 polygon; not separated from category 5 polygon by barrier
7. Reintroduction	Broadleaved or mixed woodland; coppiced; minimum area 5 ha unless part of larger complex

There are very few cases in which restoration of viable metapopulations has been achieved and adequately documented, and the underlying planning may be complex. Many insect species recovery plans include a target such as 'to establish *x* new populations by year *y*'. Selecting sites to (1) maximize chances of successful establishment and (2) assure chances of inter-colony dispersal to functionally connect those populations demands a landscape-level approach wherever feasible. In one such exercise, Holloway et al. (2003) examined the case of the Heath fritillary butterfly *Mellicta athalia* in southern England. The UK Butterfly Action Plan for this species recognized the decline from 25 to 18 populations in the county of Kent, and specified a target of returning to the former total by 2005, with due consideration to the metapopulation structure of the butterfly, so that interconnectedness was a specified condition. Holloway et al. sought to specify elements of a strategic plan to achieve this recovery target, employing GIS to produce *conservation strategy maps* (CSMs). Their procedure was anticipated to have much wider applications, and emphasized the need to integrate ecological information with landscape features (including distribution of food plants). *Habitat polygons* revealed relevant habitats across Kent, and were fitted progressively to reveal the most suitable areas based on habitat, botanical, and topographical data to show areas that may be most suitable for establishment attempts. CSMs comprise large amounts of information, and Hollway et al. pointed out that the maps can be updated easily as changes in land use occur or more information is accumulated. The seven relevant categories of high priority polygons for *M. athalia* are differentiated in Table 8.5 to indicate the kinds of information that may be relevant to a given species.

8.5 Summary

Ex-situ conservation is a last-ditch measure to save a population or a species on the verge of extinction. It involves the taking of individuals from the wild, rearing them and releasing them into a highly suitable habitat, so that there is a good chance of establishment and population increase. Some insects are suited to such an approach, but many are not, as they do not take well to captivity. For those that do breed well in captivity, there is often rapid genetic adaptation to the conditions inside cages, sometimes rendering them poorly adapted for life in the field.

Insect farming refers to the rearing of insects in captivity, while insect ranching refers to practices involving resource enhancement in the wild to improve the level of natural breeding. Sometimes it also involves augmentative captive breeding. Some birdwing butterflies have been successfully conserved in this way.

Where a population is threatened by direct human activity or by alien predators, the whole or part of a population may be translocated. Such translocations must be to a highly suitable receiving habitat for long-term survival. When this is done, great care must be taken not to stress the insects being translocated. Such individuals should also ideally be taken from a donor population of at least 1000 individuals to ensure genetic vigour.

Some captive breeding programmes have been very successful while others have failed, despite what was thought to be good knowledge of the species. Nevertheless, the evidence suggests that very good knowledge of aspects of the biology of a focal species and of its exact habitat requirements are essential for ex-situ insect conservation.

Biodiversity and assemblage studies

9.1 Introduction: defining biodiversity in space and time

Ecologists and entomologists describe patterns in insect species richness, abundance, and distribution, and seek ways to explain these. Biodiversity encompasses the variety of life, including genes, species, communities, and ecosystems (Gaston 1996). Insect biodiversity therefore includes examining the morphological (such as sexual dimorphism and differences between life history stages) and genetic variation within species, differences in characters, abundance and function between species, and patterns of species richness, composition, and structure within and between communities.

Because the task is so enormous, *species richness* (or the number of species) is generally used as a surrogate measure of biodiversity, and has in fact become the 'common currency' in much biodiversity science (Gaston and Spicer 1998). Narrowing down the issue further, individual studies often focus on describing the diversity (measuring the species richness and evaluating the species composition) of a particular, well-defined group of species, such as:

- **Assemblages**, or a group of taxonomically closely related species, e.g. an ant, dung beetle, or butterfly assemblage
- **Functional feeding groups**, e.g. detritivorous, predatory, parasitic, or herbivorous insects
- **Functional groups**, e.g. soil movers, pollinators, mud gatherers, mineral recyclers (NB: Functional feeding groups are a subset of these)
- Species **associated with a particular habitat**, e.g. epigaeic (ground-dwelling), aerial, freshwater, or cavernicolous (cave-dwelling) insects
- **Guilds**, or groups of species within the same trophic level, e.g. all the herbivore species on a particular host plant

- Species sharing a particular **life history or behavioural characteristics**, e.g. gall-forming, leaf-mining, coprophagous (dung-eating), or social insects
- **Communities** or groups of populations of insects sharing the same habitat and interacting with each other in various ways (e.g. the community of fly and beetle species that generally occupy discrete habitats such mushrooms and other fungi, or a community of herbivorous and seed feeding insects associated with a particular plant species)
- **Food webs** that include all the species involved in the transfer of energy, via feeding interactions, from its plant source to herbivores and then on to predators, parasites, and parasitoids, e.g. the suite of 3 herbivore guilds, including >90 species of aphids, flea beetles, and caterpillars, and their parasitoids, hyperparasitoids, and predators, as elucidated in an example discussed by Price (1997). Memmott (2000) and Memmott et al. (2007) provide a means of quantifying such food webs (Figure 9.1).

Because the use of *assemblages* is common in insect biodiversity studies, this is what we will refer to as the insect group of interest in this chapter. However, most of the detail applies equally well to any of the above types of insect groups. In addition to narrowing the scope by focusing on a particular group of insects,

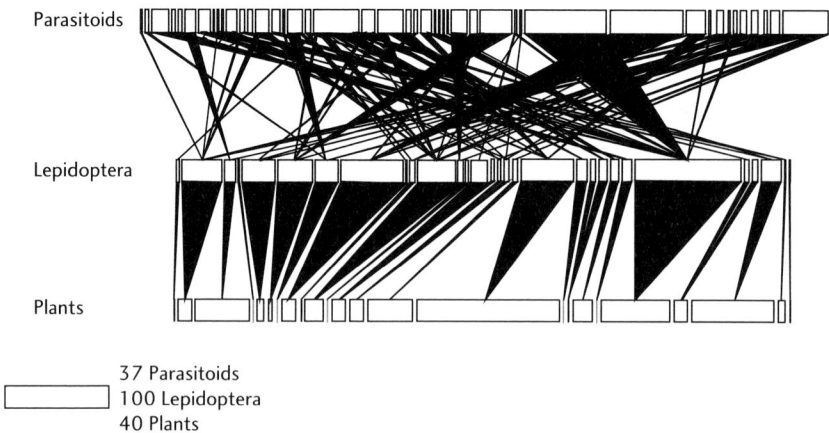

Fig. 9.1 A quantitative food web describing the interactions between plants, lepidopterans, and parasitoids. Each species is represented by a rectangle, with plants at the bottom, the lepidopterans in the middle, and the parasitoids at the top. The widths of rectangles depicting plants, lepidopterans, and parasitoids are proportional to their abundances, although for clarity the scales for the three trophic levels are different (from Memmott 2000).

in virtually all cases it is necessary to sample the group, because a complete census of its diversity is not possible. As a consequence, the outcome of such studies is a series of *estimates* of the diversity of the groups, rather than the true value of their diversity.

Sampling any of these groups (using the appropriate techniques as outlined in Chapter 3), yields a wealth of information. This includes not only the number of species caught, but also the identity of each species, the *abundance* (number of individuals) and distribution (number and locality of samples in which each species was present) of each species. Together, species richness, *species composition*, and some measure of species abundance and/or distribution, make up the most commonly used measures of biodiversity in insect studies. When considered together, these measures are often referred to as the *assemblage structure* (for an assemblage) or *community structure* (for a community). The techniques associated with measuring insect species diversity and assemblage structure therefore form the focus of this chapter.

9.2 The assemblage matrix

The starting point for quantifying assemblage structure or diversity is the *assemblage matrix* (or species by sample matrix). This matrix consists of a column for each sample and a row for each species (Table 9.1; see also Table 5.1). The body of the matrix may consist of either the number of individuals of each species recorded in each sample (abundance; Table 9.1a), or the occurrence (*presence–absence*, or *incidence*, shown as binary data) of each species in each sample (Table 9.1b). As well as being highly informative, these matrices also form the basis of all assemblage and diversity analyses (Bell 2003).

In an *abundance matrix*, row totals provide total abundance estimates for each species, and column totals provide the total abundance per sample, and the grand total (sum of row or column totals) provides the total number of individual insects caught. The variances of row, column, and grand totals can be used to assess the size of richness, range, abundance, and diversity patterns in the assemblage (see useful summary by Bell 2003). In the *occurrence matrix*, row totals provide *species occupancy* (range or distribution) data and column totals the species richness per sample.

In addition to providing a basic description of the assemblage (species diversity, abundance, and distribution), assemblage matrices are used to compile *species accumulation curves* (see below) and species relative abundance distributions, to compare species distributions (species association or co-distribution), to relate the abundances of species to each other (abundance covariation), as

Table 9.1 *Ant assemblage matrices of species abundance (a) and occurrence (presence–absence) (b) in three samples (s) in each of two habitat types (H), i.e. six sampling units (SUs), with a corresponding matrix of relational variables for each sample (c).*

Species			Samples				
(a) Abundance matrix	H1s1	H1s2	H1s3	H2s1	H2s2	H2s3	Row totals
Anoplolepis custodiens	23	33	24	1	4	0	85
A. steingroeveri	20	24	23	0	0	0	67
Camponotus angusticeps	0	5	3	0	6	0	14
Meranoplus peringueyi	1	0	7	0	1	0	9
Technomyrmex albipes	0	1	0	0	2	0	3
Linepithema humile	3	5	13	234	567	126	948
Camponotus vestitus	1	0	0	0	0	0	1
Column totals	48	68	70	235	580	126	1127
(b) Occurrence (occupancy or incidence) matrix							
Anoplolepis custodiens	1	1	1	1	1	0	5
A. steingroeveri	1	1	1	0	0	0	3
Camponotus angusticeps	0	1	1	0	1	0	3
Meranoplus peringueyi	1	0	1	0	1	0	3
Technomyrmex albipes	0	1	0	0	1	0	2
Linepithema humile	1	1	1	1	1	1	6
Camponotus vestitus	1	0	0	0	0	0	1
Column totals	5	5	5	2	5	1	23
Environmental variables							
(c) Relational matrix							
Rainfall (mm)	0	3	2	12	14	16	
Soil pH	7.4	7.6	7.8	8.1	8.1	8	
Bare ground (%)	45	50	45	20	20	20	
Vegetation height (m)	0.24	0.44	0.19	0.23	0.19	0.4	

well as to compare the assemblage structures of groups of samples, such as those from different habitat types (Ludwig and Reynolds 1988). Indeed, the types of analyses that can be performed on assemblage matrices may be grouped into those that involve comparisons between columns of the matrix (comparisons between groups of samples, also called *Q-mode analyses*), comparisons between rows of the matrix (comparisons across species, *R-mode analyses*), and those that consider both row and column patterns (quantification of assemblage structure using multivariate analyses).

While the abundance matrix is more informative than the occurrence matrix (and in fact the latter can be derived from the former), there are several instances in which occurrence data are used rather than abundance data. For example, when sampling social insects, such as the ant assemblage example used in Table 9.1, occurrence provides a less biased measure than abundance. Individuals from the same insect colony are not independent of each other, and catches are also strongly affected by the distance of the trap positions from the nests of social insects. Comparisons of abundances of social species are thus generally less meaningful than comparisons based on occurrence. Other reasons for using occurrence rather than abundance matrices include quantification of species distributions and species associations (see Ludwig and Reynolds 1988, Gotelli 2000).

In conservation-related studies we often need to know how different assemblages are related to each other. We may, for example, wish to see how the insect assemblage at a site with alien invasive plants compares with the assemblage at a natural, uninvaded site. Alternatively, we may wish to compare a burned site with an unburned one, or a polluted one with an unpolluted one, or the assemblage in a domestic garden with one in a natural area. The starting point is construction of the data matrix, as discussed in section 9.2. The replicates (sampling units, SUs), the different sites-within-treatment (to overcome pseudoreplication), e.g. the replicated burned sites, as opposed to the unburned sites, and the sites-between-treatments (e.g. burned vs unburned sites) are the columns. The species abundances are the rows, as are the environmental variables (see Chapter 5.2). These various sites can then be analysed for their similarity, i.e. we generally wish to know the extent of similarity (or difference) between the sites and to quantify this in some way. There are many similarity indices available for doing this (e.g. Bray Curtis similarity, Euclidean distance) (see Southwood and Henderson 2000), and as pointed out by Ludwig and Reynolds (1988), there is sometimes merit in using more than one form of analysis. Similarity (or difference) analysis (i.e. comparing the similarity or difference of columns in the biodiversity matrix) in fact forms the basis of multivariate analysis.

Multivariate analyses are used to compare the assemblage structure associated with, for example, different habitats, and to identify which species or which relational variables separate, or explain the variation across, different habitat types. Multivariate analysis (including ecological resemblance measures, cluster analysis, and ordination; Legendre and Legendre 1998) and a series of associated graphical techniques (e.g. dendrograms, ordinations, biplots) are commonly used to describe insect assemblages. A characteristic feature of assemblage matrices is the large number of zeros that they contain. This property of species matrices has driven the development of specialized analytical tools designed specifically to accommodate biological data. However, these techniques are not unique to entomology, and a discussion that would do the subject justice is beyond the scope of this book. We refer you rather to the excellent introductions and overviews of the use of similarity measures and multivariate analysis in ecology, provided by Legendre and Legendre (1998) and Leps and Smilauer (2003).

Finally, it is always a very good idea to maintain an electronic and hard copy *master assemblage (data) matrix*, which is the original assemblage matrix compiled, checked, and cross-checked for errors. All later modifications, exploratory data analysis, and statistical analyses are then performed on copies of this master matrix. In this way, the original matrix is always available as a reference for cross-checking, as well as a backup.

9.3 The relational matrix

As discussed in previous chapters, in most insect conservation and ecology studies one or more *relational variables* (or *environmental variables*) are measured in addition to species diversity (see Chapter 5). These variables are used either to explain patterns in insect assemblage structure or to control for environmental variation across the samples or sample groups taken. They may be compiled as a relational matrix that matches the assemblage matrix (Table 9.1c; and see Table 5.1). The relational matrix is useful in multivariate analyses, and when using some form of statistical model to explain species richness and abundance patterns.

Relational matrices differ from assemblage matrices in at least two important ways that are highly relevant when it comes to analysing your data. Relational matrices usually contain far fewer zero values than assemblage matrices. Also, column and row totals are generally not relevant because the variables encompass a range of different measurement units and repeat measurements are not additive (Table 9.1). However, this is not always the case, and an exception would include a plant species matrix that is being used as a relational data set to explain observed patterns in insect diversity. In this case, the plant data themselves form an assemblage matrix, but are not of primary interest and are merely being used to explain patterns in the insect assemblage (see Leps and Smilauer 2003).

9.4 Estimating sampling adequacy

Perhaps the most important question in biodiversity and assemblage studies is whether one has sampled the taxonomic group(s) of interest adequately. In other words, how confident are you that you have sampled all, or most, of the species in the assemblage? Similarly, how representative is the assemblage obtained from sampling of the true assemblage in the area? Without an answer to these questions, it is impossible to assess the conservation status of an area, impossible to compare the conservation values of different areas based on species richness, nor species composition, and also impossible to estimate extinction rates. Although, intuitively, it may seem valid to compare the assemblages of two areas, as long as the sample effort applied in each is equivalent, this is not in fact the case. Differences in habitat structure, environmental conditions, relative abundance distributions, and numbers of individuals collected may all result in the incorrect interpretation of assemblage comparisons based on observed species richness tallies (Gotelli and Colwell 2001).

The challenge in all insect assemblages lies largely with the rare species. It is always possible to say with more confidence that a species is present in an area than to say that it is absent. This means that a far greater sampling effort is needed to sample rare species, because the chance of capturing a rare species per sample is much lower than it is for common species (because both the abundance and distribution of rare species tends to be lower than that of common species; Gaston 1994). This is particularly relevant in insect conservation, because it is the rare species that are often of most conservation interest, and also the rare species that render particular areas biotically unique (such as *point endemic* species, although they can be locally abundant), with a high conservation status. Here we are thus concerned that we might miss particular species because the sample effort is too low, rather than not capturing them for other reasons (e.g. because the sampling technique is inappropriate or applied incorrectly).

As discussed in Chapter 3, the choice of sample size is always a compromise between practicality and sampling adequacy. Therefore, only in *depauperate* (species poor) assemblages is it possible to obtain complete species richness counts. The best we can do in species-rich assemblages is to maximize sampling effort, and at the same time, attempt to understand how representative (how complete) the species richness count obtained for that sampling effort really is. Indeed, it is no longer generally acceptable to report the outcome of assemblage or biodiversity sampling studies without demonstrating the level of sampling adequacy, or *sampling representativeness*. There are now well-established and well-tested methods for doing this (Walther and Moore 2005). Estimating species richness is a rapidly developing field of ecological analysis, and so it is well worth keeping track of advances in this area.

9.4.1 Taxon sampling (accumulation and rarefaction) curves

Sampling adequacy for species richness is commonly assessed using *species accumulation curves* and *rarefaction curves*. Gotelli and Colwell (2001) draw the useful distinction between four types of taxon sampling curves. These are (1) individual-based accumulation curves, (2) individual-based rarefaction curves, (3) sample-based accumulation curves, and (4) sample-based rarefaction curves.

- **Individual-based** taxon curves are used when the sampling literally takes place one individual insect at a time. This would include, for example, collection of individual butterflies with a net, or hand sampling of insects (see Chapter 4).
- **Sample-based** taxon curves are more common, and are used when more than a single individual is collected simultaneously, such as the contents of a pitfall trap, Malaise trap, or D-vac sample. Sampled-based curves may in fact also be used for individual-based collections (e.g. pooled across sweep net transects, or all hand sampling in a defined area).

Species accumulation curves (also known as *collector's curves*) represent the cumulative number of species obtained per individual or sample collected (Figure 9.2a) (Colwell and Coddington 1994). The shape of this curve varies depending on the order in which samples or individuals are accumulated as a result of natural variation between samples or groups of individuals. To overcome this problem, accumulation curves are generally 'randomized' to achieve a smoothed *rarefaction curve*, which provides an average representation of all possible accumulation curves (Figure 9.2a). Rarefaction curves may also be thought of as the statistical expectation (i.e. mean) of individual accumulation curves. This is achieved by randomizing the order in which samples or individuals are added to the curve. Usually, the data are repeatedly sampled a minimum of 100 times (the more the better, and 10 000 is preferable), sampling without replacement, from which to calculate the mean and standard deviation of the number of species for each additional sample or individual (Gotelli and Colwell 2001). In this way, the smooth, rarefaction curve shown in Figure 9.2a is produced from the same data represented by the uneven accumulation curve.

For a defined area, set time period, and high sample effort, an *asymptote* to species richness (Figure 9.2a) may be reached. This demonstrates that the area has been adequately sampled, and that the assemblage obtained is representative of the assemblage associated with that area. The general assumption must also be made that each rarefaction curve is produced from sampling within a fairly homogenous environment. The species × sample assemblage matrix,

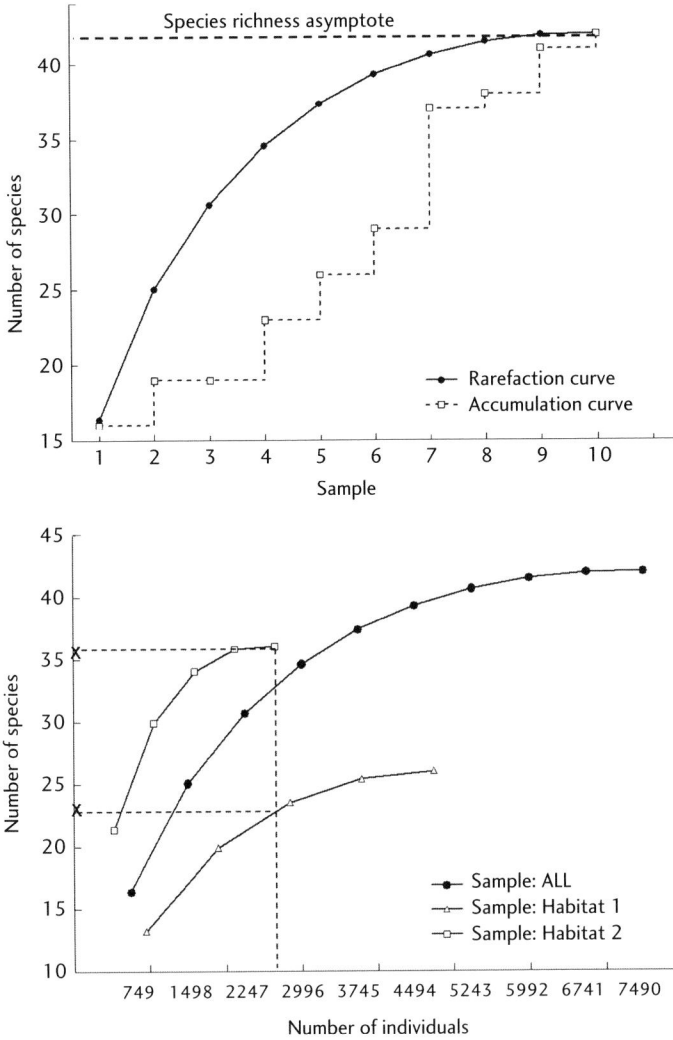

Fig. 9.2 Sample-based species accumulation and rarefaction curves scaled by sample (top) and sample-based rarefaction curves for two habitats and the combined samples for these habitats scaled by individuals (bottom).

outlined earlier, forms the basis of these calculations. Fortunately, excellent public domain software with a sound theoretical basis is available to perform these procedures and to generate rarefaction curves (*EstimateS*, Colwell 2004; *Ecosim*, Gotelli and Entsminger 2001).

9.5 Species richness

9.5.1 Comparisons of species richness

Only when asymptotes to species richness are obtained may assemblages be compared on the basis of raw data without fear of misinterpretation (Gotelli and Colwell 2001), e.g. comparing assemblages between disturbed and undisturbed sites. Unfortunately, insect assemblages seldom reach species richness asymptotes, and may never do so. However, rarefaction curves that do not reach an asymptote may still be compared after appropriate scaling by individuals, samples or area (Gotelli and Colwell 2001) (Figure 9.2b).

When *sample-based rarefaction* curves are used, valid comparisons of the species richness of two or more assemblages must be made based on the individual. In other words, the species richness of the assemblages is compared for a given number of individuals (Gotelli and Colwell 2001). For sample-based rarefaction curves, this means rescaling the curve to plot species richness against individuals (rather than samples) (Figure 9.2b). For example, in Figure 9.2b, the species richness of the two habitats is compared for the same number of individuals (approximately 2750) and, on this basis, habitat 2 has more species (an estimated 36) than habitat 1 (an estimated 23 species). In this way, the species richness of habitat 1 is *rarefied* to the same number of individuals for comparison with the species richness of habitat 2. This approach is called *scaling by individuals* because the comparison is made based on the same number of individuals for each habitat. The curve produced is therefore a 'sample-based rarefaction curve scaled by number of individuals'.

When dealing with sample-based data, it is often more accurate to refer to estimates of *species density* (number of species per unit area) rather than species richness. In this case, sample-based rarefaction curves may be rescaled by area rather than number of individuals for assemblage comparisons. Also, if sample-based rarefaction curves are not rescaled to individuals, but rather left as species richness plotted against samples, then the comparison is made in terms of species density rather than species richness. Species richness or species density values obtained in this way are obviously relative (per individual, sample, or area) and are thus not an estimate of the total number of species in the habitat or assemblage. Comparisons of the species richness or species density of assemblages are problem prone and must be conducted and interpreted with informed caution (Gotelli and Colwell 2001).

9.5.2 Estimating species richness

Observed species richness (i.e. the species richness tally obtained from a sample or individual-based study) is a biased estimator of true species richness (see Chapter 3 for definitions) in those cases where an asymptote to species richness is not reached by the rarefaction curve. In other words, all assemblages, in a defined area and for a particular time period, have a 'true' species richness that differs from the species richness that is observed by sampling (the latter is usually an underestimate) (Hellmann and Fowler 1999, Walther and Moore 2005). In studies where, based on either individual- or sample-based data, asymptotes to species richness are reached (often not the case in insect assemblages), the asymptotic species richness provides an acceptable estimate of true species richness. In many studies it may be adequate to compare assemblages across habitats or areas using rarefaction curves, and rarefying to a common number of individuals or samples, but in some instances, estimates of *total species richness* for an area or habitat are desirable. Here, a different approach is needed, that permits extrapolation from observed to expected (asymptotic) species richness.

There are three main approaches to estimating total species richness based on raw, or observed, assemblage data. These include methods based on (1) *species abundance distributions* (using parametric models such as the log-normal and Poisson log-normal distributions), (2) *taxon sampling curves* (asymptotic and non-asymptotic methods of fitting species-richness and *species-area curves* using, for example, negative exponential functions and log-log or log-linear models), and (3) *non-parametric* approaches that use the number and distribution of rare species in the assemblage (Colwell and Coddington 1994, Gotelli and Colwell 2001, Brose et al. 2003, O'Hara 2005, Hortal et al. 2006) (Table 9.2).

A good estimator of species richness should be accurate (close to the true value) and precise (showing little variation) (Walther and Moore 2005). However, factors such as *sample size, sample grain, sampling intensity, spatial coverage* (i.e. actual area physically covered by sampling), *evenness* of the assemblage, species mobility, and the size of true species richness all affect the performance of species richness estimators (Brose et al. 2003, Brose and Martinez 2004, Hortal et al. 2006). The performance of these different approaches has been extensively tested. The general conclusion, based on these numerous assessments, is that different estimators are better or worse for different assemblages as a consequence of differences in the properties of the assemblage and characteristics of the sample design. The taxon sampling curve approaches have been shown, for example, to be most useful for estimating an increase in species richness for a given increase in sampling effort or sample area (useful for decision-making when designing a study), and not for estimating total species richness (Colwell and Coddington 1994).

9.5.3 Chosing a richness estimator

This variability in the performance of species richness estimators makes life difficult when it comes to selecting a richness estimator for your study. However, non-parametric estimators appear to provide the best overall performance (Hortal et al. 2006) and are therefore the safest general option (Table 9.2). This is particularly true for insect assemblages, where a large number of rare species is the rule rather than exception. The challenge, therefore, is to distinguish between surveys that are substantially incomplete, and therefore have many rare species (many species with very few individuals), and surveys that are close to complete and where a high proportion of rarity is a true reflection of the assemblage structure (Colwell and Coddington 1994). The Chao estimators (Table 9.2), for example, estimate the true number of species in an assemblage based on the number of rare species in the assemblage (i.e. *singletons*, or those species represented by only a single individual, and *doubletons*, those represented by two individuals). Nonetheless, because different estimators provide different estimates, and are sensitive to the properties of the assemblage and the sampling design (O'Hara 2005), we recommend that several estimators are used and reported in sample-based biodiversity studies (Hortal et al. 2006). Again, it is advisable to be on the look-out for new developments in this field.

9.6 Species abundance and density

As is clear from the definition at the start of this chapter, biodiversity (or biological diversity) is much more than merely the number of species in the assemblage (although species richness is the most commonly used estimate of overall biodiversity). In addition to species richness information, abundance matrices also provide information on (1) how abundant individual species in the assemblage are (a crude estimate of species population size), (2) how abundant each species is relative to all

Table 9.2 *Generally well-performing non-parametric species richness estimators calculated by EstimateS (Colwell 2004) and SPADE (Chao and Shen 2003–05; Walther and Moore 2005; Hortal et al. 2006).*

Estimator	Use	Data
ACE	Abundance-based coverage estimator	N
ICE	Incidence-based coverage estimator	I
Chao1	Abundance-based species richness estimator	N
Chao2	Incidence-based species richness estimator	I
Jackknife1	First-order jackknife richness estimator	I
Jackknife2	Second-order jackknife richness estimator	I

(N, abundance; I, incidence.)

the others in the assemblage (relative abundance or evenness), and (3) how large the assemblage is as a whole (which may, for example, be converted using body masses to an estimate of assemblage biomass). Abundance provides valuable information on the diversity of an assemblage, allowing you to assess, for example, how many rare species there are in the assemblage, and also to make inferences about possible species interactions (using R-mode analysis, section 9.2).

Abundance data from assemblage studies can be summarized in the form of *relative abundance distributions* (also called *rank abundance distributions* or *curves,* or *dominance–diversity curves*) (Figure 9.3). Here, abundance data are usually expressed either as a percentage (Figure 9.3a), or log-transformed to accommodate large differences in the abundances of rare and common species (Figure 9.3b). Species abundances are then plotted against their abundance rank, from most to least abundant (Figure 9.3). From Figure 9.3, it is clear not only that the mite assemblage is more species rich than the springtail assemblage, but springtail abundances are also fairly evenly distributed (compared to many arthropod assemblages). A single mite species dominates the mite assemblage: the extent of its dominance is de-emphasized by the logarithmic transformation of the abundance data. Other methods of summarizing assemblage abundance data are also used, e.g. abundance frequency distributions and cumulative abundance distributions (Magurran 2004) (see, for example, Fig. 4.1).

Relative abundance distributions (Fig. 9.3) have a long history in ecology, and a number of statistical and theoretical *species abundance models* have been developed to describe and biologically explain the form of relative abundance distributions, for example geometric series, broken stick, log-series, and more recently, neutral models; see Magurran (2004) for an overview, and Tokeshi (1993) for a more detailed treatment. The interest in relative abundance distributions arises from the information they provide on the processes structuring the community or assemblage. Abundance provides a surrogate measure of how much resource a species uses (which is some function of energy availability, or productivity), and relative abundance distributions thus provide information on how species in an assemblage apportion resources (or niche space) among themselves (Tokeshi 1993, Magurran 2004). Relative abundance distributions are also used as a basis for the calculation of some diversity indices (evenness and dominance indices; see section 9.8).

Finally, in most studies, it is actually more accurate to refer to the density, rather than the abundance of a species, because sampling is usually conducted over a fixed area (grain and extent). In cases where the sampling unit (SU) is a discrete *habitat patch* or *habitat unit*, such as a plant gall or individual plant, it is most accurate to refer, for example, to the number of individual insects per gall,

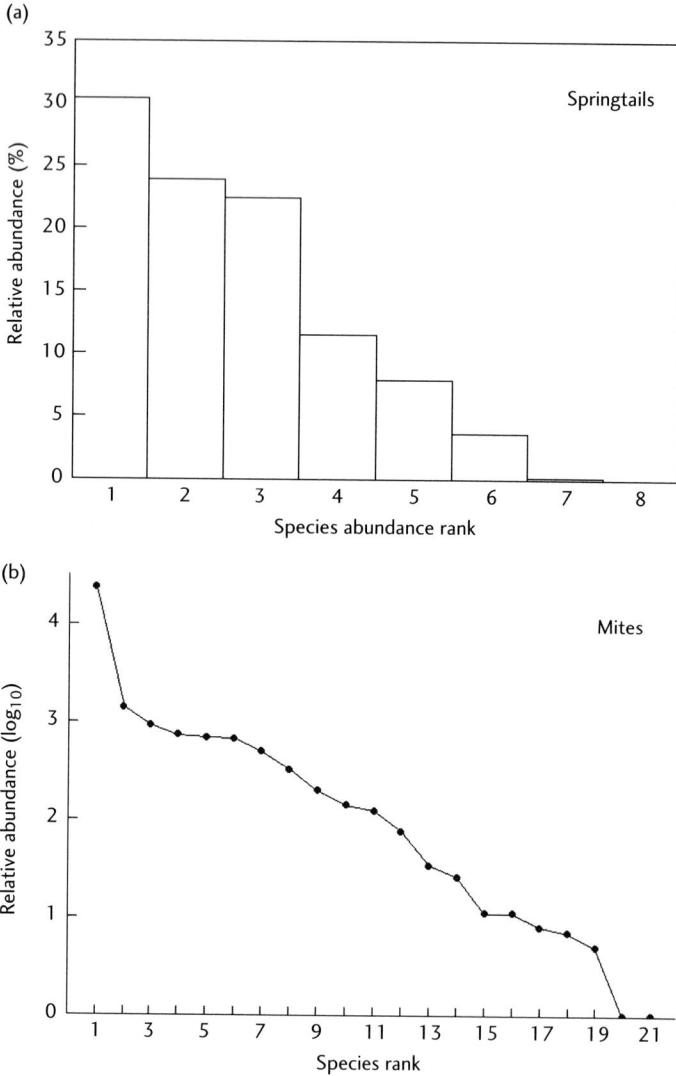

Fig. 9.3 Relative abundance distributions of a springtail (a) and mite (b) assemblage on sub-Antarctic Marion Island. In (a) the distribution is expressed as percentage relative abundance, whereas in (b) it is shown as a dominance–diversity curve using logarithmically transformed abundance values.

per plant stem, or per plant. It is appropriate to use the term abundance when the objective of the sampling protocol has been to obtain an estimate of the population size of one or more species. Sampling is usually more intensive and some information of the distribution of the population is needed, although see the discussion of *mark–release–recapture* methods in Chapter 7.4.1. Although this is usually not the case in assemblage studies, the term abundance is still most commonly used. We will use it here, but stress the importance of appreciating the distinction just described, and the consequences for data interpretation.

9.7 Species range and distribution

The presence–absence or occupancy matrix can be used to quantify patterns of species distribution (or range) in the assemblage. Here it becomes important to make the distinction between 'entire' and 'partial' studies (sensu Gaston and Blackburn 1996), where (as is usually the case) in *partial studies*, the data matrix represents only a subset of the geographic area within which each species in the assemblage actually occurs (acknowledging that assemblage composition is spatially dynamic). Unless one is working across large regional, continental, or geographic scales, the occupancy distribution provides information on the local distribution of species (Fig. 9.4). Regardless of the scale, the data are dealt with in approximately the same way, and it is only the terminology and interpretation that differ (geographic range patterns vs local distribution patterns).

Occupancy data may be expressed as *occupancy frequency distributions* (or range size distributions) to examine patterns of range sizes across species in the

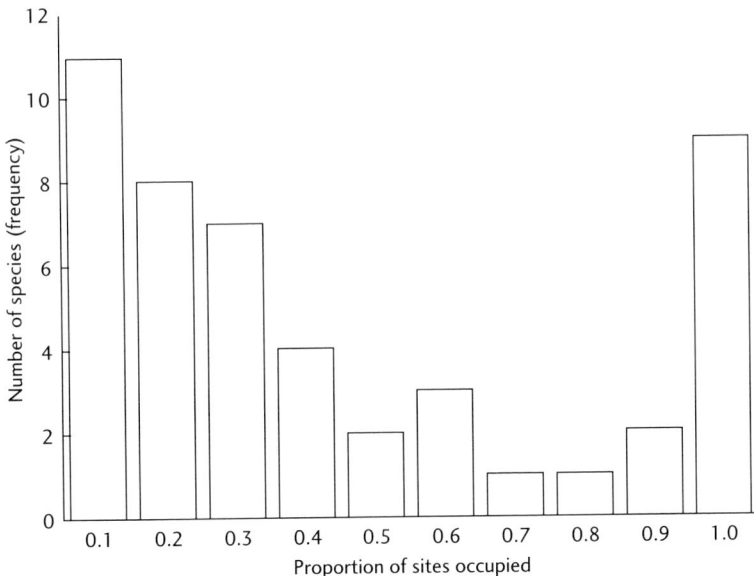

Fig. 9.4 An occupancy frequency distribution for dung beetle species in Tembe Elephant Park, South Africa.

assemblage (McGeoch and Gaston 2002). (Note that in this chapter the word 'distribution' is used in two ways: (1) to mean the physical spatial distribution (or range) of a species (the subject of this section), and (2) to mean the statistical frequency distribution of data, of which Figure 9.4 is an example.) Because species abundance and occupancy are usually positively correlated, one may ask why we bother with occupancy data when abundance data are available. One reason is that, in some cases, abundance data are not readily available or reliably interpretable (as in social insects, discussed in Chapter 3). However, the number of samples in which a species is present (occupancy) provides information in addition to abundance, i.e. on the distribution of individuals across samples (providing an estimate of species aggregation or statistical heterogeneity) and whether the distribution of a species is widespread (it is found in the majority of samples) or narrow (it is found in only a few samples).

9.7.1 Biodiversity mapping

Mapping the distribution of individual insect species, and the species richness of insect assemblages, has become an increasingly important tool in conservation assessment. Maps of the distribution of individual species not only show where a species occurs, or has occurred in the past, but are used to (1) identify species richness or *endemism hotspots*, (2) monitor changes in the geographical range size and distribution of species over time, (3) identify relationships between species distributions, climate, and other environmental variables, and (4) predict changes in species distributions as a consequence of climate change and habitat loss. For example, Parmesan et al. (1999) showed that the ranges of 63% of non-migratory European butterflies have shifted northwards over the last century, likely as a consequence of global warming. Wilson et al. (2007) showed how information on species distribution and range size could be used to predict which butterfly species were undergoing population decline.

Biodiversity mapping is useful not only for understanding the changes in distribution of rare species, but also for predicting the range expansion of invasive species. For example, locality records from across the world have been used to model the ecological niche of the invasive Argentine ant *Linepithema humile* and then to predict its spread and potential geographic distribution under current climate-change scenarios (Roura-Pascual et al. 2004).

An important use of mapped data is in conservation planning. Here, Geographic Information Systems (GIS) are commonly used to map so-called 'layers' of information and analyse the relationships between them. These layers include *spatial data* (the shape and location of places) and *attribute data* (for example, species distribution data and the distribution of habitats and other environmental variables).

These conservation assessments are possible only for those species and assemblages for which adequate distributional data are available (see Rondinini et al. (2006) for a good overview). Such data are most commonly compiled from the locality records of museum specimens, although this has the disadvantage of encompassing records often spread over several decades of collection. Particular caution may be needed when looking at common species—many museum collections of Lepidoptera, for example, are founded on hobbyist collections bequeathed in years past, when many collectors focused more on rare or sought-after species, with the result that many common species are under-represented in such accumulations. Nonetheless, museum collection data are an invaluable source of information on species historical distributions. However, current and accurate distributional data for species are also readily obtainable using a hand-held Geographic Positioning System (GPS).

A rapid and systematic first assessment of the broad distribution of a species may be obtained by (1) stratifying the region of interest into grid cells of a convenient size, (2) sampling one or more points within each grid cell, and (3) recording the presence or absence of the species at each point. Obviously, the smaller and more numerous the grid cells used, the more detailed the distributional information that will be obtained. But this also means more sampling, and so there may be a trade-off between fineness of scale and feasibility of completing the work (see next chapter, Figure 10.4).

Distributional data are valuable at several spatial scales, from mapping part of the range of particular species to mapping entire geographic ranges and species richness of assemblages globally. Occupancy data may be used, for example, in ecological landscaping to protect populations and their movement corridors, or attract individual insects to particular parts of a landscape, often to increase local diversity or for ecotourism purposes. For example, landscapes can be successfully modified to attract dragonflies by the provision of suitable water bodies, vegetation for perching and roosting, and substrates for basking.

To produce species maps, locality records (latitudinal and longitudinal point records, i.e. geographic coordinates) are required for individuals of the focal species. Importantly, absence records (localities where the individual has not been recorded) can be as useful as presence records. They can, for example, provide very valuable information on predicting where you should search for possible new localities of Red Listed species. Finch et al. (2006) give a detailed example, and Box 9.1 shows how the technique can be applied in practice. Absence records are less readily available, and not obtainable from museum collections. Another point is that absence records are generally not as reliable as presence records, because generally far greater sample effort is required to demonstrate unequivocally that a species is not present in an area than it does to confirm its presence (see Chapter 10.6 for more details; and Figure 10.5). Nonetheless, absence data are very valuable, and data sets that include both presences and absences for species are most useful. For example, they are better able to identify the environmental correlates of species distributions and for building predictive models for changes in species distributions. This approach can be expanded to include several species, and the coincidence of species on the map then indicates local area of high conservation value at the multispecies level.

Box 9.1 The use of cartographic modelling to map potential habitat for threatened endemic species

---

Populations of many threatened endemic species are declining, and these species may be known from only a few localities. An imperative is then to find other localities where the species still occur, to assist in the conservation management of the species. In cases such as this, cartographic modelling can be very useful in indicating where remnant, potentially suitable, habitat may be situated.

An example of a threatened endemic species is the Karkloof blue butterfly *Orachrysops ariadne*, which occurs only in the KwaZulu-Natal Province of South Africa. Four colonies of this species were known at the end of 2007, and two of these colonies were near local extinction. It became necessary to attempt to find other colonies, so that the viability of, and genetic diversity in, the remaining colonies could be maintained or improved through appropriate translocations or reintroductions, or through the maintenance of (the assumed) metapopulation dynamics. Because of the small number of known localities for this butterfly species, a simple cartographic model was developed to determine where suitable habitat might still remain (see figure below). Information on what factors were likely to influence the distribution of *O. ariadne* was collated from the literature. The final choice of variables to include in the model depended on what was known of the life history and ecology of *O. ariadne* at the time, and also on what coverages were available to implement the model in a GIS. Predictor values and value ranges were determined from overlays of the colony locations on the predictor variable coverages.

O. ariadne and its oviposition and larval host plant, *Indigofera woodii* var. *laxa*, are known only from untransformed grassland (including bushy grassland) from the Midlands Mistbelt Grassland vegetation type on moderate to very steep slopes with relatively moist soils (i.e. south-facing slopes). These predictors were used in the model. In addition, the mean maximum daily temperature for the hottest month and the mean minimum daily temperature for the coldest month during the flight and egg periods (i.e. the above-ground stages in the life cycle of *O. ariadne*) were used in the model as predictors. Ambient temperature influences distribution ranges, flight period and egg development time in butterflies (e.g. Lu and Samways 2001; Thomas et al. 2006; Menéndez 2007). The cartographic model was implemented in the raster-based Idrisi GIS (Idrisi Andes, Clark Labs,

Clark University, Worcester MA, USA), and the resultant predicted potential habitat is indicated in the figure below.

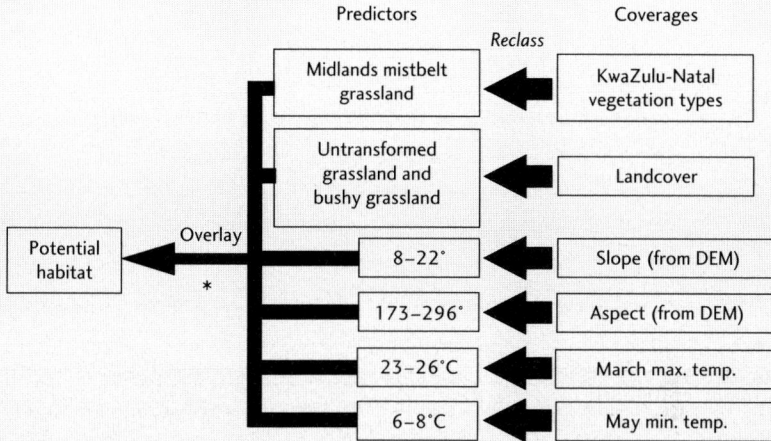

Cartographic model to predict where potentially suitable habitat for *Orachrysops ariadne* remains in KwaZulu-Natal; DEM, digital elevation model, March Max. Temp., mean maximum daily temperature for March, May Min. Temp., mean daily minimum temperature for May (Schulze 2006). *, by multiplication.

The potential habitat coverage was then overlaid on 1:50 000 scale topographical coverages to assist with the planning of the ground-truthing phase. Ground-truthing of the potential habitat map was commenced by visiting areas of potential habitat that were relatively large, especially if these areas were close to known colonies. Metapopulations may be clumped rather than more evenly distributed through the distribution range of the species (Nekola and Kraft 2002), and so searching for further colonies of *O. ariadne* was initially focused close to existing colonies (under the assumption that this species forms metapopulations). The results obtained in June and July 2008 are illustrated in the figure below. The host plant was recorded at four of the potential habitat areas visited for the first time, and oviposition was recorded at one of these. The other three areas need to be visited during the flight season. However, *I. woodii* var. *laxa* was not present on most of the potential habitat areas visited. These areas showed signs of frequent fires and/or heavy grazing, possibly explaining the absence of the host plant because it does not survive long-term annual burning. *I. woodii* var. *laxa* was, with one exception, absent from areas where it was predicted to be absent. The one exception was a small area with very few host plants near to predicted potential habitat; the scale at which the potential habitat was mapped was larger than this area of host plants and contributed to the mapping error. Collection of presence and absence data allows for the use of other distribution modelling techniques and refinement of the model at a later stage.

Adrian J. Armstrong, Ezemvelo KZN Wildlife, South Africa

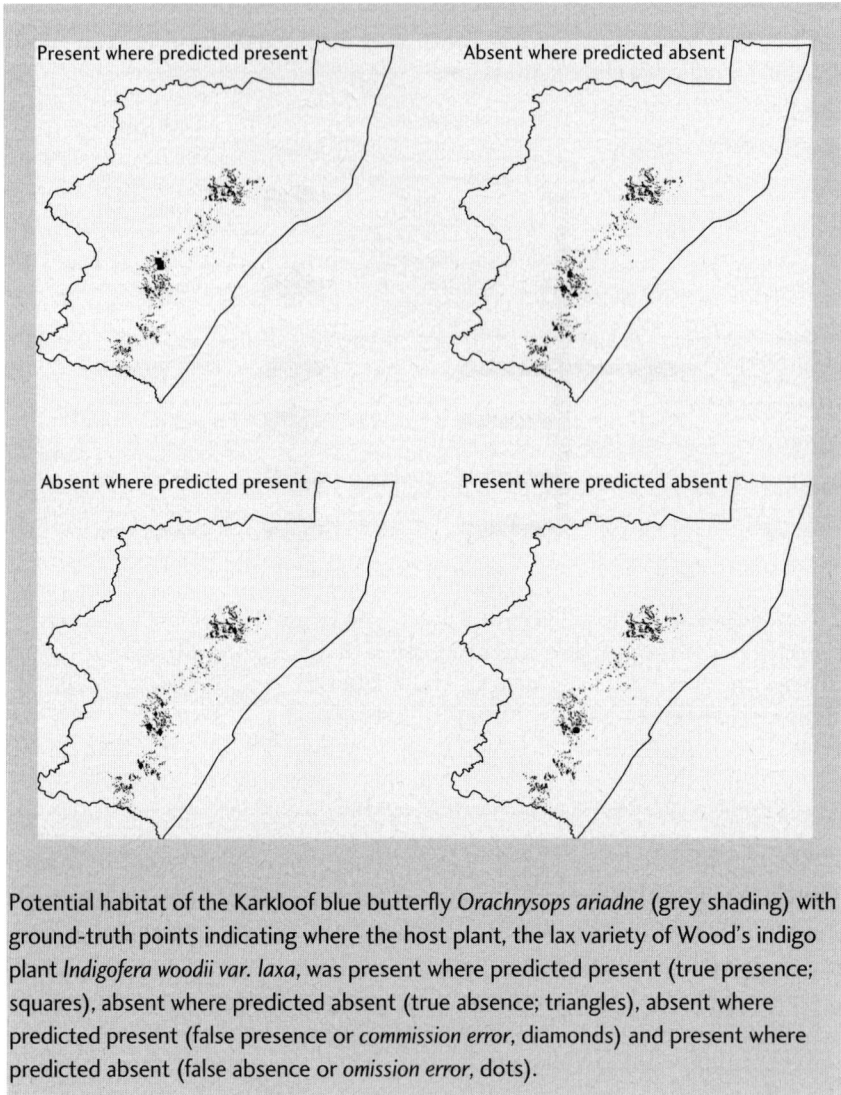

Potential habitat of the Karkloof blue butterfly *Orachrysops ariadne* (grey shading) with ground-truth points indicating where the host plant, the lax variety of Wood's indigo plant *Indigofera woodii var. laxa*, was present where predicted present (true presence; squares), absent where predicted absent (true absence; triangles), absent where predicted present (false presence or *commission error*, diamonds) and present where predicted absent (false absence or *omission error*, dots).

9.8 Diversity indices

9.8.1 Species richness, evenness, and dominance

Above we discussed measuring single components of biodiversity evident from the biodiversity matrix, i.e. species richness, abundance, and distribution. In ecology and conservation, 'diversity' is also often quantified in the form of a *species diversity index*, which provides a single value summarizing the different components of diversity, and emphasizing one or both of the species richness and relative abundance components (Box 9.2). Indices that measure the abundance of each species in a community relative to all the others are called *evenness indices* (see Figure 7.8). Closely related *dominance indices* measure the extent to which an assemblage is dominated by one or a few species (species that are highly even have low dominance and vice versa). Those indices that include both species richness and abundance components are usually better known as species diversity indices per se, or measures of heterogeneity.

Box 9.2. Diversity indices*

Diversity

Shannon index (H′)

$$H' = -\Sigma p_i \ln p_i$$

where p_i is the proportion of individuals in the ith species (or n_i/N). The higher the value of H, the greater the diversity of the assemblage or sample.

Log series α

$$\alpha = \frac{N(1-x)}{x}$$

where x is solved by iteration using $S/N = [(1-x)/x][-\ln(1-x)]$, S being the number of species in the assemblage or sample and N the total number of individuals in the assemblage or sample. However, α can also more simply be read off the table provided in Southwood and Henderson (2000) using observed values of S and N.

Simpson's index (D)

The probability that any two individuals drawn at random from a finite assemblage belong to the same species is:

$$D = \Sigma \left(\frac{n_i(n_i-1)}{N(N-1)} \right)$$

where n_i is the number of individuals in the ith species, and N is the total number of individuals (summed over all species). Thus, as D increases, diversity decreases. For this reason, the index can be expressed as $1-D$ (so that an increase in the index represents an increase in diversity).

Evenness

Simpson's evenness measure (Smith and Wilson 1996)

$$E_{1/D} = \frac{(1/D)}{S}$$

where D is the Simpson index (above), and S is the number of species in the assemblage or sample. $E_{1/D}$ lies between 0 and 1, and the larger the value the greater the evenness of the assemblage. Evenness is thus a measure of how similar species abundances are to each other in an assemblage. Ludwig and Reynolds (1988) recommend *modified Hill's ratio* (labelled *E5* by those authors) (see Figure 7.8) as it is relatively unambiguous, fairly easily interpretable, and does not require an estimate of the number of species in a community, which is affected by sample size.

Dominance

Berger-Parker index (d)

$$d = N_{max}/N$$

where N_{max} is the number of individuals of the most abundant species, and N is the total number of individuals in the assemblage or sample (across all species).
The higher the proportional abundance (**dominance**) of the most abundant species, the higher the value of d. Dominance thus measures the degree to which one or a few species numerically dominate the assemblage.

* Readers are encouraged to refer to Magurran (2004) for a detailed exposé of the advantages, disadvantages, strengths, weaknesses and variety of diversity indices, and original references.

Diversity indices have been widely used in the past. The bad news is that there is an enormous array of diversity indices to choose from (Magurran 2004), and each has its peculiarities and subtleties of meaning and application. The better news is that diversity indices are now being less used in favour of other, non- or less-aggregated measures of richness, abundance, and species composition. Nonetheless, in some cases, an index of diversity is required; for example, for comparison with previous related studies that have used a particular index, or where a single number is required to summarize overall diversity (or as much of its complexity as possible). Therefore, the advantage of diversity indices is their ability to summarize both a richness and an abundance component for the assemblage in a single value. The disadvantages are that the index values are often not intuitively interpretable, and each measure weights and combines the richness and abundance components in a slightly different, not always theoretically justifiable, way.

Magurran (2004) provides an excellent overview of diversity indices. In summary, when a diversity index is needed, she recommends the use of either *Simpson's index* or *log series* α, because both are intuitively meaningful, comparatively insensitive to sample size and their performances have been well studied (compared with the popular, widely used Shannon index, which is sensitive to sample size and more difficult to interpret) (Box 9.2). The *Berger–Parker dominance index* and *Simpson's evenness measure* are recommended as dominance and evenness measures respectively (Magurran 2004). Regardless of these useful general recommendations, it is always important to understand the performance, biases, and sensitivities of the index you are working with, and we suggest that you consult Magurran (2004), to ensure that you are able to defend your choice of diversity index. Finally, diversity indices should, wherever possible, be presented along with a measure of variation (confidence limits, standard deviation, or standard error). This can be calculated as usual (by calculating the variance in diversity across samples) for those indices that are normally distributed, or using resampling statistics for those that are not (see Manly 1997). Useful software for conducting diversity index calculations (and measures of variation associated with them) includes *EstimateS* (Colwell 2004) (see 'Useful software' at end of book).

9.8.2 Beta diversity

Species diversity is also commonly expressed in terms of three components: *alpha diversity* (local diversity), *beta diversity* (change in diversity between localities), and *gamma diversity* (regional diversity) (Whittaker 1975, Magurran 2004). The species diversity indices, discussed above, are all examples of alpha diversity measures. Measurement of alpha diversity is most common, but beta diversity is also important for understanding spatial patterns in diversity (Gaston and Blackburn 2000), and because beta diversity provides a measure of the change in diversity across space. As with alpha diversity indices, many beta diversity indices have been proposed (at least 18 distinct measures; Koleff et al. 2003).

Koleff et al. (2003) make a distinction between four categories of beta diversity measures, of which two are either widely used or recommended. These are what they term (1) 'broad-sense' turnover measures that emphasize richness differences between localities (widely used), and (2) 'narrow-sense' turnover measures that emphasize compositional turnover between localities (recommended) (Box. 9.3). As with other diversity indices, because they are interpreted differently and emphasize different components of diversity turnover, the choice of beta diversity index to use depends on the study objective. However, Koleff et al. (2003) show that in general β_{sim} (Lennon et al. 2001) performs best. Values of β_{sim} are high when the proportion of species shared between two localities is low and the proportions of species lost and gained between localities are similar (Koleff et al. 2003), i.e. compositional turnover.

Whittaker (1975) originally formulated the relationship between the three components of diversity as multiplicative, i.e. gamma diversity = alpha diversity × beta diversity. Note that the multiplicative form means that the components are not weighted equally when applied to different spatial scales (Lande 1996).

Box. 9.3 Beta diversity (β)

Compositional turnover (β_{sim})[*]

Beta diversity is usually calculated from presence–absence data and involves the measurement of three parameters, for the comparison of two sites or localities (i.e. it is a pair-wise calculation): b, the number of species unique to locality 1; c, the number of species unique to locality 2; a, the number of species shared by localities 1 and 2 (see schematic below).

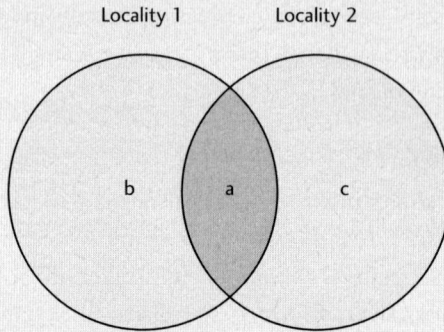

Locality 1   Locality 2

b a c

Parameters in the calculation of β (redrawn from Koleff et al. 2003).

Following Lennon et al. (2001);

$$\beta_{sim} = \frac{1}{n}\sum_{i-1}^{n}\left(1 - S_{2i}\right)$$

where

$$S_{2i} = \frac{a_i'}{a_i' + \min\left(b_i', c_i'\right)}$$

and a', b', and c' are as defined above and shown in the figure below.
β_{sim} lies between 0 (no species in common between localities) and 1 (both localities have the same number of, and identical, species).

Additive partitioning ($\beta = \gamma - \alpha$)

Following Crist et al. (2003); using any measure of diversity (such as species richness, or the Simpson or Shannon indices in Box 9.2), total diversity (γ) is partitioned into the average diversity within samples (α) and between samples (β) so that $\gamma = \alpha + \beta$. Beta diversity is estimated by $\beta = \gamma - \beta$ (Wagner et al. 2000):

$$\gamma = \alpha_1 + \sum_{i=1}^{m}\beta_i, \quad \text{given } \gamma \geq \alpha$$

where i is the number of scales in the hierarchical sample design (i = 1, 2, 3, ..., m).

The smallest sampling units are at level $i=1$ of the sample design. Thereafter the samples at higher levels of the hierarchy ($i > 1$) are formed by pooling the samples at the level below it:

α_i = the average diversity across samples at level i.

The average diversity (α_i) for samples at each level of the sampling hierarchy is

$$\alpha_i = \sum_{j=1}^{n_i} D_{ij}q_{ij},$$

where D (in this case) is Simpson's diversity index (Box 9.2) in each sample j = 1, 2, 3, ..., n_i. However, D could be any measure of diversity, or merely species richness. q_{ij} are the sample weights (the proportion of the total number of individuals found in each sample i).

Beta diversity for each sampling level i is then calculated as

$$\beta_m = \gamma - \alpha_m$$

at the highest level of the sampling hierarchy, and

$$\beta_i = \alpha_{i+1} - \alpha_i$$

at each lower level of the sampling hierarchy.

Example of hierarchical calculation of Beta (modified from Gering et al. 2003)

Level of hierarchy	Scale	Local diversity	Beta diversity	Total or gamma diversity
1	tree	α_1 (within tree)	β_1 (between trees)	γ (tree) = $\alpha_1 + \beta_1$
2	stands	α_2 (within stand)	β_2 (between stands...)	γ (stand) = $\alpha_1 + \beta_1 + \beta_2$
3	sites	α_3 (within sites...	β_3	
4	ecoregions	α_4	β_4	
5	regional	α_5		γ (regional) = $\alpha_1 + \beta_1 + \beta_2 + \beta_3 + \beta_4$

Later, an additive relationship was proposed (i.e. gamma diversity = alpha diversity + beta diversity) that weights the components equally across spatial scales (Lande 1996). This additive approach can be used in hierarchically scaled studies to assess the proportion of total diversity (gamma) found in different localities, habitats, or regions (Crist et al. 2003). For example, Gering et al. (2003)

showed that the beta diversity of forest canopy beetles was greater between forest ecoregions than between other hierarchical levels of the study (the study was hierarchical because there were sites within ecoregions, stands within sites, and trees within stands). This additive form of beta diversity falls within the 'broad-sense' understanding of turnover, as it measures the proportion of diversity explained at different spatial scales of a sampling hierarchy. Indeed, Koleff et al. (2003) recommend the use of both categories of beta diversity measures to maximize understanding of turnover in species composition across space.

9.8.3 Functional diversity

As students of biodiversity, we are all accustomed to hearing (from our less nature-enthused friends) the question, 'But why does biodiversity matter?'. *Functional diversity* provides one answer to this question, and is often measured to establish the mechanistic link between species and ecosystem functioning, and by extrapolation, our survival (Kinzig et al. 2001, Petchey 2004). For example, differences in the way that different species in an assemblage use resources (differences in either the type, quality, or quantity of resource used) imply that they each have different impacts on their environment (and cumulatively on the ecosystem), and therefore that each species has different consequences for the functioning of the ecosystem. One of the major incentives for the conservation of biodiversity is the fact that it is 'functionally significant'. Therefore, to quantify and demonstrate the functional importance of insect biodiversity (and provide policy-makers with ever more incentive to conserve insect biodiversity!), we need to be able to quantify functional diversity.

Functional diversity can be defined in a number of ways, but generally involves 'what species do' (Petchey and Gaston 2006), or the contribution that they make to one or more ecosystem processes (such as productivity, nutrient turnover, decomposition, seed dispersal, pollination). Tilman (2001) describes *functional diversity* as 'the value and range of those species and organismal traits that influence ecosystem functioning'.

The simplest measure of functional diversity is the number of trophic levels in the assemblage (e.g. herbivores, predators, parasitoids, hyperparasitoids, parasites), or the number of functional groups in the assemblage. More information is provided by adding a measure of the proportional abundance or biomass of the different functional groups in the assemblage. These measures of functional diversity are relatively easy to apply and rely only on the *a priori* existence of relevant functional group classifications (such as the classification of ant functional groups based on their response to stress and disturbance; Andersen 1995b, Majer et al. 2004, and see Table 2.1) and knowledge of the appropriate functional group for each species in the assemblage. However, this approach to measuring functional diversity is comparatively coarse, because species-level information is lost by pooling it into broad categories. As a result there is no appreciation of the contribution that individual species make to ecosystem processes.

A more information-rich and very interesting approach is to quantify the *functional structure* of an assemblage, i.e. the distribution of the relative contribution of species in the assemblage to an aggregate ecosystem function, such as pollination or decomposition (Balvanera et al. 2005). This approach is conceptually similar to quantifying assemblage structure as described above. For example, Balvanera et al. (2005) propose and illustrate the concept by quantifying the functional structure of indigenous bee pollinators of watermelon in California. The functional contribution of each species was measured by counting the number of pollen grains on a floral stigma after the visit of a single individual bee (pollinator efficiency). The pollination function of each species was then calculated as the estimated daily visits per flower multiplied by pollinator efficiency. The functional structure of the assemblage may then be plotted as pollination function against species rank (very similar to the rank abundance and dominance–diversity curves in Figure 9.3, except that the measures of pollinator function replace abundance). Functional structure thus provides information of the relative contributions of species in the assemblage to the performance of the ecosystem function of interest (pollination function in the above example).

Finally, each species in an assemblage can be classified according to a series of functionally important traits, producing a species (in rows as before) × functional trait (in columns) matrix. How one decides on what the 'important' functional traits are depends partly on the objective of the study and partly on the ecosystem function(s) of interest, as well as on the amount of biological information available (or obtainable) for species in the target assemblage (Petchey and Gaston 2006). The series of functional traits may include, for example, body size, mouthpart type (sucking, chewing, rasping), host (feeding) specificity, and seasonality. Here it is important to be able to identify those traits that do contribute to the ecosystem function of interest and those that do not, as well as those traits that provide direct measures of this function (e.g. consumption rate as a measure of decomposition function) vs those that provide only an indirect correlate of the function (e.g. body mass as a surrogate of consumption rate) (Petchey et al. 2004, Petchey and Gaston 2006).

Using multivariate analysis, the functional trait matrix can then be used to identify 'functional groups' (a more robust a-posteriori approach compared with the a-priori approach mentioned above). The functional diversity of the assemblage can also be assessed by calculating the average pairwise distances between species on the ordination (produced using multivariate analysis conducted on some measure of pairwise distance or similarity, e.g. Euclidean or Jaccard), or total branch length in the dendrogram from a cluster analysis (see Petchey and Gaston 2006 for a review of these more complex measures of functional diversity).

9.8.4 Taxonomic diversity

Taxonomic diversity measures the phylogenetic distance (or taxonomic relatedness) encompassed by an assemblage of species, and quantifies the character or feature diversity of species (this could be morphological or molecular). For example, if two assemblages have identical species richness and evenness, then most diversity measures would judge them as identical. However, one of these assemblages may contain 23 species of urban-dwelling *Drosophila* (vinegar flies, i.e. an assemblage of closely related species), whereas the other may contain 23 species of a range of Diptera, Hymenoptera, and other garden-visiting pollinators. Clearly, the species composition of the two assemblages is different, and the taxonomic diversity of the second (across several insect orders) is far greater than the first (within a single genus) because the species are more distantly related. Taxonomic diversity also draws on the concepts of richness and evenness from species diversity measures (section 9.8.1). Richness in this context refers to the number of evolutionarily novel characters or features (rather than the number of species).

Phylogenetic diversity also measures variation in characters between species. Phylogenetic diversity can be used to assign a conservation value to species and to prioritize taxa for conservation by quantifying characters (features or attributes) below species level (Faith 1994). Here, focus is placed on conserving as many different characters as possible, such as different morphological features (e.g. mouthpart types) or life history traits (e.g. bivoltine vs univoltine) of species. Characters are distributed across species, based on the phylogenetic relationships between them as a consequence of their evolutionary history. Priority species for conservation would be those species that are under-represented in the evolutionary history of the group (Faith 1994). One way of measuring phylogenetic diversity is with Faith's (1992) *PD value* (Box 9.4). This method has the advantage of few assumptions and is based on the same approach used to quantify phylogenetic pattern (Faith 1994).

The taxonomic distinctness index (Box 9.4) incorporates information on both character richness and the evenness of species abundances in the assemblage or sample. This index has the advantage that it is insensitive to sample size or effort (Clarke and Warwick 1998a). The application of this index may be useful in environmental impact assessment because its value has also been shown to be related to trophic level and functional diversity (Clarke and Warwick 1998b).

Box 9.4 Taxonomic diversity

Taxonomic diversity is based on the concept of **path** or **branch length**, or the distance (a surrogate for evolutionary time or the number of evolutionary steps) of species from each other on a taxonomic or evolutionary tree (e.g. a dendrogram based on taxonomic or phylogenetic information). Path (branch) length can be measured in several ways. Most simply, for example, using the number of taxonomic level steps between two species (so that two species in the same genus have a value of 1 and two species in the same family may be separated by a path length of 2) (Clarke and Warwick 1998). The unit measure of path length is also referred to as its weighting (or **distinctness weight** as in the taxonomic distinctness index shown below).

Phylogenetic diversity (PD)[1] (Faith 1994)

The PD of taxa in a sample or assemblage is measured by the sum of the length of the branches found along the path, along the tree connecting all taxa in the sample or assemblage (or the total path length connecting all taxa) (Faith 1992). The PD value, for any subset of taxa of size N, is the number of different nodes (branching points) on the cladogram that lie along the corresponding minimum spanning path;

PD = $(N-1)$ + number of internal nodes (branching points) on the minimum spanning path.

The best subset of N taxa (i.e. the subset inclusive of the highest phylogenetic diversity) is the one that spans the most nodes on the cladogram. The best addition to a subset of taxa (i.e. the taxon that adds most to the overall phylogenetic diversity of a group) is the taxon that adds most nodes to the minimum spanning tree.

Taxonomic distinctness index[1] (Clarke and Warwick 1998)

This index has three forms (from Clarke and Warwick 1998):

1 Taxonomic diversity (Δ) is the average path length between two randomly chosen individuals from the same assemblage or sample. Species abundance data are required and this index thus incorporates information on both measures of character richness and the evenness of species abundances in the assemblage.
2 Taxonomic distinctness (Δ^*) is the average path length between two randomly chosen individuals from different species.
3 Taxonomic distinctness based on presence-absence data (Δ^+) is the average path length between two randomly chosen species.

$$\Delta^* = \left[\Sigma\Sigma_{i<j} w_{ij} x_i x_j\right] / \left[\Sigma\Sigma_{i<j} x_i x_j\right]$$

where x_i $(i = 1, 2..., s)$ is the abundance of the ith species, $n(= \sum_i x_i x_i)$ is the total number of individuals in the assemblage or sample, and ω_{ij} is the **distinctness weight** given to the path length linking species i and j.

Thus, the larger the value of Δ^*, (or Δ^+, as in figure below) the greater the degree of taxonomic distinctiveness encompassed by the taxa in the assemblage or group. The histograms below are the values of the taxonomic distinctness index for 1000 random selections from offshore nematode assemblages at three sites in Britain (from Warwick & Clarke 1998).

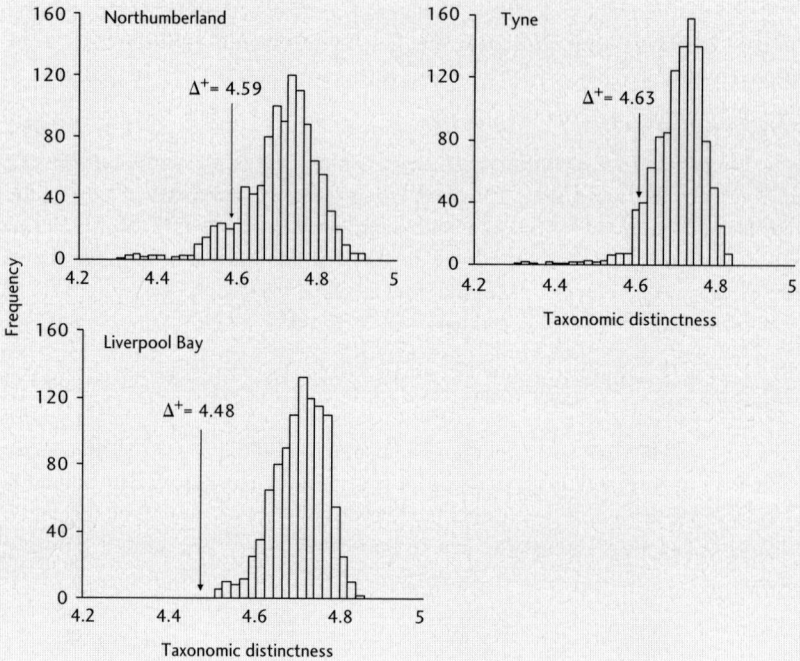

[^1]Both PD and Taxonomic distinctiveness may be calculated using the software PRIMER.

9.9 Phylogenetic comparative methods

Any one species has adapted to, or is in the process of adapting to, its local environment, which is constantly changing. This environmental change may be natural, in both the short term (over days) and the long term (over millennia). Overlying these natural changes are anthropogenic stresses, which may be such that some species are able to survive while others not. Survival of a species in these changing conditions depends on a species' characteristics or traits. It can be very informative to know what traits enable one species to survive rather than another, given particular natural and anthropogenic environmental conditions.

Yet each species has an evolutionary history which inevitably constrains the nature of the traits. Comparing species must take cognizance of this phylogenetic pedigree. In the past, many comparative studies have been done from a statistical point of view, as if all species are completely unrelated, when in reality species have evolved from a time-based array of common ancestors. Thus comparative studies should use information on the evolutionary relationships between organisms. This approach, known as the *phylogenetic comparative methods* (PCMs) was developed by Harvey and Pagel (1991) (although based on concepts going back to Aristotle), and is now adopted as a standard in comparative species studies. However, the effectiveness of the approach depends on the phylogeny being thoroughly resolved (Freckleton 2000), which of course is often not the case with insects, although attempts have been made to overcome this shortcoming (Quinn and Keough 2002).

PCMs are used to test for correlated evolutionary changes in traits or to determine the extent to which a trait is a signal of its evolutionary past. Indeed, evidence for adaptation to local environmental conditions only becomes clear after allowing for traits attributable to phylogeny. So, from a conservation perspective, we recommend studies comparing species traits be done on closely related, *sibling species*, or, where this is not possible, only after factoring in the phylogenetic constraints. Furthermore, it is also important, when focusing on differences in traits, to ensure that the environmental conditions are comparable, which is equivalent to similar selection pressures. Thus we recommend that for understanding the survival chances of one species as compared to another, that the first comparisons are done with *sympatric* sibling species. Samways and Lu (2007) did this for a common species and a highly threatened sympatric sibling species of butterfly. The outcome was that despite the threatened species being a much stronger flier than the common species, it was constrained by requiring a particular form of its host plant species. This then enabled a conservation strategy to be put into place, involving encouragement of this particular host plant. This study also illustrated another issue; that it is important to pinpoint the key threat(s) (as opposed to general threats, such as 'landscape fragmentation') for effective fine-filter conservation measures to be put in place.

9.10 Summary

Because of the very high biodiversity of insects, ecological and conservation studies most often involve working with large groups of insect taxa, such as guilds, assemblages, or food webs. The biodiversity measures used for such groups of taxa include their species richness, abundance, biomass, range size, distribution, species composition, diversity indices, and measures of functional, taxonomic and phylogenetic diversity. The analytical stage of quantifying biodiversity begins with the assemblage matrix, and usually also with the *relational variable matrix*. Because sampling (rather than a census) is generally involved, estimates (rather than true or complete values) of the above variables are obtained. As a consequence, it is extremely important to understand how adequate or representative these estimates of biodiversity are, using for example taxon sampling curves and a suite of species richness estimators. Care should be taken in interpreting diversity differences between sites of samples when asymptotes to species richness are not achieved. Many of the approaches to measuring biodiversity outlined above are developing rapidly, with newer methods providing both novel insights and improved the accuracy and precision of biodiversity estimation.

10

Studying insects in the changing environment

10.1 Introduction: drivers and methodology

The major drivers of global environmental change, and which affect biodiversity, include habitat loss and alteration, invasive alien species, pollution, and climate change. The catalyst for much of today's insect ecology and conservation research is rapid global environmental change, combined with the recognized importance and functional significance of insect biodiversity.

In the context of global change, insects are studied to understand (1) the impact of environmental change on insect biodiversity, and (2) the consequences of a change in insect biodiversity on overall biodiversity and ecosystem functioning as we know it, as well as (3) to examine the possibility of using insects to provide some sort of early warning system or indication of the extent, rate, and nature of environmental change. Studies must also examine ways of (1) effectively preventing insect species loss, (2) managing and maintaining current levels of insect biodiversity, and (3) restoring insect biodiversity to areas that have suffered losses. Methodological approaches for doing so for different forms of environmental change are discussed in this chapter. In many ways, studying insects in a changing environment will involve the application and integration of techniques, methods, and approaches discussed in the previous chapters.

10.2 Environmental change in context

The basis for studying the environment, and its intrigue for scientists and entomologists, is that neither the biotic nor abiotic components of the environment are static; both are characterized by their variability in time and space. Weather has daily and seasonal cycles, and climate has long-term trends. Similarly, individual activity patterns of insects vary according to time of day and season, and populations fluctuate across seasons and years. Perhaps the most important challenge in studying the impact of human-induced environmental change on insect biodiversity is separating out which components of variability in the system are human-induced and which are not. Baseline variation must therefore always be considered in studies of environmental change.

The appropriate space–time frame of the study must also be considered (see Figure 1.3). The spatial and temporal scale of biodiversity patterns and the processes that drive them tend to scale together such that environmental impacts small in spatial scale (e.g. accidental point source pollution, such as a pulse of effluent) tend to have impacts over comparatively short time frames. By contrast, the environmental impacts of climate change are not only global but are driven by processes that operate over long periods of time (Figure 1.3). The implication of space–time scaling is that whereas the impacts of point-source pollution may be detected fairly rapidly, studies of climate change impacts (whether retrospective or predictive) must necessarily be long term. Studies conducted at inappropriate spatial or temporal scales relative to their relevant environmental change process are unlikely to yield useful or significant results.

10.3 Approaches to studying environmental change

10.3.1 Methodological approaches

There are at least four broad methodological approaches to studying the impacts of environmental change, and the responses of insect biodiversity to such change. These approaches include (1) modelling (from conceptual to mathematical), (2) mensurate field studies (observational or collection/ sample-based), (3) experimental (manipulative) field studies, and (4) laboratory experimentation (see also Chapter 3). This book focuses largely on the field-based component of insect conservation studies. However, it is important to remember that other approaches are available and that these approaches are often complementary.

The most comprehensive understanding of insect biodiversity, and its response to environmental change, is likely to arise as a consequence of using more than a single methodological approach. This is because all approaches have specific strengths and weaknesses, and complementary approaches may lead to the greatest advances in understanding.

Diamond (1986) provides an excellent overview of the relative strengths and weaknesses of laboratory vs field experiments and so-called 'natural experiments' (i.e. non-manipulative surveys of the environment and communities in them). We cannot conclude that one approach is generally superior to another, because the best approach must be decided based on the biological system, species, processes and sites involved, as well as the research question being posed. Nonetheless, when selecting an approach for a particular study, it is important to appreciate the advantages and drawbacks of each approach.

Laboratory experiments (LEs) are superior in enabling the researcher to regulate or control external influences, or independent variables. LEs are usually set up to explicitly test the effect of one or a few independent or explanatory variables. They provide the most confident conclusions about cause–effect relationships. However, the drawback of LEs is that relationships are tested in biologically unrealistic settings and thus may not apply under natural conditions. In practical terms, LEs are also restricted in the time frame over which they can be run, as well as the spatial scale of processes and interactions that can be tested. Maintaining species in the laboratory also generally presents challenges and financial costs, and LEs tend to use a single or a few species, rather than examine multispecies responses and interactions.

Mensurate field studies (MFSs) (or 'natural experiments') are studies where the environment is not manipulated, and habitats or sites are sampled and replicated across space in an attempt to encompass the natural variability of interest. However, natural experiments also include before–after comparisons, where species and communities may be compared before or after a natural of human-induced change in environmental quality. Methodology for these before–after situations is discussed by Green (1979) and Underwood (1991, 1992, 1994). MFSs have all the advantages that LEs do not, and vice versa. MFSs can encompass broad spatial and temporal scales, and are, as a result, biologically realistic. However, independent variables cannot be controlled, and the confidence one is able to place in identified causal or explanatory variables is often weak compared to LEs.

Field experiments (FEs) are intermediate between LEs and MFSs because they allow the regulation of at least some independent variables. Their spatial scale and temporal duration are, however, also usually limited by logistic and financial constraints. Nonetheless, they provide a more realistic biological context than LEs, and a more powerful approach than MFSs.

10.3.2 Modelling

Models are an extremely useful heuristic tool in biology that can complement any of the above approaches. The word *model*, or some 'representation of reality' (Starfield and Bleloch 1991), is fairly all-encompassing, and may range from something as simple as a flow diagram to more complex statistical or mathematical models. Modelling can also be applied to aid virtually any research question, including those involving populations, communities, or broadscale biodiversity-level problems (Gillman and Hails 1997).

There are several reasons for model building: (1) to define a problem, (2) to organize one's thoughts, (3) to understand one's data, (4) to communicate and test understanding of these data, and (5) to make predictions (Starfield and Bleloch 1991). A simple *conceptual model* is, in fact, very useful for understanding different types of modelling problems (Figure 10.1). Areas 3 and 4 in Figure 10.1 are commonly encountered in biology and this is certainly true for entomology and many of the challenges for insect conservation. The objective of modelling in these situations is primarily to define the problem, organize one's thoughts and plan the data that should be collected. In situations such as this, it is easiest to start with simple descriptive models (of which Figure 10.2 is an example). Any approach to model design may benefit from incorporating the best available biological information as realistically as possible.

The first step is to identify and list the variables and parameters of the problem. *Variables* take on different values for each case or instance, whereas *parameters* in the model are those measures that are unlikely to change between cases (Starfield et al. 1990). The next step may be to categorize variables into groups that are most closely related, and then to identify how each variable and parameter may be related to all or some of the others in the model (or problem). Most

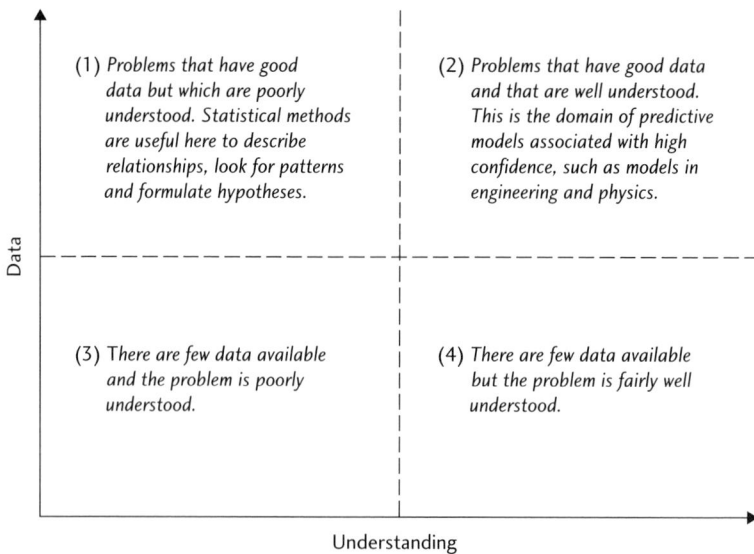

Fig. 10.1 Categories of modelling problems as explained by Starfield and Bleloch (1991).

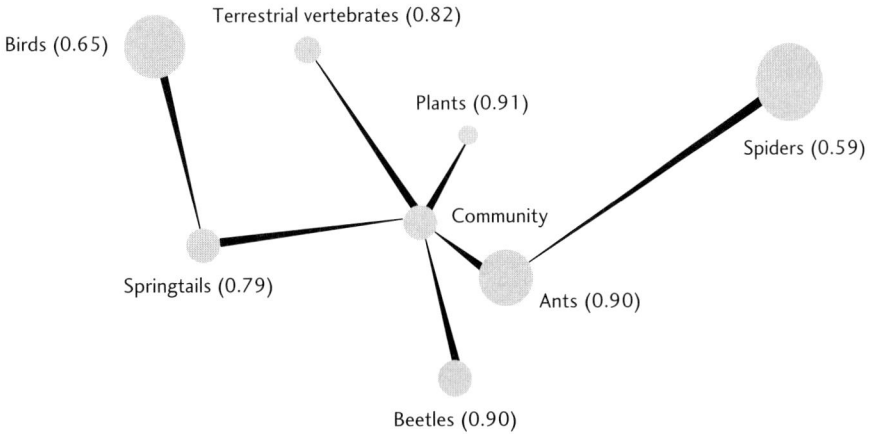

Birds (0.65)

Terrestrial vertebrates (0.82)

Plants (0.91)

Spiders (0.59)

Community

Springtails (0.79)

Ants (0.90)

Beetles (0.90)

Fig. 10.2 Comparison of the community patterns derived from seven taxon subsets, and from the total data set. A matrix of correlation coefficients was compiled from a pairwise comparison of the relevant transect similarity matrices. Results are displayed in three dimensions using semi-strong hybrid scaling (stress = 0.07). Minimum spanning tree linkages are superimposed to indicate nearest neighbours in community space. The correlations of each taxon with the community are shown in brackets. The use of plants, ants, and beetles is a particularly good reflection of the community as a whole (from Bisevac and Majer 2002).

simply, these can be denoted with unidirectional or bidirectional arrows and with arrows of different widths (to denote the putative strength of the relationship between two variables). A model, in the form of a flow diagram, graph, or even a set of formulae, may emerge from this process, depending on the type and complexity of the problem. Examples of more complex models that are useful in insect conservation include population dynamics, predator–prey, and meta-population dynamics models (often represented as differential or deterministic equations). The dynamics of ecological communities (also often modelled with differential equations and community or stability matrices), such as competitive and mutualistic interactions, can also be examined with such models (Gillman and Hails 1997).

This section has merely scratched the surface of what is possible with modelling (see Starfield and Bleloch 1991, Starfield et al. 1990, Gillman and Hails 1997). Formalizing your thinking about a problem in this way is extremely valuable, and will lead to better construction of hypotheses and better-designed studies. For an excellent introduction to some underlying principles and applications of theoretical ecology towards describing the real world, see May and McLean (2007).

10.4 Bioindicators

A *bioindicator* is a species or group of species that (1) readily reflects the abiotic or biotic state of an environment, (2) represents the impact of environmental change on a habitat, community or ecosystem, or (3) is indicative of the diversity of a subset of taxa, or of wholesale diversity, within an area. A *bioindicator system*, on the other hand, is a tool to extract biological information from an ecosystem and to use this information for making scientifically based management decisions (van Straalen and Krivolutsky 1996).

Our often poor knowledge of insect biodiversity, especially in species-rich parts of the world, along with high levels of species extinction, is a major challenge to conservation. This challenge can only be met with efficient approaches to gathering information with limited resources. Bioindicators that both readily reflect and represent the *state of the environment* (SoE) are one approach (Table 10.1), while bioindication has become an essential component of conservation strategies aimed at addressing a wide array of biodiversity threats. The use of selected, suitable species, or species groups, to represent some component of their environment or broader biodiversity is one of the tools available

Table 10.1 *Definitions and categories of bioindication.*

Bioindicator categories	Definition	Examples and related terms
Environmental indicator	A species or group of species that responds predictably, in ways that are readily observed and quantified, to environmental disturbance or to a change in environmental state	Related terms: Sentinel, exploiter, bioassay, accumulator, biomarker
Ecological indicator	A species or group of species that demonstrate the effects of environmental change (such as habitat alteration and fragmentation and climate change) on biota or biotic systems	
Biodiversity indicator	A biodiversity indicator is a group of taxa (e.g. genus, tribe, family or order, or a selected group of species from a range of higher taxa), or functional group, the diversity of which reflects some measure of the diversity (e.g. character richness, species richness, level of endemism) of other higher taxa in a habitat or set of habitats	Related terms: Surrogate, umbrella, flagship, focal species or taxon

for addressing the biodiversity crisis. Insects make promising bioindicators because of their high species richness, large biomass, and the importance and diversity of their functional roles, as well as their responsiveness to environmental change.

10.4.1 Bioindication terminology and development

The terminology associated with bioindication is varied and extensive, with many synonyms. Bioindicators are different from abiotic indicators, such as soil quality, landscape structure, or potential evapotranspiration. Bioindicators are also different from composite or indirect indicators, such as land-use metrics, environmental diversity, or policy indicators.

There are three categories, or types of bioindication: environmental bioindication, ecological bioindication, and biodiversity indication (Table 10.1). Although these categories are not always mutually exclusive, they do successfully distinguish the vast majority of bioindication studies. The distinction is also useful because the categories have very different objectives, and subsequently different approaches, methods, and necessary conditions that the bioindicator should fulfill (McGeoch 1998, 2002).

In many cases, in both environmental and ecological bioindication, the objective involves identifying species that are both sensitive to environmental quality and conspicuously responsive to a change in that quality. The bioindicator response is generally quantified using measures of species abundance and distribution. Once the bioindicator objective has been determined, the process of developing a bioindicator system involves the following general stages (1) a-priori selection of one or more potential bioindicator taxa, (2) identifying sensitive and suitable species from the assemblage of potential taxa using field sampling and quantitative techniques, (3) testing the proposed bioindicator on an independent data set, and finally, (4) further developing the bioindicator for use by managers and practitioners. The following points are important to consider during the first three stages of the process: (1) using several, or a 'basket', of species, to improve the reliability of the bioindicator (Hammond 1994); (2) separating stochastic abundance fluctuations from those associated with the environmental change of interest; (3) being able to associate a quantitative indicator value and associated level of significance with the bioindicator species or assemblage (ideally this should be a linear relationship, Duelli and Obrist 2003), and (4) ensuring that species selected are recognizable, readily sampled and quantified, and likely to remain so. Table 10.2 outlines in further detail each of the nine steps that should be followed when identifying and testing a bioindicator.

Table 10.2 *Nine steps involved in the identification, testing, and eventual adoption of a bioindicator (from McGeoch 1998).*

Step	Description
1	Determine broad objective (environmental, ecological or biodiversity indication)
2	Refine objectives (by making them scenario specific) and clarify end-point (by identifying the final desired outcome of the process)
3	Select potential indicator based on accepted a-priori suitability criteria (see Table 10.3)
4	Accumulate data on the proposed bioindicator using appropriate sampling or experimental design protocols (see Chapter 3)
5	Collect quantitative relational data (weather, habitat quality, etc., Chapter 3)
6	Establish statistically the relationship between the indicator and the relational data (information on the environmental stressor of interest)
7	Based on the nature of the relationship, either accept (preliminarily) or reject the species, higher level taxon or assemblage as a potential indicator
8	Establish the robustness of the indicator by developing and testing appropriate hypotheses under different conditions (see Fig. 10.3)
9	If the null hypotheses are rejected make specific recommendations, based on the original objectives, for the use of the (now realized) bioindicator and further development of the bioindicator system

10.4.2 Taxa that make good bioindicators

Bisevac and Majer (2002), in terms of assessing mine site rehabilitation, suggest that trends in species richness and community composition were better represented by certain invertebrates than by vertebrates, and that these were highly cost-effective indicators. They found that plants, ants, and beetles were the taxa with the highest correlation with the overall community composition (Figure 10.2). Kremen et al. (1993) even suggested that terrestrial arthropods could be used for virtually any monitoring programme, so long as the goals are well defined. Invertebrates in general have some attributes which lend themselves to biomonitoring and biodiversity assessments

- Being small, they are often highly sensitive to very local conditions
- Most being mobile and reactive to changing conditions, they often respond by moving away from, or towards, adverse or optimal conditions respectively

- With their short breeding rates and short generation times, they are often highly responsive numerically to changes
- With their great variety of growth rates, life history styles, body sizes, food preferences, and ecological preferences, they can, at the species level, often be linked to specific environmental variables
- Their abundances fluctuations provide an extra sensitivity layer over that of species richness for indicating subtle changes in environmental conditions,
- Many are relatively easy to survey.

These advantages are offset by the taxonomic challenge, especially at the sensitive, species level and particularly in the great biodiversity hotspots in the world. Nevertheless, there are certain taxa and functional groups, and combinations, that have great potential, as discussed by McGeoch (2007).

Maes and Van Dyck (2005) provide a useful stepwise decision-making framework for the compilation of a set of taxonomically diverse indicator species, based on Hilty and Merelender (2000). There are five steps in the process:

1) Decide which ecosystem attributes the indicator taxa should reflect. This involves deciding on which *biotope* attributes of the focal ecosystem are suitable for a variety of specialist species particularly appropriate in information rich regions. This means that the multispecies group should contain species that are sensitive to ecosystem change, such as fragmentation, overall environmental change, and microclimate change, and that are dependent on one or more of the typical biotope attributes, and preferably all biotope attributes, i.e. they should be sensitive to changes away from the 'ideal' conditions.

2) List all species or taxonomic groups that meet the baseline information criteria. This means that the taxonomy should be clear, biology and life history well known, and the species' distribution is also well known, as well their tolerance levels to environmental pressures being known, especially in relation to specific changes in environmental conditions.

3) Use only intermediately rare and easily detectable species, which are evenly distributed in the focal area. This means determining the most apparent stage, which may be the eggs in the case of lycaenid butterflies (see Box 9.1) but adults in the case of satyrid butterflies. Good field guides are of course also essential to this step. In the case of threatened species, which may be locally common at one particular site (e.g. the Yellow presba dragonfly in Figure 6.5), they must not be sampled in a way that causes death or even stress.

4) List available information on niche and life history, as well as sensitivity to environmental stressors: niche and life history criteria in relation to trophic level, reaction time to environmental changes, mobility, minimum area requirements, detailed niche information, fine structural, but essential, aspects of the biotope, and sensitivity to different environmental stressors.

5) Compile a set of complementary species from taxonomic groups to satisfy every criterion in steps 1–5.

Although Maes and Van Dyck (2005) found this approach using these attributes to be effective for the well-known biota of wetlands in Belgium, all this information is simply not available for most parts of the world. Thus the search for bioindicators for the biodiverse parts of the world usually involves some sort of compromise (Spector and Forsyth (1998). Nevertheless, there are some insect taxa, and combinations of them, that have proved to be suitable in multispecies decision-making frameworks worldwide (Kremen et al. 1993; Pearson 1994; Cranston and Trueman 1997; Yen and Butcher 1997; Ward and Larivière 2004; McGeoch 2007). Briefly, Coleoptera, especially ground beetles (Carabidae) (Niemelä et al. 2000), tiger beetles (Cicindelidae/-inae) (Pearson and Cassola 1992) and dung beetles (within the Scarabaeidae), and some other beetle groups (Michaels 2007) are well recognized as ecological bioindicators, and have been tested in biodiversity assessments. Dung beetles have been extensively used in studies as indicators of disturbance and habitat quality, particularly in the tropics and subtropics. Ground beetles have been applied in similar contexts. The use of Hymenoptera in bioindication includes largely ants, but also honeybees (particularly as environmental indicators of pollutant levels in agroecosystems) and other apidoid communities. Beneficial organisms, including parasitoids, have been proposed as indicators of sustainability (Thomson et al. 2007), although the use of predatory arthropods as useful indicators, at least in Australian agroecosystems, has less value as they are mostly polyphagous generalists (New 2007). Ants have been strongly promoted as bioindicators (Alonso 2000; Kasperi and Majer 2000; Andersen et al. 2002), mostly of land-use and restoration (Majer 1983), because of their high diversity and functional importance, especially in the southern hemisphere. Greenslade (2007) has developed the potential of Collembola to assess the ecological status of a range of ecosystems in Australia. Butterflies have been widely used in the tropics for indicating the impacts of logging on the ecosystem, while grasshoppers are particularly good at reflecting changes in savanna conditions. Blood worms (Chironomid larvae) are commonly used as both environmental indicators of freshwater pollution. They often increase under eutrophic conditions while at the same time, the commonly used EPT taxa (Ephemeroptera/Plecoptera/Trichoptera; mayflies/stoneflies/caddisflies) decrease. These are commonly used in freshwater health assessments (Rosenberg and Resh 1993; Spellerberg 2005; Walsh 2006). There are of course many taxa to choose from, and as pithily summarized by Andersen (1999) 'My bioindicator or yours?', simply recording that a certain insect group responds to changes does not mean that it qualifies as a bioindicator; as all the extra steps outlined in Table 10.2 would have been left out of the selection process.

When choosing a potential bioindicator taxon, it is important to make a choice that is appropriate for the bioindication objective (Duelli and Obrist 2003), the geographic area, and habitat within which the study will be conducted (Debuse et al. 2007), and that uses standardized methodology (Neville and Yen 2007). To satisfy these stipulations, it may be necessary to use a range of different taxa, which may represent different functional groups, as well as aiming to satisfy some of the decision framework mentioned above. Just because a particular taxon has been shown to be useful on one continent for one particular bioindication objective, does not mean that it will necessarily be suitable for any other. Ground beetles, for example, appear to be of more overall value in the northern hemisphere than in the southern hemisphere, but this may be more a question of our state of knowledge rather than a true reflection of hemisphere differences (Horne 2007). Nonetheless, there are some guidelines for selecting potentially good bioindicator taxa that can then be tested in the field (Table 10.3). Not all these characteristics are essential for a bioindicator and which characteristics are relevant can only be determined by the rationale and objective of the bioindicator.

One other point bears consideration. Some taxa, such as grasshoppers and ants, can be very responsive to changed conditions, such as grassland disturbance or fire, yet return quickly after these impacts. This is in contrast to harvestmen and terrestrial crustaceans, which are not only sensitive but also can take a long time (many years) to recover. Thus, in a conservation assessment of a landscape, one taxon might not necessarily be a surrogate for another.

Finally, there is the issue of practicality. Bioindicators may simply be research tools for exploring the value of a particular area, as opposed to candidates for use on a regular basis by managers (see also 10.4.4). This is an issue that needs to be decided upon before starting a research programme on the choice of bioindicators. On the one hand, the aim may be to search for biondicators which are primary tools, say, for selecting sites for conservation priority. On the other, the aim may be to search for bioindicators which can be used by managers, for example, to test whether a particular management activity (e.g. burning regimes) has been

Table 10.3 *Suggested useful characteristics for selecting potential bioindicators prior to testing (modified from McGeoch 1998)*

- Cost efficient and effective (time, funds, personnel)
- Sampled and sorted easily
- Adequate representation in samples (abundance)
- Ease and reliability of storage
- Taxonomically well known, readily identified, taxonomic expertise available
- Sampled individuals expendable
- Spatial and temporal distribution predictable to ensure long-term continuity
- Relatively independent of sample size
- Baseline data on biology and population dynamics available
- Low genetic and functional variability
- Sufficiently sensitive to provide early warning
- Able to differentiate between natural cycles and trends and those produced by anthropogenic stress factor
- Representative of critical components, functions and processes
- Show a well defined response, i.e. either die or decrease, change or mutate, replace or be replaced by other species
- Readily accumulate pollutants
- Easily cultured in the laboratory
- Capable of providing continuous assessment over a wide range of stress
- Recognized importance to agriculture, environment, etc.
- Economic importance as a resource or pest
- Representative of all trophic levels and major functional guilds
- Representative of related and unrelated taxa
- Include a broad range of body sizes and growth forms
- Tend to be distributed over a range of habitats or environments
- Representatives from low, medium and high diversity groups
- Wide range of host specificities

effective. While ants can be good indicators of changing conditions, they generally have to be sampled with pitfall traps and examined under the microscope. This can be very impractical in practical reserve management, for example, and it may be more practical to use more highly visible taxa such as butterflies or grasshoppers, which can be assessed visually or with close-focus binoculars (see

Chapter 6.5.7). Likewise, for freshwater assessment, adult dragonfly species give parallel results to a suite of benthic macroinvertebrates, and yet provide more sensitivity at the local spatial scale as the taxonomic resolution is at the level of species and not at higher taxonomic levels as with benthic macroinvertebrates (Smith et al. 2007) (Box 10.1).

Box 10.1 The Dragonfly Biotic Index

The Dragonfly Biotic Index is an index of the ecological integrity and health of a freshwater system, and complements indices based on benthic macroinverte-brates (Smith et al. 2007). Overall, it is more sensitive than macroinverebrate scores as it operates at the species level rather than the higher taxonomic level. It is also easy to use, as all it requires is close-focus binoculars.

Each dragonfly species is given a score (see table below), which is the sum of (1) geographical distribution (scored 0–3), (2) Red List status (scored 0–3), and (3) sensitivity to disturbance (scored 0–3). The scores for any one species thus range from 0 to 9, with a common, widespread generalist that is highly tolerant of habitat change scoring 0, while a narrow-range endemic, which is threatened and highly sensitive to habitat change scores 9. Every South African odonate species has been assigned a score (Samways 2008). The total DBI of a water body (stream, river, or pool) reflects the total odonate assemblage, thus enabling water bodies to be compared and restoration to be monitored (Simaika and Samways 2008). The index is robust and has great value in conservation planning.

| | | Sub-indices | |
Score	Distribution	Threat	Sensitivity
0	Very common throughout South Africa and southern Africa	LC (GS)	Not sensitive; little affected by habitat disturbance and may even benefit from habitat change due to alien plants; may thrive in artificial water bodies
1	Localized across a wide area in South Africa, and localized or common in southern Africa; or very common in 1-3 Provinces and localized or common elsewhere	NT (GS) or VU (NS)	Low sensitivity to habitat change from alien plants; may occur commonly in artificial water bodies
2	National South African endemic confined to 3 or more Provinces; or widespread in southern Africa but marginal and very rare in South Africa	VU (GS) or CR (NS) or EN (NS)	Medium sensitivity to habitat disturbance such as from alien plants and bank disturbance; may have been recorded in artificial water bodies

		Sub-indices	
Score	Distribution	Threat	Sensitivity
3	Endemic or near-endemic and confined to only 1 or 2 Provinces	CR (GS) or EN (GS)	Extremely sensitive to habitat disturbance from alien plants; only recorded in undisturbed natural habitat

Abbreviations: IUCN species threat status, LC = Least Concern, NT = Near Threatened, VU = Vulnerable, EN = Endangered, CR = Critically Endangered. GS = Global status, NS = National status.

10.4.3 Quantifying the indicator value of a species

An excellent method for quantitatively identifying potential bioindicator species from an assemblage of species is the *indicator value* (*IndVal*) method (Dufrêne and Legendre 1997) (Box 10.2). This method combines measurements of the degree of specificity of a species to an ecological state (such as habitat type or disturbance level), and its fidelity (or frequency of occurrence) within that state. It was first applied to an assemblage of 189 ground beetle species across 69 localities and 9 habitats in Belgium. The method is based on an assemblage matrix (see Chapter 9.2). Species with a high specificity and high fidelity within an environmental state will have a high indicator value for that state. High fidelity (frequency of occurrence) of a species across sample sites is generally associated with a large abundance of individuals. Both these characteristics aid sampling and monitoring, and both are important requirements for a useful bioindicator. The *IndVal* method has several advantages over other indicator measures used for ecological bioindication (McGeoch and Chown 1998). For example, the *IndVal* is calculated independently for each species, and there is complete flexibility with regard to the state (site, sample or habitat) categorization on which the *IndVal* measures are based.

The *IndVal* method allows the identification of both 'characteristic' species (i.e. with high specificity and fidelity to a state and thus a high %*IndVal*), and 'detector' species (i.e. species that span a range of ecological states and have intermediate specificity) (Box 10.2). McGeoch et al. (2002) demonstrate this using dung beetles as indicators of habitat conversion from closed canopy forest to open, mixed woodland. Detector species may be more useful indicators of the direction of change than highly specific (characteristic) species restricted to a single state (Box 10.2). This is because the abundances (and thus the fidelity) of characteristic species may decline rapidly under changing environmental

Box 10.2 The Indicator Value (IndVal) method

Concept

This method is based on the measurement of how specific a species is to a particular ecological state as well as its fidelity to that state (how frequently it is found, or its occupancy) (Dufrêne and Legendre (1997). By combining specificity and fidelity measures for each species, an indicator value is obtained that is independent for each species in the assemblage. The higher the indicator value, the more characteristic a species is of a particular ecological state and the better environmental or ecological bioindicator it will make.

Calculation

1 The calculation of the indicator value is based on the assemblage matrix (see Chapter 9). Before calculation of indicator values it is thus important to have the assemblage matrix complete, including row (across sites) and column (across species) totals.

2 Important for this method is that a separate assemblage matrix must be constructed for each 'group' of sites. Here the definition of the 'groups' will be determined by the objective of the study.

3 For example, groups may be different habitat types or structures (grassland, woodland, forest), disturbed vs undisturbed habitat (nature reserves vs harvested forest patches), or different stretches of a river (upper vs lower reaches). In other words, these groups may be determined a priori by the objective of the study.

4 Alternatively, sites may be assigned to groups objectively with the use of analytical techniques such as clustering or ordination (Legendre and Legendre 1998). This approach is used, for example, when the environmental state or condition of sites is poorly understood and the site cannot readily be grouped into sites of comparable state.

5 Next, the indicator value (*IndVal*) is calculated in the following way:

Specificity measure:

$$A_{ij} = Nindividuals_{ij}/Nindividuals_{i\cdot}$$

where $Nindividuals_{ij}$ is the mean number of species i across sites of group j, and $Nindividuals_{i\cdot}$ is the sum of the mean numbers of individuals of species i over all groups.

Fidelity measure:

$$B_{ij} = Nsites_{ij}/Nsites_{j}$$

where $Nsites_{ij}$ is the number of sites in group (habitat) j where species i is present, and $Nsites_{j}$ is the total number of sites in that group.

The indicator value (expressed as a percentage) for species i in group (habitat) j value is then:

$$\%IndVal_{ij} = A_{ij} \times B_{ij} \times 100.$$

6   Species with high and significant percentage *IndVals* may be considered bio-indicators of the group of sites in question (70% has been used as an arbitrary lower threshold in some studies).

7   Use Dufrêne and Legendre's (1997) random reallocation procedure to calculate the significance of each *IndVal* (and in fact to complete all *IndVal* calculations; see 'Useful software' at end of book).

Detector species

The approach described above identifies so-called 'characteristic' species, i.e. a species with both high specificity and fidelity in a particular habitat type, environmental state or group of sites (lower right hand cell in the figure below). However, many species have other combinations of specificity and fidelity values and can be classified differently based on these characteristics (as characteristic, detector, tramp, rural or vulnerable species (McGeoch et al. 2002)). For example, species with intermediate specificity levels but a definite preference for a particular habitat type (detector species) may in some cases provide bioindication information complimentary to that provided by characteristic species. Because characteristic species are highly specific to a particular habitat or environmental state they may disappear (or become locally extinct) relatively rapidly. They would of course no longer be of value to bioindication. The fidelity of detector species can then be monitored across habitats to detect the existence and direction of environmental change (McGeoch et al. 2002). The *IndVal* method can also be used for identifying detector species. Both characteristic and detector species are thus of potential value as indicators.

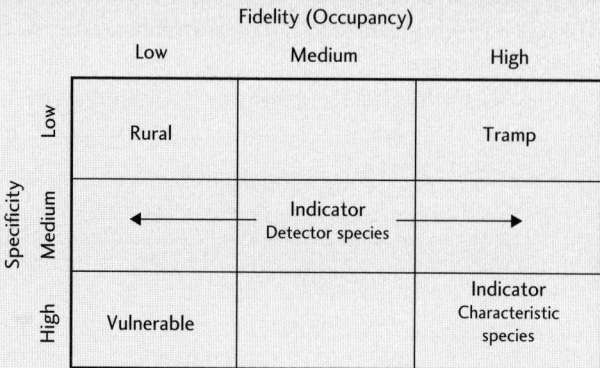

conditions to the point where they are regarded as vulnerable. These species will become increasingly difficult to sample, and may disappear rapidly with no further value for monitoring thereafter.

Characteristic indicator species also provide no information on the direction of ecological change (although changes in their abundance may remain useful for monitoring within the habitat to which they are specific), because they are highly specific, and thus restricted to a single ecological state. By contrast, species with moderate specificity levels (detector species, Box 10.2), may be more useful for monitoring change.

10.4.4 Testing the proposed bioindicator and developing the bioindicator system

Once one or more species have been identified in a quantitative way as being potential bioindicators (e.g. using the *IndVal* method described above), these species must be tested (Figure 10.3). Testing is important to determine the

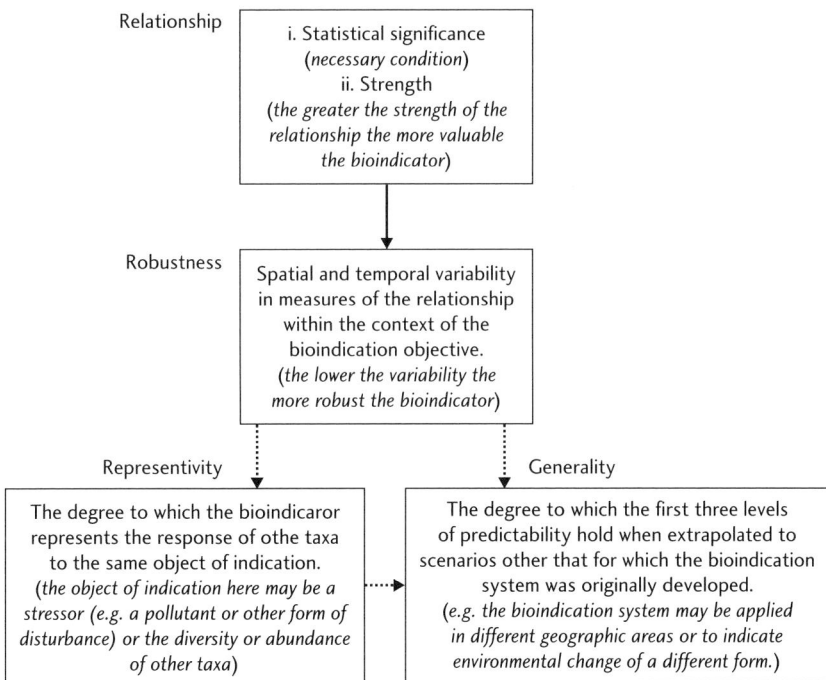

Fig. 10.3 The predictability hierarchy in bioindication. As a bioindicator moves along the hierarchy of levels of predictability (from the significance of a relationship to the generality of the bioindicator) it gains value for biodiversity assessment and conservation management (from McGeoch 2007).

reliability of the response of the bioindicator (or how robust). This testing needs to be carried out using an independent data set, i.e. using a data set different from the one that was used to identify the indicator species in the first case. This data set may already be available from another study, or will have to be collected by sampling different sites and different years. This repeated sampling will generate confidence intervals around the response of the bioindicator, and, in so doing, significantly improve the value of the bioindicator (Figure 10.3). In a worst-case scenario, the indicator selected may be shown to respond inconsistently, and therefore cannot be used as a reliable bioindicator. However, the careful a priori selection of taxa and further quantitative initial testing (e.g. using the *IndVal* method) should reduce the chance of this happening.

Once the bioindicator has been tested and found to be robust, further development of the bioindication system involves making it user-friendly as a tool for conservation managers, practitioners, and non-experts. This may include the compilation of illustrated guidelines to the application of the bioindication system, as well as illustrated keys to the taxa involved or an electronic expert system. For a bioindicator to move beyond the realm of primary research to application, it must be practicable to the end-user. Further development of the bioindication system may also involve linking the insect bioindicator into a broader *state of the environment* (*SoE*) system or monitoring programme which may involve several different bioindicators and biotic, abiotic, political, and economic measures, e.g. the *2010 Target* of the Convention on Biological Diversity to reduce the rate of biodiversity loss (Dobson 2005).

10.5 Biodiversity assessment

Biodiversity indication has been driven by the urgent need to prioritize land areas for protection and to identify regions of the world with unique and high levels of biodiversity. Because our knowledge of the taxonomy and distribution of invertebrate species in particular is poor, comprehensive biodiversity assessment is only just beginning, with the Global Dragonfly Assessment being the first one. Generally however, it is necessary to use some surrogate of biodiversity as the basis for decision-making, both to overcome the *taxonomic challenge* and to save the time and expense required for comprehensive biodiversity surveys.

Biodiversity indication therefore involves the use of either (1) individual species, (2) the species richness of target taxa, (3) higher taxon richness, (4) functional groups, or (5) levels of rarity, endemism, or threat, to estimate a generally broader component of biodiversity. Although abiotic characteristics of the environment are often easier to measure than species diversity, the use of abiotic measures as a surrogate for biodiversity has not always been effective (Araujo et al. 2001; Brooks et al. 2004). Landscape planners have emphasized that the best way forward is to use both *landscape surrogates* in addition to *species surrogates*. This, in turn, enables selection of reserve areas. The shocking revelation, however, is that when such thorough and insightful selection is done, about half the land area needs to be conserved if regional biodiversity is to be maintained (Reyers et al. 2002).

Unfortunately, although a wide range of approaches have been used, and there are many case studies, the search for biodiversity indicators has not been very successful. Most studies find no, or only very weak, significant relationships between elements of biodiversity and proposed surrogates (i.e. biodiversity indicators). Some approaches are more promising than others. For example, lower-taxonomic levels (e.g. genera) tend to predict species richness better than higher taxonomic levels. Also, the relationships between the richness of the biodiversity indicator taxon and some other measure of biodiversity tend to be much stronger at large scales (across continents) than at fine scales (e.g. within protected areas).

Despite the growing acceptance that different aspects of biodiversity (taxonomic, biogeographic, and threat status) are more often than not weakly correlated (Orme et al., 2005), the use of *presence–absence data* for selected species to model assemblage species richness has met with more success (MacNally and Fleishman 2002, 2004). The method was developed by modelling the species richness (based on a general linear model) of butterflies in the central Great Basin of the USA as a function of the occurrence of a subset of selected, proposed indicator species. Very widespread and rare species were excluded from the initial indicator species set on the basis that neither group is likely to serve as an effective indicator of spatial variation in biodiversity. The best set of indicator species (explanatory variables in the model) was then determined. A subset of 4–5 (<10%) of the butterfly species were found to explain a significant proportion of the variation in species richness (77–88%) (MacNally and Fleishman 2002). Not only was the explanatory power of the models strong, but the models proved robust when tested using a formal validation process, including a test of the models using a spatially and temporally independent data set. Furthermore, a model incorporating the occurrence patterns of six butterfly species predicted 82% of the variation in combined butterfly and bird species richness (>130 species) (Fleishman et al. 2005). Finally, these models are also particularly valuable because they were developed and tested at a scale at which conservation management decisions take place. This species-occupancy modelling approach is certainly worth exploring further in different geographic regions and with different taxa.

Threatened insect species as a group have been found to be useful for conservation planning. Lawler et al. (2003), for example, showed that sites selected with single taxonomic indicator groups provided between 61 and 82% of all other species, although no taxonomic group provided protection for more than 58%

of all other threatened species. However, threatened species overall performed well as an indicator group, covering an average of 84% of all other species. Maes and Van Dyck (2004), in a very penetrating paper, showed that the threatened Alcon blue butterfly *Maculinea alcon* was a very useful indicator for the quality and area of wet heathlands in Belgium. Interestingly, they found that complementary information of the species in a multispecies, multitrophic indicator group (nine species: two birds, two butterflies, two plants, two dragonflies and a grasshopper, importantly, after using a stepwise decision-making framework) usefully signalled distinctions in biotope area and configuration, vulnerability to fragmentation, eutrophication, and desiccation. Nevertheless, much more work is needed in this field, particularly on various combinations of indicator and surrogate groups in different geographical areas, because in Norway, certain Red Listed saproxylic beetles showed a negative correlation with a group of indicators (Sverdrup-Thygeson 2001) The risk of using rare species as indicators is that the investment in developing them as indicators may come to naught should populations further decline (the species becomes too difficult to locate to be useful as an indicator) or become extinct.

With the shortage of time for conservation action to avert the decline of biodiversity worldwide, short-cut measures are required to expedite the acquisition of land areas and to assess the value of transformed or restored areas. The idea of conserving certain species, and in doing so, many other species as well as communities, has great appeal. Such *umbrella species* (Simberloff 1998; Fleishmann et al. 2000; 2001; Roberge and Angelstam 2004; Betrus et al. 2005) are usually seen as being iconic and ideally need to be highly *apparent* and easy to sample. On this basis, it is tempting to use, for example, charismatic vertebrates as surrogates for insects. But this approach risks missing many insects with restricted geographical ranges and special habitat requirements. Besides, many such vertebrates are under pressure and may well have already disappeared from parts of their range.

While early reference to umbrella species was based more on intuition than on science, the field has developed to an extent where an *umbrella index* (*UI*), which quantitatively defines a potential umbrella species, is now available (Fleishmann et al. 2000). This index rates the trade-off between the level of protection given by the potential umbrella species and the financial resources available, so that conservation action focuses on where the umbrella species are present. If conservation of all the areas where the umbrella species are present turns out to be too expensive, then the aim would be to conserve an affordable subset of all these areas. The umbrella index selects the areas based on occurrence rate of potential umbrella species, co-occurrence of other species, and sensitivity to human disturbance. However, there is little point in having umbrella species which are so generalist and widespread that their umbrella value is meaningless. Rather, the aim is to employ species which are highly relevant to the conservation goals, especially particular ecosystems. Bried et al. (2007), for example, tested the concept using dragonflies and plants as umbrella species for the conservation of wetland impoundments. Although they found that the performance of the umbrella index was questionable, this did not detract from their recommendation that the UI warrants more testing, a conclusion with which we concur.

10.6 Biodiversity mapping

Quantifying the distribution of species across and within their geographic ranges, as well as monitoring changes in the distribution of species, provides extremely valuable information for insect conservation (see sections 6.4, 6.5, 9.7 and Figs 6.10 and 7.18). Maps of species distributions are important not only for detecting range shifts induced by climate change, but are also useful for examining range contractions by species threatened by other forms of environmental change, as well as for monitoring range expansion by invasive species. For example, Roura-Pascual et al. (2004) used historical (mainly museum and literature) presence records of the distribution of the invasive Argentine ant *Linepithema humile* to quantify potential geographic distribution of the species and to predict how this might change under projected climate change scenarios. Initially, they correlated the presence records of the species in its native range (north-eastern Argentina and southern Paraguay) with environmental data (topography, climate, and normalized difference vegetation index, NDVI). They then used the relationship between the presence of the Argentine ant and environmental variables in its native range to predict its likely potential distribution elsewhere in the world.

However, before being able to make any assessments of species distributions, past, present, or future, one must have data. In some cases, historical data are available in the form of museum collections or literature data on past species distributions. These data form an extremely valuable baseline for assessing changes in species distributions and, for example, attributing range shifts to climate change. For example, data from 1967–1973 on butterfly communities across an elevational gradient in Spain could be used to demonstrate uphill shifts of almost 300 m in butterfly species richness and composition by comparison with current data (Wilson et al. 2007).

While predictive models have been widely used in the field of pest control, using models such as CLIMEX (Sutherst and Maywald 1985), they have been little used in conservation. One of the problems with such modelling approaches is that they place the emphasis on climate, when in reality there are a whole host of interacting variables, biotic and abiotic, often at the scale of microhabitat, which determine the actual and potential range of a species. In a study of 15 ladybird species, the actual and potential geographical range could be predicted with 100% certainty on climate alone for only 4 species, with the accuracy being between 33 and 89% for another 10, and 0% for 1. Various biological features such as host availability, varied diet, presence or not of natural enemies, and presence or not of roosting sites all played a role in reducing the effect of climatic predictive modelling (Samways et al. 1999).

Data for mapping are usually available in the form of point locations, or latitude and longitude (x, y, and also sometimes z or elevation) coordinates associated with each insect specimen. Point locality data are most valuable for studying species distributions and allow distributions to be plotted and examined at a variety of spatial scales. It may be necessary to scale up point records (decrease the data resolution by pooling data into grid cells of a particular size) for comparison with other taxon data available at coarser resolution, or for matching with environmental data such as rainfall, land use, etc. The locality information obtained with the handheld GPS should be supplemented with information on date and time, as well as weather conditions and habitat information. Where the data are to be used in large-scale spatial modelling, other variables, such as elevation and exact microhabitat for each species, should also be recorded, as well as supplementary information on, for example, abundance and proportion of life stages present. Where the focus is on Red Listing a species, these additional data are very important, including assessment of level of threat and even effectiveness of any conservation management activity.

One problem with historical data is usually a lack of absence records. Records of species absence, although usually somewhat less certain than presence records, can be as valuable as records of species presences when mapping the distribution of a species (Finch et al. 2006). Wherever possible, studies should thus record the geographic coordinates of sites at which species were not found as well as those at which they were found (Figure 10.4). Databases also usually include so-called *omission* and *commission errors*. Omission errors are sites or localities where the species has not been recorded, but where it is in fact present. Omission errors can be made, for example, as a result of low sample effort, particularly for rare species. Commission errors are localities where a species has been recorded as present, but where it is in fact absent or at least not resident (Figure 10.4). When gathering distributional data and compiling species distribution databases it is important to consider the rate of such errors and possible ways of minimizing them.

Use of a BIOCLIM-type approach for modelling distributions of insects is particularly useful as it requires minimal data for building the model. Using a sensitivity analysis, Finch et al. (2006) investigated the approach on a set of 160 dragonfly species, and showed that the models overstated the distribution

Stage 1

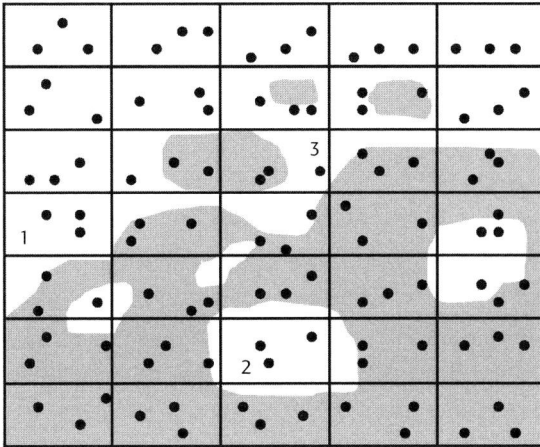

Stage 2

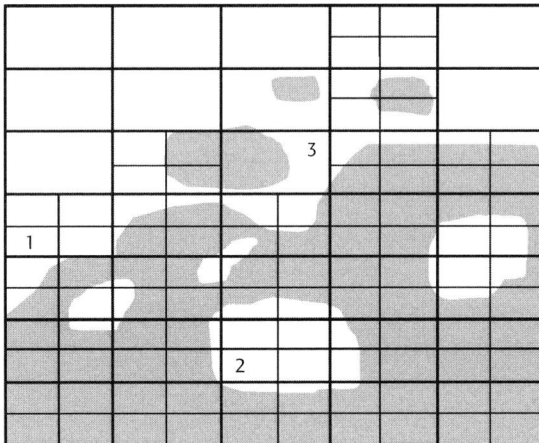

Fig. 10.4 Staged, grid cell-based approach to quantifying the distribution of a species. Shaded area represents true species distribution (species present) with unshaded area denoting species absence. Large grain (grid cell size) is used in stage 1 to cover the extent of the putative distribution of the species within the area of interest. Dots denote randomly positioned sampling points in each grid cell (three in this case). In stage 2, the grain is decreased by subdivision of original grid cells in those cells in which the species was recorded as present during the first stage. Either all fine-grain cells are sampled in stage 2, or cells may be sampled selectively, for example, to refine the distributional edge of the species. 1. Example of a commission error. A migrant or transient individual was sampled in this grid cell even though the species is not resident in the area. 2. Example of an omission error. A species is present in this grid cell but was not sampled in stage 1. 3. Even though the species was not sampled in this grid cell in stage 1, the cell is included for fine grain sampling in stage 2 to ensure that this was not an omission error.

of habitat generalists and species with distinct outlier records. However, what is interesting from a conservation perspective is that the model accurately predicted the restricted ranges of habitat specialist species (Figure 10.5) (see also Box 9.1 Wilson et al. (2004)). The value of such an approach is that the model can also be used with respect to changing environmental conditions, noting the

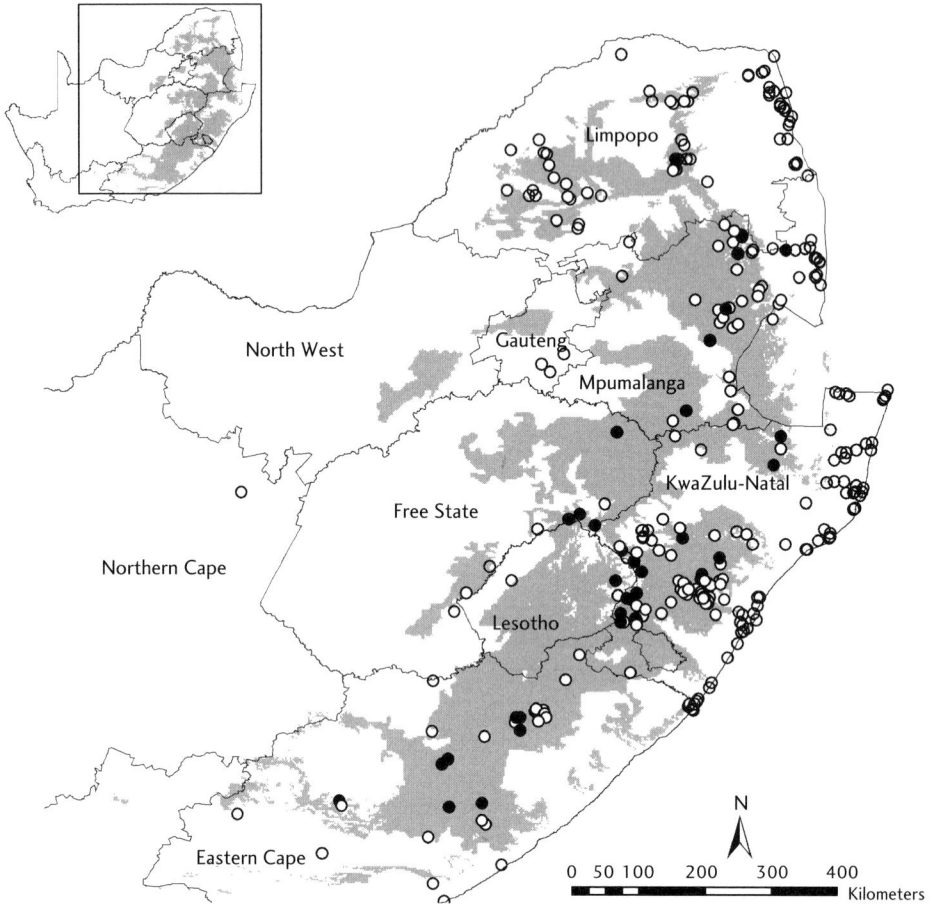

Fig. 10.5 Application of predictive modelling for insects in conservation planning. The specialized Mountain malachite damselfly *Chlorolestes fasciatus* is highly sensitive to stream disturbance and is characteristic of relatively good-quality upland streams. The point distribution records are given as black circles and the 'absence records' as open circles. Using a BIOCLIM-type modelling approach, the species' predicted geographical distribution can be drawn (shown here in grey), which provides target areas where the species can be expected to occur, assuming the habitat is in good condition (from Samways 2005).

reservations we had earlier in this section, that climate alone is not necessarily a good predictor of where a species will actually occur, as there may be various, sometimes subtle, biotic and abiotic variables which also influence geographical distribution.

Historical data must be supplemented with current distribution data, and it is usually necessary to survey the distribution of the species to quantify its present geographical or elevational range. Indeed, this is essential when undertaking Red List assessments, which must be redone at least every 10 years. In cases where no historical data are available, comprehensive surveys of current species distributions must necessarily form the baseline for future monitoring of distributional changes. When surveying a species to establish its distribution (or *extent of occurrence* or *area of occupancy*; see Figure 6.10) a *stratified approach*, and sometimes also *staged approach*, is recommended (Figure 10.4). Each stage is a compromise between data coverage (having information across the full extent of the species range or across the area of interest) and data resolution (having data at fine scales). When species have clearly defined habitat requirements, and the distribution of such habitat is well known, it is comparatively straightforward to design the survey. However, when a species' habitat requirements are poorly known, or it is known to be a generalist species with a fairly continuous distribution, the design of the survey is likely to be more complex.

One approach to gathering information on the distribution of a species is to place a grid over the extent to be surveyed (Fig. 10.4). Each grid cell is then visited and the presence or absence of the species in that cell determined. Any one of the sampling techniques discussed in Chapter 4 may be used, depending on the taxon of interest, and the focus could be a single species or an entire assemblage. The centre of the grid cell, grid-line intersection points, or a randomly selected site inside the grid can be used to survey the cell. Importantly, the exact x, y (and elevation wherever possible) locality coordinates are taken at each sample site, although the inference made is that the species is either present or absent in the entire grid cell. More than a single randomly selected site within the cell should be surveyed if logistically feasible.

The survey site(s) within each grid cell can also be chosen based on knowledge of suitable habitat availability within the cell. The grain, or grid cell size, may be determined by what is logistically possible to achieve in terms of the ability to visit each cell to determine whether the species is present or absent. When the extent is very large and there is virtually no information on the distribution of the species, then it may be advisable to start with a fairly coarse grain (large cell sizes). After these comparatively few cells have been visited, increasingly finer grid cell sizes can be used to hone in on the finer scale distribution of the species across the range. For example, particular focus could be given to delimiting the range edge of the species (Figure 10.4).

When historical information is available, designing the survey becomes slightly easier, because particular parts of the distribution can be targeted rather than surveying the entire extent of the range. For example, (1) localities where the species has previously been recorded, (2) localities where the species has previously not been recorded, and (3) localities at and beyond the historical range margins will provide information on any change in the distribution of the species since it was first surveyed. Repeatedly sampling the same localities over time, where possible, is important to separate possible climate-related range shifts from other causal factors, e.g. habitat destruction.

Where the elevational range of a species must be quantified (or any other linear habitat, such as species with riverine or coastal distributions, or distributions along road verges), a transect-based, rather than grid-based, survey may be used. The linear extent of the area is divided into one or more transects, each with several subsections as, for example, described in Chapter 6.5.7.

An alternative approach to quantifying species distributions are atlas projects, such as *The Millennium Atlas of Butterflies in Britain and Ireland* (Asher et al. 2001). This includes a map for each butterfly species, determined during three survey periods, i.e. pre-1970, 1970–1982, and 1995–1999. The *Millennium Atlas* includes the presence or absence of butterflies at a resolution of 10 km squares, although some data are also largely available at finer resolutions of down to 1 km square. Atlases are, of course, long-term projects, and involve large collaborative networks of researchers and volunteers, as well as good coordination of activities and data management. In such projects, it is important to record sample effort (recording effort) for every grid cell (such as the number of recorders who provided check lists for that cell and number of visits to the cell).

Finally, for representation of geographic ranges of insects, we do not recommend using blocked-in ranges (as often found in bird field guides), unless the aim is to explicitly illustrate the *extent of occurrence*, which is based on the minimum convex polygon of point recordings (which is *not* the geographical range). For insects, it is much more meaningful to use only points where the species has actually been recorded (with a historical tag where appropriate). The reason for this is that many insect species occur only at specific geographical locations, which may, for example, be characterized by moist conditions, with wide areas of unsuitable dry habitat in between. Blocking in the range would imply that a particular species occurs across all the in-between areas, grossly over-representing its overall density, especially for rarer species (Jetz et al. 2008).

10.7 Summary

Methodological approaches for assessing and responding to environmental drivers of change in the world around us, include *modelling*, *mensurate field studies*, *experimental field studies*, and *laboratory experimentation*. Measuring change in environmental conditions can be done with *bioindicators*, of which certain insect groups, or a combination of them, can be very effective. Procedures are available for quantifying and testing the proposed bioindicators, and for developing a bioindicator system using insects. Insects are also being increasingly used in biodiversity *assessments* and biodiversity mapping, and can be used as complements to other organisms for overall biodiversity assessments. Mapping insect distributions is an increasingly valuable tool for (1) monitoring changes in distribution (e.g. for both rare and invading species), as well as (2) understanding the drivers of changes in insect distribution and abundance.

11

Key questions for insect conservation in an era of global change

11.1 Forms of environmental change

It is beyond the scope and objective of this book to provide the theoretical background to each form of environmental change and our current understanding of their impacts on insect biodiversity (see work by T. Tscharntke and colleagues for extensive work on insect conservation in fragmented agricultural landscapes, e.g. Tscharntke et al. 2007). Needless to say there is a vast literature and well-established theory on topics such as habitat fragmentation, biological invasion, and climate change. Rather, the aim of this section is to provide a brief overview of some of the important questions and variables, sampling, and measurement considerations in studies of insect and; environmental change. Several of the techniques mentioned in this chapter as being appropriate for application to study a particular form of environmental change are dealt with in detail in previous chapters.

11.1.1 Habitat loss, fragmentation, and isolation

One of the consequences of human population growth is increasing loss of habitat. As a result, remaining patches of natural and semi-natural habitat are shrinking and becoming more and more isolated from each other (Fahrig 2003; Lindenmayer and Fischer 2006) (Chapter 7). Tools such as aerial and satellite imagery, remote sensing, and geographic information systems (GIS) make it possible to quantify the extent of habitat loss and changes in habitat quality, and the size, isolation, connectivity, and spatial configuration of habitat patches. This information becomes the relational or explanatory variables in studies of the impact of habitat change on insect populations and communities.

Measures of habitat connectivity that include biological information, such as the dispersal distance of the species, can be better measures of connectivity than those based purely on landscape metrics (such as nearest neighbour analysis) (Hanski and Pöyry, 2007). However, this view must be framed in terms of the conservation objective because any one *landscape element* may provide *connectivity* for one species (or life stage) but not for the next, i.e. the landscape is *variegated* (a *differential filter*). This emphasizes the point we made in Chapter 7 — that insect conservation is about suitability of the landscape for the species of concern. It is also important to remember that the species occupancy of a patch is determined by both local factors (such as the presence of a host plant) and regional factors (such as the presence of enough source habitat in the region to maintain a metapopulation and long-term species survival).

Not all landscapes and habitats lend themselves to fragmentation studies. If a landscape is only weakly fragmented, or if the transitions in habitat quality are not marked, insect populations and communities are unlikely to demonstrate a response. The history of the area must also be considered because the age of current patterns of disturbance and fragmentation will determine the likelihood of *extinction debt* and the status of population recovery.

Detailed knowledge of the biology of the species is of utmost importance in population-level studies of habitat change and fragmentation. This includes its specific habitat requirements, and the recognition that different life history stages may have different habitat requirements (e.g. larval host plant may differ from adult nectar sources) (Dennis et al. 2007). Information on the movement behaviour, dispersal ability and migration pattern of the species is needed to appropriately relate the research question to (1) the extent of the study and (2) the definition of patch isolation and connectivity (see Chapter 7). Dispersal distance frequencies across individuals in the population are also important. Just because an individual is able to disperse a particular distance does not necessarily mean that it will, with most individuals typically dispersing a short distance and only a few dispersing a long distance (see Figure 7.12). Nevertheless, this occasional long-distance dispersal of a few individuals also has important outbreeding value. Adverse conditions may stimulate such long-term dispersal, with the Silhouette dropwing dragonfly *Trithemis hecate* in Botswana dispersing over 20 km, even before it has fully matured. Thus, it is essential to be aware that dispersal of just a few individuals, which may not appear to be numerically significant, can nevertheless be of conservation significance, with the occasional individual transferring genes to neighbouring populations and countering the genetic effects of isolation.

Species also respond differently to habitat edges and to *ecotones*, and some species actually have higher abundances along fragment edges than in the interior of patches (McGeoch and Gaston 2000). This information will be critical for deciding on the placement of *sample units* in patches (see Chapter 3.3.6).

The better known the species biology, the better will be the study design and the stronger will be the inferences drawn about the effect of habitat loss and fragmentation of the population dynamics of the species. This is one reason why many of the insect fragmentation field studies have been on Lepidoptera rather than on less well-known and less conspicuous taxa.

Studies of the impacts of habitat change and fragmentation on insect assemblages rather than populations are complicated by the fact that the habitat requirements and dispersal distances of each species differ, and the effects of habitat change may be masked by considering community-level properties such as species richness. Nonetheless, community-level fragmentation studies ask questions such as what is the relationship between species richness, composition, species nestedness, and habitat patch size and isolation.

Variables typically measured in fragmentation studies include species population size, the lifespan of individuals, variation in population size within patches over time, rates of immigration and emigration per patch, and the identification of source and sink habitat patches (Hanski and Pöyry 2007). Few generalizations are possible about the response of insect species to fragmentation, and as a result most studies are likely to motivated by the need for management recommendations to conserve a particular rare and threatened species. Studies on the impacts of fragmentation and habitat change provide valuable information for reserve design, including minimum habitat requirements, effective patch sizes, and spatial configuration necessary to conserve rare and threatened species.

11.1.2 Disturbance, decline in habitat quality, and the loss of critical resources

The loss of a critical resource independently of any change in habitat patch area or isolation can of course also have devastating consequences for insect populations and communities. For example, the removal of dead and decaying wood from boreal forests during traditional timber harvesting practices is considered to be the primary mechanism responsible for the decline in saproxylic beetle diversity and the large number of Red Listed species in boreal forests globally (McGeoch et al. 2007). Lewis and Basset (2007) compiled a 'wish list' of methodological approaches for tropical forest insect biodiversity studies (Table 11.1). Researchers complying with this list will ensure that studies are comparable and that over the longer term, generalities can be made about disturbance effects on tropical insect biodiversity. All of the points in this list are as relevant to virtually any disturbed insect community as they are to tropical forest insects,

Table 11.1 *A 'wish list' for studies of the effects of disturbance on tropical insects (from Lewis and Basset 2007).*

- Take into account the geographical distribution and endemicity of taxa, rather than focusing solely on overall species richness or diversity values

- Report both species richness and diversity measures, and control for the critical influence of sample size on species richness values by using rarefaction

- Be explicit about the nature of replication in the investigation

- Document clearly the forms of habitat disturbance, and the time since disturbance events

- Document the history of human and natural disturbance in the studied areas

- To avoid publication bias, publish negative results (where no significant disturbance effect is found), as well as positive ones

- Consider employing or exploiting experimental protocols, such as before–after control impact (BACI)

- Use sound concepts of taxonomy (where morphospecies correspond to unnamed species, rather than fuzzy groupings of unidentified specimens)

- Use a multitaxon approach to reach more general conclusions about the impacts of disturbance on diversity; where this is not possible recognize clearly the limitations associated with individual study taxa

- Include summary data on numbers of individuals of each species recorded from individual sampling locations (perhaps as electronic appendices), to facilitate subsequent meta-analyses

---

and reinforce points made in earlier chapters. Again a fair understanding of species biology and habitat requirements is necessary to identify how a change in resource availability impacts on particular components of insect biodiversity.

11.1.3 Impact of invasive alien species

Biological invasion is recognized as one of the primary threats to biodiversity worldwide, along with habitat loss and climate change. The accidental (and sometimes deliberate, such as for agriculture or biological control) introduction of alien species, including vertebrates, plants, microbes, and other invertebrates, is also of major concern to insect conservation. Alien invasive plants, for example, may impact negatively on insect biodiversity by changing habitat quality, outcompeting native host plants, and interrupting vital ecological interactions. Social insects, in particular ants and wasps, have established themselves worldwide as successful alien invasive species. A range of life history traits associated with sociality in insects appear to facilitate successful invasion (Moller 1996).

Studies of biological invasion relevant to insect conservation thus include (1) monitoring ports of entry and other introduction pathways for alien species introductions; (2) examining the distribution, range expansion and the invasion process by invasive insect species; (3) understanding mechanisms limiting or facilitating the distribution, and retarding or enhancing the spread, of alien species; (4) studying the impact of invasive alien species (insect or other alien species) on insect biodiversity; and (5) establishing the consequences of biological invasion for ecosystem functioning (e.g. impacts of invaders on pollination, dispersal, and decomposition processes).

A combination of experimental and mensurate studies are useful for understanding the disruptive effects of biological invasion on insect biodiversity. Paired comparisons of communities between invaded and uninvaded areas with otherwise similar habitat and climate conditions are a mensurate approach to understanding the impact of invasive species. Removal or insertion experiments involve the deliberate manipulation of the invasive species and the measurement of its impacts on local populations or communities. Introducing an invasive species into a previously uninvaded habitat for the purpose of understanding its impact is, of course, ethically questionable and should preferably be avoided. However, the before–after control experiment (BACI) provides a powerful alternative in situations where the species is spreading rapidly and currently uninvaded areas will soon be invaded (Underwood 1994). An example of this is South African dragonflies, where certain rare endemic species had not been seen for many years (one for 83 years!) but reappeared once alien invasive trees had been removed. In this example, the alien trees shading out their habitat were the *key threat*, and removal of the trees, after identifying them as indeed the key threat, led to remarkable recovery of the species (Samways et al. 2005). Shade alone was suspected as the *key adverse factor*, which was subsequently confirmed experimentally (Remsburg et al. 2008).

An approach which needs further investigation is developing methods to predict the potential invasibility of any particular insect species. Climatic modelling has had a long history of relevance in the realm of pest control, and is now growing in importance in assessments for biological control agents (Boivin et al. 2006). Experimental approaches on habitat tolerance are also being developed in an effort to predict the risks associated with particular species that have congenerics which are known to be notorious invasive species (Addison and Samways 2006).

11.1.4 Climate change

In addition to habitat destruction and biological invasion, climate change is one of the major forms of environmental change impacting biodiversity. Some of the earliest substantial evidence for the impact of climate change on the geographic range and conservation of species comes from insect studies. Insects have in fact become important model organisms for understanding the impact of climate change on biodiversity (Box 11.1) (Wilson et al. 2007).

The effects of climate change on insects will differ between species depending on their biology, current environment, and geographic distribution. As a result, some species are likely to be more susceptible to climate change than others. Given our fragmentary understanding of the direct, indirect, and likely interactive effects of parameters of climate change on invertebrates, as well as the large taxonomic, life history, and habitat diversity in the group, identifying those species at greatest risk will be difficult.

One of the important characteristics of the response of biodiversity to climate change is that species tend to respond individualistically. In other words entire assemblages do not respond in a similar way; rather, each species in the assemblage tends to respond in a more or less unique fashion (see for example McGeoch et al. 2006), as they have done in the prehistoric past (Ponel et al. 2003). This characteristic has important implications for climate change studies because it means that assemblage-level changes (such as changes in species and total abundance) are unlikely to be informative about species-level responses. Nonetheless, there is a variety of ways in which climate change affects insects, and the study of each of these requires a somewhat different approach.

Climate change has altered and will continue to alter the geographic and local distributions of species. For example, in the northern hemisphere a northward range expansion of many butterfly species has been observed over the past few decades, with southern range limit contraction in some cases. Many species are also shifting upwards in elevation as temperatures at lower elevations become increasingly warmer (Wilson et al. 2007). Studying the distribution of insect species is discussed in Chapter 10.

Mean temperature increases, and higher winter and evening temperatures, can have profound effects on invertebrates. Although increased temperatures in general result in increased development rates and a forward shift in phenologies, survival, distribution, migration, dispersal activity, range size and position, activity periods, and food consumption rates will also be affected (Hughes 2000, Malcolm et al. 2002, Walther et al. 2002). For example, using a database of historical sightings of dragonflies, Hassall et al. (2007) showed that dragonfly flight dates in Britain are approximately 12 days earlier now than they were in the 1960s. However, the predicted impact of such changes on invertebrates is geographically variable, and high-latitude and -elevation species may be most seriously affected (Kennedy 1995).

Changes in the carbon-nutrient balance in plant tissue as a result of increases in carbon dioxide will reduce the nutritional quality of plant tissues and alter production of secondary compounds. Predicted effects for insect herbivores, particularly chewing insects, include increased first-instar mortality, increased development time and consumption, and decreased digestive efficiency and performance. Lower development rates may also increase herbivore mortality from natural enemies and result in asynchronous plant–insect life cycles (Bale et al. 2002). Interactions between species are predicted to be significantly altered by climate change, particularly when interacting species respond differently to the change. It is the interaction between the effect of climate change on insect life history and the resulting changes in phenology (timing of life history events) that will be important in determining how insects respond to climate change.

Water availability is one of the most important determinants of the distribution and abundance of insects. Because of their small body size, insects are particularly vulnerable to water loss, and water loss rates are positively related to precipitation levels (Chown and Nicolson 2004). Increased frequency of extreme events, such as floods, droughts, and fires, will increase mortality and may result in extinction of restricted-range species.

Finally, climate change will interact with other components of environmental change. For example, species in and adjacent to agroecosystems may be threatened by increases in the use of pesticides that would occur if predictions of increased pest damage as a result of climate change were realized. Species in habitats with heavy invasive loads may be particularly threatened as invaders alter resource availability and are generally superior competitors (Chown et al. 2007). Any positive response of species to climate warming (higher population levels and range expansion as a result of increasing temperatures at higher latitudes) may be outweighed by negative impact of habitat loss.

Therefore, climate change is likely to bring about changes in insect life history, physiology, mobility, population dynamics, synchrony with their host plants and other resources, interaction strengths and complexity, range size and isolation, geographic distribution, and range. Because of the multitude of ways in which climate change impacts on insects, studying these effects in the field will require a broad range of approaches and techniques. Regardless of the approach adopted, it is important to remember that insects will respond to microclimate conditions more clearly than to macroclimate. Because of their small size, insects experience, and are able to sometimes actively select, microhabitats and microclimate conditions that can be quite different from the climate information provided by standard weather stations (usually positioned 1.5 m above the ground).

Measuring microclimate, or the climate near the ground or plant surface, or within the microhabitat to which the insect is directly exposed, is therefore an important part of insect–climate change studies (Chapter 5).

Field-based approaches and techniques for studying climate change impacts on insects include:

- Mapping and monitoring changes in species distributions and activity periods as discussed above. Because climate change has been shown to change both the latitudinal and elevational ranges of insect species, quantifying such change and monitoring it over time forms an important component of insect–climate studies. Quantifying the relationship between the current distribution of a species and climate (constructing a *climate envelope* for the species) is also an approach used to project future possible geographic range shifts in species (Hijmans and Graham 2006; Pearson et al. 2006). However, it is important to base models on the whole, and not just part, of the geographical range of a species, otherwise there is likely to be an increase in the estimate of extinction risk (Akcakaya et al. 2006).

- Quantifying the distribution of species along natural climate gradients, such as temperature (e.g. elevational, latitudinal) or aridity gradients. This approach provides fairly large-scale spatial information. This information can be used as a spatial surrogate for predicting species responses to change over time, such as increasing temperature or increasing or decreasing aridity (a so-called 'space-for-time' substitution). However, the approach depends on the assumption that climate will change over time along the same trend that is currently observed over the spatial climate gradient (Dunne et al. 2004). The usefulness of this approach also depends on the relative importance of climate in determining the species distribution along the gradient. Other factors such as habitat type or biotic interactions can confound the interpretation of species distribution patterns along the gradient.

- Field experiments in which the microclimate associated with particular habitats is manipulated. For example, patches of habitat can be irrigated, desiccated, warmed or cooled (both soil and air temperatures), or the quantity and quality of light or carbon dioxide concentrations altered. Rain-out shelters, or cloches, are one such technique that can be used to both reduce precipitation experienced and warm the environment under the shelter (Figure 11.1). Of course, as in all experiments, it is important to consider controls and unwanted treatment effects (see Chapter 3). It is also very important in microclimate manipulation experiments to directly and simultaneously measure the change in microclimate being induced by the treatment. The advantage of field experiments is that they have the power to directly

Fig. 11.1 A rain-out shelter (or cloche) made from transparent polycarbonate and used to dry and warm the local environment over cushion plants on sub-Antarctic Marion Island, and to quantify the response of the mite and springtail species inside the plant clump to climate change-related drying and warming (Le Roux et al. 2005; McGeoch et al. 2006).

identify mechanisms driving species' responses to change. However, field experiments on climate change tend to be criticized for being too short term or with treatments insufficiently replicated. Short-term species responses may also differ from those observed over longer periods. Climate change experiments are usually also expensive and time consuming to construct and run. It is also a challenge to select and adequately manipulate the appropriate direction and degree of change (Dunne et al. 2004). Such decisions are usually guided by climate projections. Field experiments usually reflect average changes over time, whereas extreme weather events, or temporal shifts in weather patterns, may be more important drivers biotic changes.

• A combination of modelling, field sampling, and field experimental approaches along with physiological and laboratory experimentation is advocated to achieve a comprehensive understanding of climate change impacts on insect biodiversity. Dunne et al. (2004) show how powerful a combination of field experiments and environmental gradient study information can be for predicting the response of species to climate change. In this approach, the climate change experiment is replicated along a natural climate gradient. For example, the experiment is repeated at low, medium, and high elevations and both experimental and gradient survey data used in combination to infer species responses to climate change.

Box 11.1 The importance of maintaining accurate spatial data and recording details of habitat for climate change studies: dragonflies in Europe

---

Geographical range extension of dragonflies in Europe

Northward extensions of dragonflies in Europe were already being recorded by the mid-1990s (Ott 1996). An example is given here of the northward expansion of the Mediterranean species *Crocothemis erythraea* in Germany in response to elevated temperatures.

The adult male of *Crocothemis erythraea* is a very striking, habitat-tolerant yet responsive species, making it a good candidate for assessing how insect species move in response to climate change.

On the basis of good historical records dating from the 1980s (mapped in grid cells 10 × 10 km) from the German Federal State of Rhineland-Palatinate, it was possible to track geographical range expansion accurately over time. This study emphasizes the value of storing accurate, spatially explicit, baseline data for studies on the effect of climate change on insects.

About a decade after this initial study, new data and publications (regional species lists, atlas programmes, and data from federal mapping programmes) on this species for the different federal states were analysed, and a new map was produced, showing the subsequent distribution of this species in Germany (Ott 2007). The figure below shows the range expansion as of 2008, and illustrates that the species had expanded its range from southern Germany right to the north of the country, expanding 700 km northwards within only two decades. This change in geographical range correlates with temperature changes across the country over the same period.

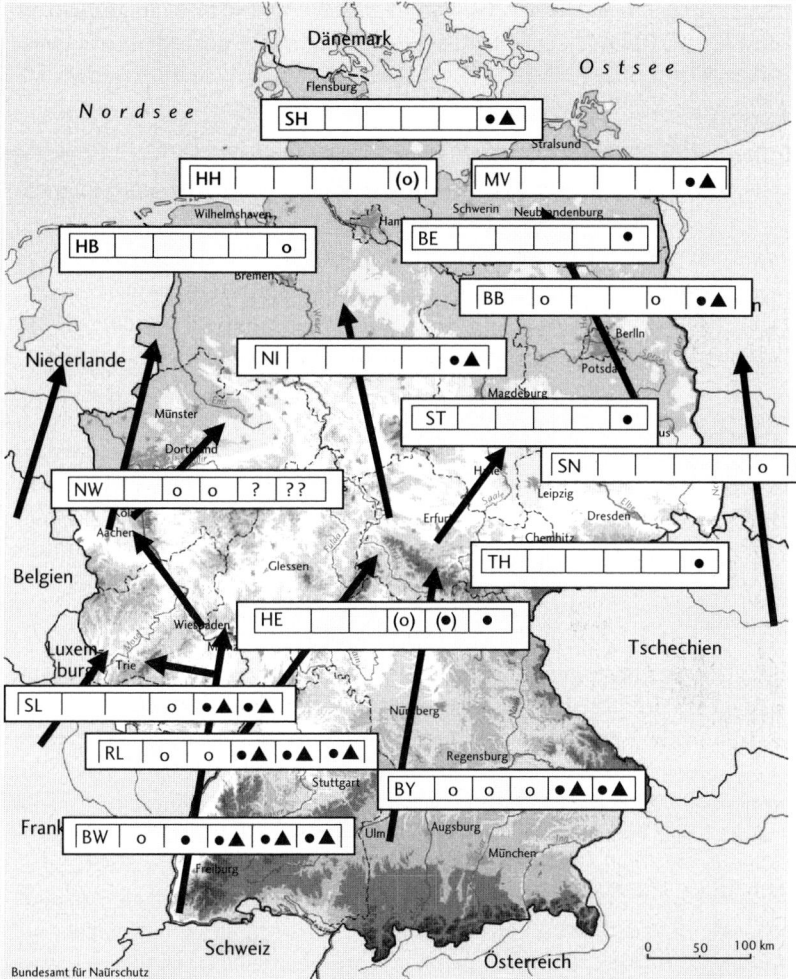

Quelle: Bundesamt für Naturshutz (BfN), 2006

BW	o		•	•▲	•▲	•▲

Geographical range expansion of *Crocothemis erythraea* in Germany. In two decades the whole of Germany was colonized by this species, traditionally considered to be 'Mediterranean'. (Key: first block: abbreviation for particular German Federal States; second block: presence status prior to 1970; third block 1970–1979; fourth block 1980–1989; fifth block 1990–1999; sixth block 2000–2008. o = species present, • = species present and also breeding, ▲ = population levels increasing, () = probably increasing).

Similar range expansions in Germany were also documented for other odonate species formerly considered of Mediterranean origin (see compilations in Ott 2001, 2008): *Erythromma viridulum* also reached northern Germany; *Anax parthenope* has become more common in many regions (e.g. Rhine Valley, Brandenburg); *Aeshna affinis* now breeds in many parts of Germany; invasions of *Sympetrum fonscolombii* are now a regular occurrence, with the appearance of second-generation individuals. Similar 'Mediterranization' of the dragonfly fauna has also occurred in other European countries, with *E. viridulum* even colonizing the UK.

The importance of monitoring and the value of atlas programmes

This case study clearly shows that such an expansion can be documented only by good record-keeping. The more accurate the basic data, the more detailed the information we can obtain, and, in turn, the more conclusive will be the evidence. Where only basic information is available (e.g. that a species was present in a certain area at a particular time or during a certain time period), the results will have limitations. But where there is more detailed information (e.g. on day *a* at *b* hours the species was present at pond *x* (elevation, coordinates, detailed documentation of water quality) with *y* males and *z* females, pairing and egg laying was recorded, exuviae and/or larvae were found, and *n* other species were also recorded) we then have all the information necessary for detailed comparison and analysis. Where this type of information is then available for many water bodies across a region over several years or even decades, we can then obtain a clear picture of the changing character of the insect assemblage in response to changing conditions, such as effects of climate change and its synergism with other impacts.

This temporal comparative study demonstrates the importance of thorough monitoring and mapping, over a certain time span, to achieve a picture of the dynamics of range change. This is conceptually important. Geographical ranges are traditionally represented as being static, which is generally not the case, especially in these times of rapid climate change. This point is emphasized by the European Commission's Habitats Directive, where the quality of biotopes is listed in Annex I, and the status of populations is reassessed after 6 years.

However, it must be emphasized that the monitoring of only a few water bodies only gives part of the overall picture. Full clarity only comes about after analysis of data over longer time periods and for a large area, covering at least a region or state, and even larger area where possible. This is being done for the SLL-plus-Atlas project, led by B. Trockur, which encompasses Belgium, France, Germany and Luxembourg, 65 401 km^2 and 117 050 data points in all. The data are consists of all verified information from published and grey literature, as well as new mapping data sets. This approach will feed into other planned projects for parts of the region (e.g. for the atlas of dragonflies in Rhineland-Palatinate). The main objective of this atlas project was in fact to produce an overview of the dragonfly fauna of the region (description of all species, distribution, ecological analysis, etc.), with climate change not originally central to the project. Yet, arising out of the

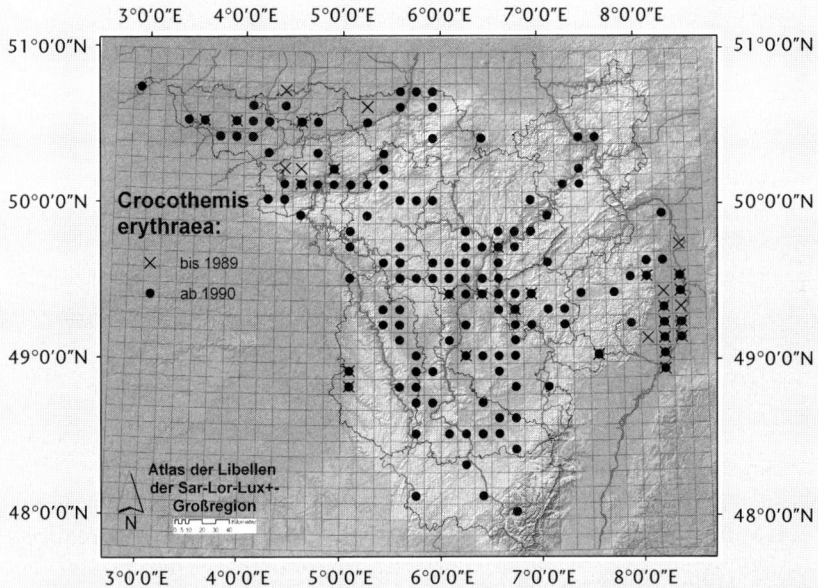

Range expansion of *Crocothemis erythraea* in the SLL-plus-region (map from the Trockur et al. working group). (X = the species present in a grid cell of about 10 × 10 km up to and including 1989; ● = present 1990 or later).

Increase in number of grid cells populated by various odonate species in northern Europe in the two time periods prior to 1990, and 1990–2006

Increase in occupied grid cells (absolute)	Species	Total number of grid cells with the species present	Percentage of grid cells with the species present
+109	Crocothemis erythraea	149	25.3
+107	Erythromma viridulum	212	36.1
+101	Calopteryx virgo	409	69.6
+87	Sympetrum sanguineum	354	60.2
+81	Cordulia aenea	306	52.0
+79	Calopteryx splendens	421	71.6
+74	Platycnemis pennipes	422	71.8
+74	Gomphus pulchellus	296	50.3
+73	Aeshna mixta	273	46.4
+72	Anax imperator	421	71.6

high quality of the data, it has also been possible to track some of the influences of recent, rapid climate change. This new spin has also led to only two time periods being finally chosen for making temporal comparisons with respect to the tipping point of the effects of recent rapid climate change: before and after 1990. The figure below shows the expansion of *Crocothemis erythraea* before and after 1990. It was at about this time that the species expanded its range dramatically.

The value of this binary time period is emphasized in the table above. *Crocothemis erythraea* was the species showing the greatest increase in the area, but it was followed by *Erythromma viridulum* and three other species of Mediterranean origin: *Aeshna mixta*, *Anax imperator*, and *Gomphus pulchellus*, which are now present in half to more than two-thirds of the grid cells.

Jürgen Ott, L.U.P.O GmbH, Germany

11.1.5 Biotechnology, biological control, and sustainable harvesting

11.1.5.1 Biotechnology

Although modern, biotechnological pest management tools are designed partly to reduce environmental contamination and the non-target impacts of traditional pest management schemes, there remain concerns about the impacts of these strategies on insect biodiversity (Hill and Sendashonga 2006). The use of crops that are genetically modified to be insect resistant should reduce the quantity of pesticide usage and in this way also reduce the impact on non-target insects. Nonetheless, these genetically modified (GM) crops may still have an impact on non-target species that feed on the crop and that are also susceptible to the strain of *Bacillus thuringiensis* (Bt) used, e.g. currently mostly Lepidoptera and Coleoptera. The use of herbicide-tolerant GM crops may also reduce the plant and weed diversity in agroecosystems and reduce the insect biodiversity associated with the agroecosystem. The use of GM crops may thus also alter agricultural management practices, and in this way indirectly impact insect biodiversity patterns.

To date, little evidence has been found for significantly negative impacts of GM crops on non-target insect diversity. However, because the technology is fairly new and development ongoing, as well as because of the broadly agreed philosophy of case-by-case biosafety assessments for genetically modified organisms (GMOs), research in this field (van Wyk et al. 2007) is likely to remain active (as specified in the Cartagena Protocol on Biosafety (CBD 2002; Hill and Sendashonga 2006). This philosophy recognizes the large variation in the biotechnology itself (such as different pest targets, varying levels of expression and expression in different plant parts), as well as the importance of geographic context in determining possible unwanted effects of the application of GMOs.

Biosafety assessments therefore examine and monitor the likelihood and extent of non-target and other potential unwanted impacts of GMOs (Box 11.2). Non-target studies must examine (1) the spectrum of non-pest herbivore species exposed to the GM crop in the field, (2) impacts higher up the trophic system (predators and parasitoids), as well as (3) the impact of the use of herbicide-tolerant crops on insect biodiversity in agroecosystems.

The impact may differ markedly across the geographic range within which the GM crop is grown, as well as between seasons with different weather conditions. Impacts may also accumulate slowly over time. Combined with laboratory toxicity and behavioural choice tests, farm-scale and medium to long-term field trial assessments within the region in which the crop will be, or already is, planted will provide the most relevant information on any unwanted impacts of GMOs on insect biodiversity (Squire et al. 2003). A clear distinction is made between field trials designed to test potential impacts (these are usually conducted over smaller spatial and temporal scales) and long-term monitoring of

---

Box 11.2 A model for non-target risk assessment of genetically modified crops

---

Any cropping system, although generally less biodiverse than natural habitat, may still host up to several hundreds of species. These non-target species, while not crop pests, will be exposed to a greater or lesser extent to crop management practices, including pest management. This includes exposure to GM, particularly insect resistant, crop plants. It is of course not possible to assess the impact of the GM crop on all non-target species. It is therefore important to use a scientifically defensible and transparent strategy for selecting relevant species on which to conduct toxicity testing (Birch et al. 2004). Non-target risk assessment should preferably be conducted on the unmodified crop before the GM crop is released into the environment. However, the process described below is equally valuable to develop post-release monitoring programmes. Birch et al. (2004) have developed a five-step non-target risk assessment model which they illustrate using Bt maize in Kenya as a case study. This model was essentially developed to select, from the numerous species present in an agroecosystem, those species on which detailed impact assessment studies will be conducted (usually involving laboratory-based toxicity testing). The model as outlined by Birch et al. (2004) is summarized below.

Step 1: Listing of non-target species and categorization of functional groups

- During this step all the insect species (and usually other invertebrates such as spiders and earthworms) associated with the crop in the region of interest are listed. This process will usually involve a combination of expert knowledge, literature information, and hands-on biodiversity assessment. The

non-target species list will differ between crops and for the same crop in different agroecological settings and different geographic regions.
- Once the list has been compiled, the functional group, trophic level, and trophic position of each species is specified and the list categorized according to trophic position and function.
- Andow and Hilbeck (2004) identify two broad types of function:
  - *Anthropogenic function*, or functions of direct and immediate concern to humans; e.g. pests (negative function), natural enemies (positive function), or rare and endangered species (conservation priorities or species of cultural significance). In the case of insect resistance GM, several pest species may be present that are not the actual target of the GM crop.
  - *Ecological function* includes those functions important to ecosystem functioning; e.g. pollinators of wild plant species, decomposers, seed dispersers.
- Several functional groups will thus be identified under each of the above types. There will also undoubtedly be species for which the function is unknown. It is important to explicitly include an 'unknown' category to ensure potentially important species are not forgotten about entirely.

Step 2: Prioritization of species and functions
- In this step, each species is the list is assigned a priority based on the likelihood and consequences of its being negatively affected by the GM crop.
- In Birch et al.'s (2004) model a selection matrix is compiled in which each species is given a score for each of three criteria. The criteria are:
  - *Maximum potential exposure* (MPE). This is based on the abundance of the species in association with the crop and the crop environment in the region of interest. It also includes how frequently (temporal component) and how broadly (across fields) the species is found on the crop and how specific it is to the crop and the crops associated habitat (termed linkage). The greater any of these factors the greater will be the MPE.
  - *Potential adverse effect* (PAE). This is a score based on the significance or impact of the adverse effect of the GM crop on the species should there be one. A key predator of an important pest would, for example, score highly, as would a host-specific pollinator of a rare plant species.
  - *Potential likely exposure* (PLE). The PLE is based on the likelihood of direct exposure of the species to the transgene product or metabolite. If, for example, the transgene is expressed in the leaf material of the crop and the species is a leaf-chewing herbivore then PLE will be high for that species. By contrast, an aphid may bypass exposure to the transgene on plant cell material by directly accessing the phloem.
- Once all species have been scored based on these criteria, the species are ranked with the species scoring highest on all three criteria given a rank of 1. At this point a precautionary approach should be adopted. For example, if a species has a high MPE, but the PAE for the species is unknown, then

it would be safest to include such a species in the final list of candidates for testing.

- The final list of candidates for testing could include one species from each functional group within each of the two function types (see above), or several species from each functional group considered to be most important. The final selection, while guided by the scores and ranking process, will be based on expert opinion and therefore to some extent subjective. The length of the final list of candidates for testing will also be determined by financial and logistical constraints. The decision making process, even if not entirely quantitative, must be defendable and transparent.

Step 3: Analysis of exposure pathways

- The objective of this step is to describe the one or more pathways by which each candidate species may be exposed to the plant and the transgene product.
- The insect could come into contact with the transgene product in a variety of ways, e.g. via direct contact with plant residues, root exudates, pollen, dead leaves, or other plant tissue. Indirect contact is also possible for predators and parasitoids that feed on insect herbivores that have consumed the GM crop.
- This step could further refine the candidate list of species, but is also important for designing toxicity tests. In laboratory toxicity experiments insect species must be exposed to the transgene product in a way that is ecologically realistic.
- Because there are so many potential exposure pathways, and because different pathways are relevant for different species, this step is an important exercise for ensuring relevance of the outcome of toxicity testing.

Step 4: Hazard identification and hypothesis development

- During this step hypotheses are developed for each species on the final candidate list to test possible hazards.
- For example, if a species feeds on a particular plant tissue will it be directly or indirectly affected? Will the effect be lethal or sublethal? Will reproductive performance and mating success be affected and thus population size in future generations? Will there be indirect effects as a consequence of higher trophic level impacts?

Step 5: Protocols, parameters to be measured and experimental endpoints

- Finally, methods are developed and experiments designed to test the hypotheses defined in the previous step and based on the identified exposure pathways.
- Laboratory methods can be of two types: (1) ecotoxicological where the insect is exposed directly to the transgene product extract, and (2) whole-plant methods where the insect is exposed to the entire plant or parts thereof (Andow and Hilbeck 2004).

- Controls in these experiments should be genetically identical to the GM crop with the exception of the inserted transgene.
- Experiments should examine longer-term effects and not just immediate mortality levels, i.e. they should follow individuals through the life cycle and quantify factors such as stage-specific mortality and female fecundity.
- In general, experiments should be designed so that they are as ecologically realistic as possible and as far as possible consider longer-term and higher order impacts.

For further details refer to the CABI Publishing series Environmental Risk Assessment of Genetically Modified Organisms, edited by Kapuscinski and Schei, including Hilbeck and Andow (2004) and Hilbeck et al. (2006).

impacts of GM crops that have already been commercially released (here the design of sampling programmes that are well-replicated, broad scale and long-term is possible).

Woiwod and Schuler (2007) give an excellent overview of GM crops and insect conservation. They conclude that the scientific evidence suggests that the GM crops grown to date do not appear in themselves more hazardous to the environment than those produced by conventional breeding. Indeed, it appears that Bt crops are of overall benefit to non-pest insects as they result in reduced use of wide-spectrum insecticides. However, there is another side to the increased expansion of GM agriculture. Destruction of the Amazonian rain forest is currently 25 000 km² per year, principally for the production of soya and beef. One of the drivers of this massive landscape disturbance is the demand in Europe for GM-free soya. As Woiwod and Schuler (2007) point out, this will inevitably have a major impact on many insect species, and will no doubt result in huge numbers of *Centinelan extinctions* (see p. 92). These issues illustrate that the conservation of insects and other biodiversity is very much a large-scale socio-political issue.

A related matter in agriculture is the demands being made on natural capital by the production of biofuels. Large cassava biofuel fields in Brazil have pest problems that smaller fields surrounded by natural vegetation do not. The reason for this is that the natural borders around the small fields are a reservoir of natural enemies which then move on to the cassava and control the pests (Samways 1979). Bickel et al. (2006) showed a high degree of ant population subdivision in Borneo, possibly through disruption of metapopulation dynamics from oil palm planting. These studies are simply a glimpse of the potential disturbance biofuels can have on insect and other biodiversity, and it is an area which urgently requires much more research.

11.1.5.2 Biological control

Classical biological control (CBC), in which biocontrol are agents deliberately introduced into an area outside of their native range for use as a pest management tool, was formerly considered to be an environmentally sound form of pest management. Although CBC has certainly reduced dependence on certain chemical control measures, concerns have been raised about the higher trophic level and *food web* consequences of the introduction of exotic species to control pests and weeds (Howarth 1991; Simberloff and Stiling 1996; Memmott et al. 2007). However, the issue is really about what level of risk we are prepared to accept, given that pest control using biological methods is not likely to stop (Samways 1997b, 2005). Considering this in perspective (Samways 1988a), although CBC does carry risks, they are relatively small in terms of their impact on insect diversity compared with the major environmental drivers of insect loss (habitat loss, landscape fragmentation, threats from invasive alien organisms, climate change, and even the large-scale side effects of GM crops; see 11.1.5.1). Nevertheless, there is one very significant issue with CBC—once a biological agent has been introduced and has established, it almost certainly can never be recalled. In short, CBC is irreversible, and an irreversible violation of *sense of place* (Lockwood 2001). Nevertheless, there are instances where CBC can benefit conservation (e.g. Zwölfer and Zimmermann 2004), and these mostly pertain to the biological control of invasive alien plants.

Risk analyses of a range of natural enemies (and of pathogens, which present some of the same risks as biocontrol agents) was undertaken by van Lenteren et al. (2003, 2006a) with the conclusion that generalist insect predators carry particularly high risks to non-target organisms (indigenous organisms that were not the intended targets for the agents). The same group (Van Lenteren and Loomans 2006; Van Lenteren et al. 2006c) have reviewed the risks of releasing alien biological control agents, and conclude that CBC introductions to date have been relatively safe with less than 1% having caused population-level effects on non-target species. It should be borne in mind, however, that only a minority of introductions have included a careful evaluation of non-target effects, and also considering the point discussed in section 11.2.3, that species with high connectance in food webs can have a major effect on community stability.

While stringent pre-release risk assessments provide considerable insight, they may not present the full picture as the assessments are done in the artificial conditions of the laboratory. However, one area where insect conservationists can positively engage the practice of CBC is to examine consequences of biological control agents on the broader food web. For example, a herbivore insect introduced to control an alien invasive plant species may itself be adopted as host by a range of indigenous and non-indigenous predators and parasitoids. The consequences of the shift in balance between the number and strength of links in the food web that results is currently largely unknown. Network theory provides a powerful new approach for quantifying and understanding food webs, where species are considered nodes in the network and interactions between them become links in the network (Memmott et al. 2007). Food web studies in association with biological control agents are thus of interest to the field of insect conservation and the conservation of ecological interactions (Memmott et al. 2007) (Chapter 9).

Van Lenteren et al. (2006) give an excellent review of procedures for undertaking risk assessment of potential natural enemies, with reference to the international guidelines developed through the International Standard for Phytosanitary Measures No. 3 (https://www.ippc.int/IPP/En/default.jsp). Guidelines and methods are currently being developed further, with ecological factors for determining the environmental impact of an alien natural enemy being increasingly used in evaluations. Host range information is being used to reject or accept introductions, alongside determination of the potential establishment of potential natural enemies based on their thermal biology and their potential to establish in a new area. These approaches are being combined with assessments of dispersal ability, as well as direct and indirect effects on non-targets. These considerations are leading to a point where ecological and behavioural data can be used in a quantitative, stepwise way as part of a risk assessment procedure. Quick-scan approaches are also being considered, which may result in 'white lists' of supposedly safe species for specified geographical areas with specific climates. The aim is to have a procedure in place which will be effective and efficient for biological control practitioners while at the same time having minimal environmental risks.

11.1.5.3 Sustainable harvesting

Insects themselves are sometimes considered agricultural produce. For example, in Africa some species of edible caterpillars and wild silk-producing moth species are harvested as an important part of local diets and economies. These species are commonly harvested from natural populations rather than actively cultivated. In this case, studies of the maximum level of harvesting possible to ensure sustainability of the resource may be necessary. Such species may become of conservation concern as a consequence of either over-harvesting or habitat loss combined with sustained levels of harvesting. Studies concerning the sustainable use of insect species must examine spatial and temporal characteristics of the population dynamics of the species, including density-dependent and -independent impacts such as weather conditions, natural enemies, and harvesting levels. Sustainable harvesting is, of course, also relevant in situations where an important resource for insects is harvested or destroyed in the process, such as the example of timber harvesting in boreal forests mentioned above.

11.1.6 Aquatic and terrestrial systems contamination

Because of human reliance on fresh water, global investment in monitoring water quality is enormous. Aquatic insects and other invertebrates have a long tradition as indicators of the quality of freshwater systems and pollution levels (Bonada et al. 2006). The classic pepper moth case and the increase in melanism in moth populations as a consequence of industrial-era air pollution levels provided an early demonstration of terrestrial insect responses to atmospheric pollution. Despite the apparent importance of pollution, including that of pesticides, there is much we still do not know on how pollutants and synthetic compounds affect insects and food webs in general, despite the wealth of knowledge coming from applied entomology, and the projected increase in pesticide usage by 270% compared with present levels by the year 2050 (Tilman et al. 2001). A salient reminder of the many subtle impacts such foreign chemicals may have on ecosystems comes from the findings of Cilgi and Jepson (1995) that butterfly larval and adult fitness can be reduced by dosages only 1/640 of the field dose rate. Furthermore, such chemical impacts are likely to be synergistic with other impacts, such as habitat fragmentation, with detergents, for example, being a major synergistic impact over alien plants in reducing rare, endemic, and threatened damselflies on a tropical island (Samways 2003). In view of this triple concern (increased pesticide usage/subtle effects/synergisms), we recommend that much more research be done in this area.

In addition to effects on numbers, the quality of individuals may be influenced by chemicals, the effects of which may manifest as greater incidence of deformities. The topic of *fluctuating asymmetry* (FA) has been advanced as one possible way to measure this in insects. It is defined as 'small, random departures from expected bilateral symmetry in an organism' (Markow 1995). In the past FA has been measured and used as an indicator of genetic and environmental stress in insect populations (as well as other taxa). High levels of FA, relative to a control population, were seen as evidence for environmental stress (such as pollution or the existence of a population bottleneck). For example, FA in wing length and wing cell size in damselflies has been examined as a possible response to parasite load or water pollution. However, current consensus suggests that

FA varies naturally between populations and over time in a complex way and that FA has an epistatic genetic basis (Leamy and Klingenberg 2005). As such it is unfortunately of little, or at least not immediate, value to environmental monitoring.

11.2 Consequences of species loss for ecosystem function

11.2.1 Importance of diversity

Maintenance of species diversity (not just the number of species per se, but also their variety and their various interactions; i.e. *ecological integrity*) of ecosystems is important for conservation of natural ecosystem function, as it leads to important issues of how an ecosystem can be managed to be stable, and whether a particular ecosystem is sensitive to a particular type of disturbance (i.e. the extent of its *resilience*) (Loreau et al. 2002, Weisser and Siemann 2004, Ives 2007). This view stems from an understanding of the role of species diversity for maintaining ecosystem function and evolutionary processes. We may also argue this issue from another perspective. The conservation of ecological integrity is essentially the scientific underpinning of the *precautionary principle*.

Wardle et al. (2000) drew attention to the fact that there are principally two schools of thought with regard to this diversity–function issue. On the one hand, there are clear, causative relationships between diversity and ecosystem function (Tilman 1999). The alternative view is that ecosystem services are not necessarily driven by species diversity per se, but rather that the main drivers of ecosystem properties are the key functional attributes or traits of the dominant species present, as well as the composition of functional types (Hooper and Vitousek 1998). The issues surrounding this debate are extremely important because, as Wardle et al. (2000) point out, it focuses attention on whether or not we should really consider rare species where the goal might otherwise be to conserve ecosystem function.

11.2.2 Temporal perspectives, adverse synergisms, and discontinuities

Any discussion on the diversity–function debate requires some perspectives on time, both from the past (Graham et al. 2006) and into the future. In the past, relationships have been decoupled and re-assembled with changing global climate (Ponel et al. 2003). Throughout the Quaternary, geographical ranges of species shifted back and forth, tracking optimal environmental conditions. This emphasizes that, in terms of evolutionary conservation, the *sense of place* is largely an artefact of time. While rapid evolutionary change in insects may occasionally occur (Mavárez et al. 2006), either to these historical climate-driven events or even to current anthropogenic landscape change, in general, the modern mix of current anthropogenic pressures on many species will be too much for them to survive. Not only will this arise from population susceptibility to environmental change, but also from genetic inbreeding (Saccheri et al. 1998). This is largely because many impacts are speculated from models to be synergistic, with, for example, the joint effect of fragmentation and global climate change being described by Travis (2003) as a 'deadly anthropogenic cocktail'. This seems indeed to be the case, with habitat loss and global climate change being responsible for huge declines in some British butterflies (Warren et al. 2001). Furthermore, it is beginning to appear that reserve networks in Europe will be unable to cope with the dynamic geographical range shifts necessary for the long-term survival of many species (Kuchlein and Ellis 1997).

The crucial question then hinges on the extent to which there will be decoupling of interactions (e.g. pollination, herbivory, parasitism) (Lindborg and Eriksson 2004), with cascade effects as the climate changes across a world dominated by human landscapes. These mosaics inhibit movement of many species and prevent them from finding suitable source habitat conditions. This, combined with deterioration of patch quality, is causing considerable local extinction, especially among habitat specialists (Kotze et al. 2003; Valladares et al. 2006), and leading to ecosystem *discontinuities* involving insects (Samways 1996). Much more research is required in this area to identify where there are likely to be *catastrophic regime shifts* (where changes accumulate to a point where there is a radical change to a new ecological state), or whether opportunities for restoration exist (Scheffer and Carpenter 2003). For example, the insect fauna on Easter Island have undergone an irreversible change in composition and abundance. Having said that, by far the most morally sound, sensible, expedient, and cheapest route to conservation is to prevent landscape deterioration and biodiversity erosion in the first place.

11.2.3 Species numbers, rarity, and ecosystem function

Among plants, functional diversity (see Chapter 9.8.3), as measured by the value and spectrum of species traits, rather than by simply species numbers, can strongly determine ecosystem functioning (Diaz and Cabido 2001). In turn, insect species interact within and across various trophic levels, and they are thus likely to have multiple effects on food webs. Dunne et al. (2002) have shown that food-web structure can affect compositional biodiversity, leading to local extinctions. The removal of a *number* of species affects ecosystems differently depending on trophic *functions* of species removed, with food webs being more robust to random removal of species than to selective removal of species which have the most trophic links to other species. In turn, the robustness of a food web is improved with increased connectivity within it, although this may be independent of species richness per se. Nevertheless, in changing environments, a large number of species is probably needed to reduce temporal variability in ecosystem processes. However, on balance it is the organisms with longer lifespans, bigger bodies, poorer dispersal capacities, more specialized resource use, lower reproductive rates, and some other traits that make them particularly susceptible to landscape change (Diaz et al. 2006). For approaches in insect conservation, this emphasizes the importance of maintaining as much diversity as possible, in the largest areas possible (including a biodiversity-rich agricultural landscape), to buffer changes that are likely to occur in the future (Tscharntke et al. 2007).

Removal of highly connected species, in contrast to removal of species with few or weak trophic connections, can have major consequences for biodiversity throughout the food web. For example, Sole and Goodwin (2000) showed that removal of only 5–10% of these highly connected species can lead to major ecosystem change. Furthermore, loss of only a few important predators of grazers can have a disproportionately large effect on ecosystem diversity, which may involve subtle effects with time delays (Duffy 2003; Lindborg and Eriksson 2004). This means that for maintaining food-web stability, and hence the conservation of functional diversity, it is essential to maintain diversity of highly connected species. Changes to food-web stability can be very subtle, and may depend on genetic diversity as well as species diversity (Reusch et al. 2005). What all this amounts to is that we need to avoid homogenization of habitats and loss of mobile predators from high trophic levels, because, as McCann (2007) puts it, according to this view, nature is a balance between bottom-up (driven by habitat heterogeneity) and top down (driven by predators) forces—a challenging research agenda for insect conservationists.

Changes in biodiversity can be both the cause and the result of changes in both stability and productivity. This two-way effect leads to feedback loops as well as other impacts which influence how communities change with biodiversity loss. Food webs mediate these interactions, with consumers modifying, dampening, and even reversing these biodiversity–productivity linkages (Worm and Duffy 2003). Indeed, Fontaine et al. (2006) show that plant communities pollinated by the most functionally diverse pollinator assemblage contained about 50%

more plant species than did plant communities pollinated by less-diverse pollinator assemblages. Furthermore, the positive effect of functional diversity was explained by a complementarity between functional groups of pollinators and plants. Thus, the functional diversity of pollination networks may be critical to ecosystem sustainability.

A further consideration is of interactions across ecosystems as well as within them. Knight et al. (2005) show how trophic cascades can occur when fish consume large numbers of dragonfly larvae, which reduces pressure on pollinators, which, in turn, are normally eaten by the adult dragonflies. As a result, plants near ponds with fish receive higher levels of pollination than plants away from ponds. We now need further information on various interactions across ecosystems and landscape elements to make more informed decisions at the level of the whole landscape.

11.2.4 The importance of food-web connectance

It is important to conserve whole ecosystems, with all levels and types of food web *connectance* intact. This immediately strikes a positive chord for insect conservationists, where the huge diversity of insect life involves considerable connectance, albeit that individual species in many cases play only a minor role in food-web structure. But this is where we need to be very careful: what may appear to be a rare species may only be so because it is being suppressed by natural enemies and pathogens, as well as other possible factors, the principle of which underpins classical biological control. Another interesting facet comes from studies on plant communities where the collective effect of rare species increases community resistance to invasion by aliens and minimizes any impact (Lyons and Schwartz 2001), while maintaining rare species in the community also helps maintain ecosystem function (Lyons et al. 2005).

When there is gradual loss of rare species and specialists, the result can be catastrophic regime shifts. If this line of thought is pursued further, it would suggest that transformation of formerly extensive ecosystems into remnant patches will leave the resultant fragments not simply as smaller reflections of the whole, but rather as impoverished areas subject to adverse environmental conditions and *ecological relaxation* over time. Each patch will then also go on its own trajectory, leading to new food-web structures and even different species components as new, mobile, and resourceful species enter the system.

An additional component to consider is that of spatial scale (Peterson et al. 1998). At the small, local scale, there may be an increase until all available niches are filled. Then, with a further increase in species diversity, especially of effective competitors, functioning decreases. However, at the larger spatial scale, regional species can complement each other, such that ecosystem functioning increases with increased species diversity. This emphasizes the importance of maintaining large complementary patches and networks across wide areas to maximize biotic diversity (Bond and Chase 2002) (see below, Figure 10.7).

11.2.5 Rare and threatened species as flagships

Explicit population models suggest that prediction of the effects of fragmentation requires a good understanding of the biology and habitat use of the species in question, and that the uniqueness of species and the landscape in which they live confound simple analysis (Wiegand et al. 2005). Yet for practical large-scale conservation, and given the regular shortage of resources, clearly we cannot know all the insect and other species, even in a small area, especially in the species-rich lower latitudes and other biodiversity hotspots. This is further highlighted by the fact that some, and probably many, putative species are *species complexes* (Hebert et al. 2004). So, to get a measure of the value of a landscape, we choose a subset of taxa. This subset may be certain taxa which are well-known, or a size class, particular functional types and even Red Listed species (Lawler et al. 2003). In terms of practical conservation, rare and threatened species can be selected at least as icons, and to some extent surrogates, and used as one barometer of changes in ecosystem form and function.

Thus, to sum up, current evidence is suggesting that to maintain ecosystem health and resilience, we need to conserve the range of food-web connectance, even involving rare species, throughout a wide variety of landscapes. The challenge facing us is that much of the impact has momentum that is not stoppable for a long time, with the inevitability of *extinction debt* (Tilman et al. 1994). While in the past populations could shift across natural landscapes on a regional scale unimpeded by anthropogenic obstacles, the situation is not the same today, with the myriad of human-transformed land mosaics. To maintain any semblance of the past, we need to think ecosystem health and vigour in terms of massive linkages between large, good-quality landscape patches in addition to maintaining a landscape mosaic that is minimally disturbed. Embedded in this approach is the knowledge that all species have a role to play and deserve conservation attention in their own right, and as a means of ensuring some ecosystem resilience. Iconic, rare and threatened species can thus play a role as flagship species, signifying the conservation significance of the landscape.

11.3 Lessons learnt from diversity value: towards a synthetic management approach

After systematic reserve selection (of which the approach by Reyers et al. (2002) is an excellent example where insects were included, which normally happens very rarely), management of the landscape should optimize insect and other biodiversity conservation. But first we must be very aware that overlying the *coarse filter* landscape approach is the *fine filter* species approach. In other words, the coarse filter approach is a catch-all to engage the precautionary principle and to conserve as many species as possible, and their myriad of interactions, based on general principles of effective conservation. This approach is simply a way of dealing with such complex compositional and functional, and changing, biodiversity 'in the hope' of conserving as much as possible. Where we have found out about the specific details of the biology of species of conservation concern and we wish to build in that knowledge to conserving that species, we can do so in addition to the landscape ecology approach. In other words, the landscape approach does not exclude the species approach. This two-layered approach may also include a third, in-between layer, called the *mesofilter*, which includes structures and features such as hedgerows, fallen trees, muddy patches, etc. (Hunter 2005). For insects, one structure of great importance is that of decaying and dead wood for the saproxylic insect guild.

The findings emerging from insect conservation biology are leading to the emergence of a *synthetic management approach* which is a foundation for managing the landscape to maximize the opportunities for insect and other biodiversity conservation. It is a conceptual way forward based on research findings available to date. The synthetic management approach is summarized in Box 11.3, with more details being given in Samways (2007a, b). Of note is that the approach is based on six principles and a central tenet (Box 11.4) The central tenet refers to a population level theme, that of the *metapopulation trio*, and involves a combined approach which maximizes patch size, maximizes patch quality, and reduces patch isolation (see Box 11.4). More research is now required to test this conceptual way forward.

Box 11.3 An integrated approach to insect conservation

The six principles of coarse-filter insect conservation at the landscape level, combined with a metapopulation 'golden thread', as well as an overlay of fine-filter, species conservation.

Six principles of insect conservation

For effective conservation at the landscape, coarse-filter level, there are six interrelated principles that are pivotal for success:

1. **Maintain natural reserves.** Reserves should be as large as possible and as many as possible. 'How large' depends on the organisms in question, and the availability of land. Important for specialist species.

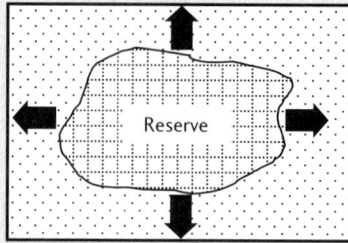

Maximise protected area

2. **Maintain quality habitat heterogeneity.** The aim is to maintain as much natural variety as possible at various spatial scales. Important for maintaining a diversity of opportunities for as many indigenous species as possible. This involves removal or suppression of invasive aliens.

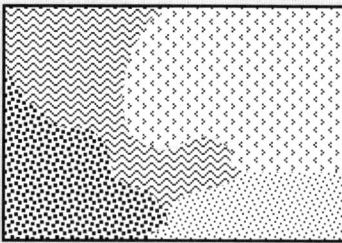

Four diverse habitats   BETTER THAN   Two habitats

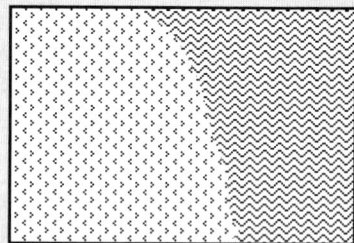

3. **Reduce contrast between disturbed areas and adjacent natural areas.** Softening hard, contrasting edges allows for more variety of microclimate and microhabitat conditions, especially for small ectothermic organisms like insects. Soft edges, i.e. gentle ecotones, also improve connectivity within the landscape.

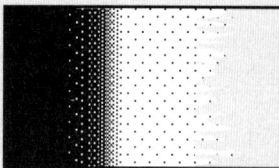

BETTER THAN

4. **Outside reserves, maintain as much undisturbed or minimally disturbed land as possible**. Land sparing or set-aside is effectively instigating conservation headlands, conservancies, agri-environment schemes, and other management

activities that provide undisturbed, or at least, less-disturbed areas. In this way, the area of occupancy and connectivity are improved, so improving chances of survival.

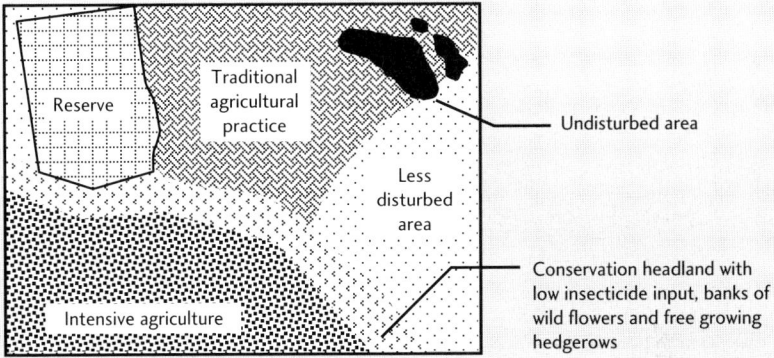

5. **Simulate natural disturbance.** Most landscapes experience some disturbance, whether abiotic, such as fire or flooding, or biotic, such as grazing impact from large herbivores. These disturbances may be patchy and at small spatial scales (e.g. trampling around waterholes) or large ones (e.g. extensive grassland fires). The aim is to simulate these natural conditions, with particular attention to extent, intensity and timing of management activity.

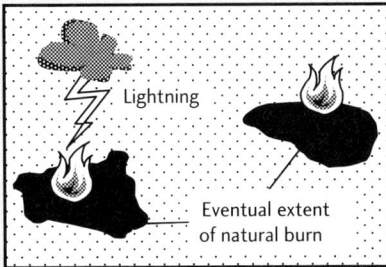

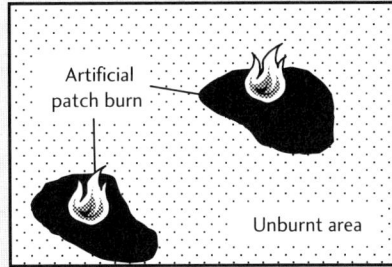

Natural disturbance   Management activity

6. **Link patches of quality habitat.** It is essential to maintain the ecological status quo and evolutionary potential of landscapes. This means allowing movement of organisms and maintenance of high population levels to buffer adverse conditions and promote genetic vigour. Full corridors may not always be feasible, and *stepping-stone habitats* become a second choice.

Patches of high quality habitat

Corridor or linkage, also of high quality habitat

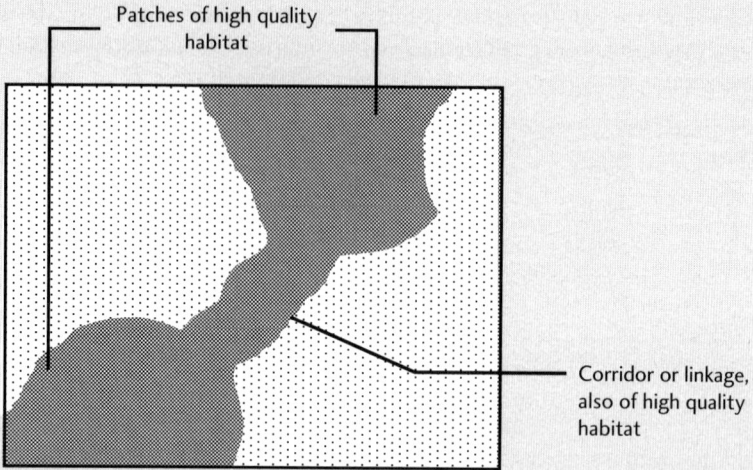

The central tenet

Maintain the metapopulation trio of large patch size, good patch quality, and reduced patch isolation.

The six principles above are a landscape level approach. At the population level there is the principle of maintaining the *metapopulation trio*. To maintain metapopulation viability, it is essential that a large population is maintained through good quality habitat and sufficient habitat size. These large, good-quality habitats also need to be close together to ensure movement back and forth between them. This population level approach is the essence of single-species, or fine-filter conservation.

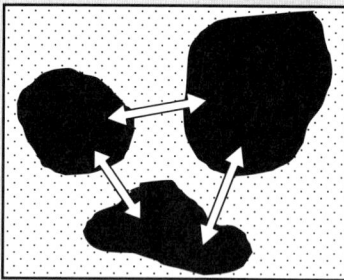

BETTER THAN

Large, high quality patches close together

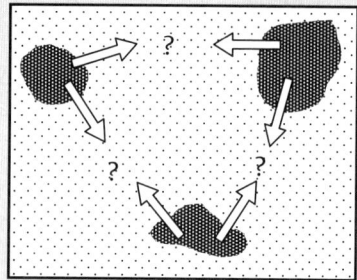

Small, low quality patches, far apart

Box 11.4 A conceptual model of ecosystem restoration triage

There is a gradient from ecosystems with intense-and-frequent, mostly anthropogenic, disturbance and low ecological integrity, to ecosystems with intense-and-very infrequent or mild-at-any-time natural events and high ecological integrity. Urbanization is at one end of this spectrum and the 'original' state at the other.

'Original' is in parentheses because (a) it depends how far back we go in time, and (b) because we can never be completely sure what all the original structural, compositional, and functional biodiversity was at any one time in the past. The lost 'original' state is the pristine state, which no longer exists, as anthropogenic impacts reach all parts of the world. Restoration here is a biocentric, deep-ecology view, where there is a genuine aim to bring back all aspects of ecological integrity.

The starting point in the decision process is whether to restore or not restore ecological integrity. It does not invoke decisions on whether to regreen, ecologically landscape, or rehabilitate. These three have various cultural, aesthetic and engineering components, and not just a biocentric one. There are two extremes of 'doing nothing': (a) where ecological integrity is irretrievably lost (e.g. a harbour for large ships), and (b) where ecological integrity is intact. Where ecological integrity is irretrievably lost, only regreening, rehabilitation, or ecological landscaping, but not restoration, is possible. The third prong of triage is the one where ecological restoration is restorable, and is the highest level of biocentricity.

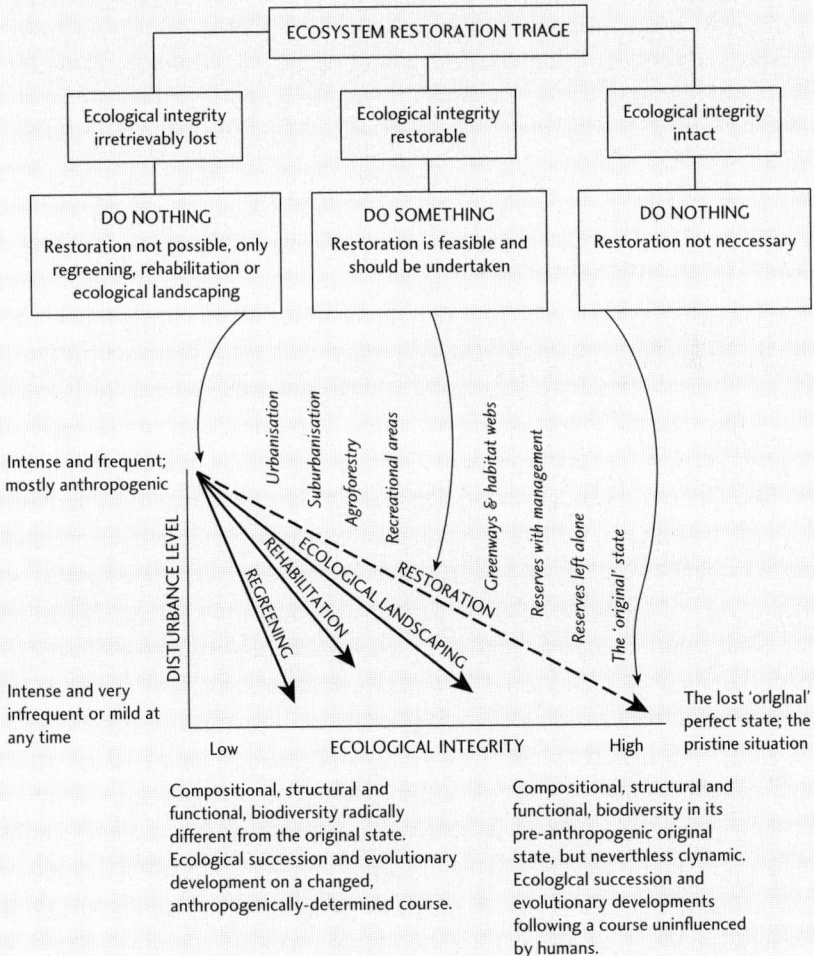

Compositional, structural and functional, biodiversity radically different from the original state. Ecological succession and evolutionary development on a changed, anthropogenically-determined course.

Compositional, structural and functional, biodiversity in its pre-anthropogenic original state, but neverthless clynamic. Ecologlcal succession and evolutionary developments following a course uninfluenced by humans.

Regreening is simply putting back a vegetation cover with more consideration for aesthetics and engineering value than for ecological integrity (e.g. grass cover of road cuttings). The maximal ecological integrity value for regreening is roughly at the level of recreational areas, with disturbance ranging from intense and frequent (e.g. mowing) to infrequent and mild. **Rehabilitation** aims to recover some ecological integrity but has a major aesthetic and/or human cultural components combined with ecological considerations (e.g. mine dump rehabilitation, removal of pollutants from a stream). Like regreening, the maximal ecological integrity value achievable through rehabilitation is low. This contrasts with **ecological landscaping** which deliberately aims to restore the historic; 'natural' ecosystem, which may be aesthetic (deliberately or inadvertently anthropocentric) or not (purely biocentric). Carefully planned planting of indigenous trees along roadsides is an example of ecological landscaping. Researched well, ecological landscaping can have great ecological integrity value, at least over time after indigenous biodiversity returns. Ecological landscaping is also of value to greenways, ecological networks and reserves with management. We are then finally left with restoration, which can normally only be done on minimally degraded ecosystems (hence the dashed line in the central area of the figure). **Restoration** aims for the historic 'original' state, but this is rarely actually achievable (because of, for example, invasive aliens) (hence the dashed line in the lower right of the figure) (from Samways 2000).

11.4 Triage and restoration

As Rolston (1994) has pointed out in his penetrating and thoughtful book, real natural value comes with wild nature. Restoration, he says, connotes the idea of putting something back. Yet we need to be clear on what we are putting back and what we are not, and that restoration is a forward-looking activity to rehabilitate for the future: 'We cannot go back in history and undo the undoing we humans once did'. Indeed, although restoration sounds noble, it is in effect a last-ditch approach to patch up natural systems, or at lease steer them into a trajectory where they then take over and develop their own character. However, such a restored system is never really the genuine article, unsullied by humans. It is much better, and indeed even much cheaper, to prevent ecosystem degradation and loss of biodiversity in the first place than to attempt to restore ecological integrity. Nevertheless, we must try to help nature recover arguably even if the financial cost is great. There is nothing uglier or sadder than a filthy and polluted river running through a town. Restoration is out of respect for ourselves as well as for nature.

Restoration is primarily the domain of botanically minded ecologists simply because plants form the main biotic architecture of the would-be restored area. Insects usually come along in tow, pulled by the attraction of the plants and the various resources supplied by them. As with all conservation-related activities, in restoration there needs to be a very clear idea of what is the goal, bearing in mind that restoration is a starting point rather than an end point. A useful conceptual starting point is to think in terms of *triage*, the principle of which is to decide which areas and to what extent an area is restorable, i.e. where to do and put effective restoration (Box 11.4, p. 362). On the one hand, an area may be so degraded that restoration to some semblance of the original ecosystem is not possible, although *regreening, rehabilitation*, or even *ecological landscaping* may be so (Box 11.4). On the other hand, the system may be in such basic good shape that if left to its own devices it may well recover close to its original character, as with abandonment of allotments in forest clearings. For other situations, and this is the tour de force of restoration, it is essential that the interventionist prong of triage is put in place. In other words, the aim is to kick-start the system; for example, reinstating the natural hydrological process by removing alien plants, and planting appropriate indigenous plants so as to give nature the framework it needs for then going on to heal itself from the ravages of human impact. Insect conservation, while generally being a passive partner in this activity, is nevertheless an important facet, as the insect fauna then becomes a good indication of whether and to what extent ecological integrity is being restored. Ecological monitoring (see below), using insects, and—not to be too entomologically chauvinistic—other biodiversity too, such as lichens, then becomes an important exercise for determining how well we are doing in letting nature regenerate itself. At times a helping hand may be required, such as the thinning of planted trees. This is in effect the adjustment of the trajectory of the restoration process.

The aim may not necessarily be to let ecosystem succession run its full course; for example, the intention might be to allow sunny areas to remain in afforested areas so as to provide a *mesofilter* for certain species which require sunflecks or sunny, sheltered locales. There may also be restoration in one area (e.g. letting forest re-establish) yet ecological landscaping (e.g. the building of a pond) in another, with the aim of creating as much quality habitat heterogeneity as possible, as mentioned as part of the synthetic management approach outlined in section 11.3. Whatever the goals, there is an enormous amount that insect ecologists and conservationists can do by way of engaging restoration activities.

11.5 Ecological monitoring

Ecological monitoring involves the 'systematic collection of ecological data in a standardized manner at regular intervals over time' (Spellerberg 2005). Importantly, monitoring is always motivated by a particular objective that has to do with detecting shifts in baseline conditions, identifying trends or assessing compliance with certain standards (Spellerberg 2005). It is a way of keeping an eye on the continuing state of the environment, of ascertaining the success of a management activity (as in section 11.3) and determining the success of a restoration activity (as in section 11.4).

Ecological monitoring has a poor reputation in some quarters because in the past it has been associated with significant long-term financial investment with little return. This unfortunate turn of events must not be used to deny the critical role that monitoring plays in our ability to understand and manage the effects of global environmental change on biodiversity. Rather, it should be used as a valuable lesson to ensure that monitoring programmes are designed well and every effort made to ensure that the data gathered will achieve the monitoring objective.

Ecological monitoring is a very necessary activity, for several reasons (Spellerberg 2005). First, many ecosystems and communities remain poorly understood and data gathered in the process of monitoring can add substantially to our knowledge of these systems. Long-term data are needed to understand how human-induced change is affecting insect diversity and the environment. Also, information gathered from monitoring can be used as an early warning to detect the types and sizes of change of significant concern and to catalyse political and management action to retard or prevent such change. Monitoring also provides information that can be used to better understand the implications of a changing environment for the things we care about, i.e. insect biodiversity and our quality of life.

Monitoring has a long tradition in entomology, with monitoring programmes used to develop forecast systems for insect pest populations in insect pest management. However, more recently monitoring has become necessary to understand (1) the impact of climate change on insect species distributions and phenologies, (2) the level and impacts of environmental pollutants, and (3) as bioindicators to assess the state of biodiversity and ecosystem functioning. Insects are also sometimes used in State of the Environment (SoE) reporting by nations that use such reports to monitor the state of environmental quality and of their natural resources. Here monitoring is often divided into the so-called *pressure–state–response framework*. This framework divides the monitoring objective into three components, namely (1) monitoring the driver of change, environmental stressor or 'pressure' (e.g. climate change or the distribution and abundance of an invasive species); (2) monitoring the state of the component of the environment to be protected (e.g. native insect biodiversity, the population size of a rare or threatened species); (3) monitoring society's response (such as the effectiveness of management interventions) (Spellerberg 2005).

In designing a monitoring programme, all the basic principles of designing a sampling programme apply (see Chapter 3). Again, for example, it is critical to

Table 11.2 *Seven characteristics of highly effective monitoring programmes (from Lovett et al. 2007).*

- Programme designed around clear and compelling scientific questions
- Review, feedback, and adaptation included in the design
- Measurements chosen carefully and with the future in mind
- Quality and consistency of the data maintained
- Long-term data accessibility and sample archiving planned for
- Monitoring data continually examined, interpreted, and presented
- Monitoring included within an integrated research programme

establish clear objectives and scientific hypotheses at the start of the process. The objective and hypotheses will direct the choice of biotic and abiotic variables as well as the frequency and duration of sampling events, and number, position, and extent of monitoring sites. Bioindicators are often used in monitoring programmes once they have been tested and shown to be robust indicators of the environmental stressor of component of biodiversity of interest.

Monitoring does, however, require a number of additional considerations to ensure its success (see Table 11.2). Perhaps the most important element of a monitoring programme is standardization of sample design and sampling methods for the duration of the programme (see Chapter 3). While standardization is important in any sampling programme, it becomes of critical importance in a long-term programme because of the enormous investment over long periods of time. A long time series of data can easily be devalued or rendered useless as a consequence of a change in sampling methodology (see Spellerberg 2005 for a comprehensive treatment of environmental monitoring). Importantly, because monitoring is inherently long term, plans have to be developed to ensure effective data management and knowledge transfer. The monitoring programme may well live to experience significant information and other technological advances, as well as turnover in principle investigators and personnel. As a result, the programme has to be designed to accommodate such change without compromising the quality and value of the programme.

An exemplar insect monitoring programme involves the four complementary schemes for assessing changes in butterfly biodiversity in the UK. This includes the combined use of information from (1) Red Data Books on butterfly species conservation status, (2) multiscale atlases and mapping schemes to monitor changes in species distributions, (3) transects that generate population time series data, and (4) occasional surveys that quantify population characteristics of selected species across their range (Thomas 2005). Other well-developed monitoring schemes for insects include those for freshwater macro-invertebrates (Dallas and Day 1993; Wright et al. 1993; Revenga et al. 2005; Thomas 2005).

11.6 Engaging citizen scientists in insect conservation

Not everyone loves insects, and most people do not see that they have any relevance in their lives, despite the fact that Losey and Vaughan (2006) have estimated that the net value of insect services in the USA alone is of the order of $57 billion dollars per year. This lack of appreciation, known as the *perception challenge*, is an area in which insect ecologists and conservationists can do much more (Lemelin 2007). Nevertheless, there are some societies which are superb ambassadors for the field, and of particular note is Butterfly Conservation based in the UK and the Xerces Society in the USA. Both produce magazines with glorious colour pictures which cannot fail to appeal—hopefully to everyone!

Some countries have a traditional connection with certain taxa, especially the Japanese culture with dragonflies (Primack et al. 2005). In the UK, there is a site guide for butterflies and dragonflies (Hill and Twist 1998). Suh and Samways (2001) give a procedure for developing an insect awareness trail. Important considerations are having various trail lengths to suit different people with different levels of motivation, with sensitivity to the needs of parents with small children, older people, and the physically challenged. Insects are, after all, one group of animals readily accessible to these three groups of people, and encouragement of their appreciation is something that insect conservationists can engage with to a much greater extent. There are other important issues with such trails, such as making identification boards and trail leaflets available. This information builds of course on good field guides. Stopping points to view the insects are also important, and the provision of seating in at least some of these areas is also a consideration. Some government-sponsored conservation agencies are also engaging the citizen scientist, even in very rural areas (Box 11.5).

Box 11.5 Community involvement in the conservation of threatened endemic invertebrate species

Threatened endemic invertebrate species can benefit greatly from human community involvement in their conservation. Indeed, without such cooperation from people living in areas where the species occur, some populations of these species may go extinct.

An example relates to the Karkloof blue butterfly *Orachrysops ariadne* colony near the Nkandla Forest Reserve in KwaZulu-Natal, South Africa. This butterfly is endemic to the province of KwaZulu-Natal and was known from only four sites in 2007. The number of eggs recorded during monitoring at the Nkandla colony site declined greatly from 250 in 2002 to 57 in 2003, and the decline in egg numbers continued until 2005 when only 11 eggs were recorded. The reasons for the decline in recorded egg numbers were the uncontrolled herbivory from domestic livestock and burning at the monitoring site.

In response to this, an open day was held with members of the Chube community, on whose land the Karkloof blue butterfly colony occurs, at the Nkandla Forest Reserve in November 2004. The Chube community members were informed by staff of Ezemvelo KZN Wildlife (EKZNW), the provincial conservation organization, of the important biodiversity asset on their land. Discussion at the meeting centred around the life history of the Karkloof blue butterfly, the egg-monitoring programme, and the impact of uncontrolled fires and herbivory on the oviposition and larval host plant (the lax variety of Wood's indigo plant *Indigofera woodii* var. *laxa*). The actions identified for the monitoring site by the participants at the meeting were firebreak establishment, fencing, and alien plant control. The Chube community members were then taken to the colony site to observe the monitoring site and identify the host plant. Afterwards, these Chube community members held discussions with others who did not attend the meeting to obtain consensus on the way forward. This led to further joint discussions with EKZNW to consolidate the outcomes of the discussions and to decide on co-management actions for the colony site.

A Karkloof blue butterfly task team was established for the co-management of the monitoring site in 2005. However, certain members of the Chube community were apprehensive about EKZNW's intentions. Their fears were allayed when EKZNW agreed to a memorandum of agreement in respect of the monitoring programme and the co-management of the site with the Chube community. After this, the fence line and firebreak were demarcated. The memorandum of agreement was signed in 2008. Prior to the management interventions, community members who were interested in learning about the monitoring technique were encouraged to participate. Community members and members from the

Members of the Chube community and Ezemvelo KZN Wildlife staff at the open day at Nkandla Nature Reserve in 2004 (left) and monitoring egg numbers of the Karkloof blue butterfly *Orachrysops ariadne* in 2008 (right).

was established around the monitoring site after the egg monitoring in 2006. In 2007, an uncontrolled wildfire burnt through the monitoring site after the egg monitoring but before the firebreak was established. In 2008, the firebreak was established in early May, prior to the annual egg monitoring in early June.

Fencing of the monitoring site and alien plant control measures commenced in 2007. Some members of the Chube community were trained in alien plant control and firebreak burning, and were employed to undertake these activities at the site with the assistance of EKZNW. The entire Karkloof blue butterfly colony (including the monitoring site) is now incorporated into EKZNW's Invasive Alien Species Programme. EKZNW's Community Conservation Division is actively involved with the Chube community. The Karkloof blue butterfly management guidelines brochure was translated into isiZulu, the language of the Chube community, and environmental awareness campaigns commenced with learners in the Nkandla area, focusing on key biodiversity assets in the area. The Department of Transport has agreed to change the alignment of the road running past the monitoring site to ensure that its widening for tarring does not affect the monitoring site. The number of Karkloof blue butterfly eggs recorded at the monitoring site between 2006 and 2008 has been more than double the number recorded in 2005. With the completion of the fencing in 2008, the egg numbers should increase further and through continued monitoring and careful management, the site should eventually attain the numbers recorded in 2002.

Sharon Louw and Adrian J. Armstrong, Ezemvelo
KZN Wildlife, South Africa

Connection with nature is important for our well-being (Rolston 1994), with the concern that there is currently an *extinction of experience*—a disconnect between us and nature (Miller 2005; Cheesman and Key 2007). A positive response has been an effort to consider *rescuing the extinction of experience* (Samways 2007c) through an ethical progression of valuing from the utilitarian through intrinsic to feeling and loving nature (Fox 1993; Stokes 2006).

This progression of ethical value is strongly linked to ensuring resources are available to future generations through children. Indeed, in a public survey in a botanical garden with a view to the development of a dragonfly observation trail, it was children (and older people) who were most interested in the insects, which were very tangible to them. Thus insects in an urban park can be part of appreciation of sustainability and a source of positive feelings and beneficial services (Chiesura 2004). Yet we have to be careful to draw a clear distinction between inculcating an appreciation of nature per se (which can nevertheless lead to the development of conservation awareness) and the targeted conservation of rare species and their habitats. Urban parks usually provide habitats only for the common and widespread generalist species, with the rare specialists often being restricted to the wilderness areas (Ferreira and Tidon 2005). Indeed, few Red Listed insect species occur in urban parks, although they may nevertheless be in urban areas and under threat from building developments. Thus, rescuing the extinction of experience may not necessarily be in step with the exercise of *complementarity*—the setting aside of reserve areas based on their irreplaceable biota. This echoes Usher's (1986) point that we must also consider conserving *typicalness* as well as rarity.

Rescuing the extinction of experience links well with butterfly gardening (or the wider concept of insect gardening), where plants are established to attract insect species into close proximity with people (Xerces Society 1998). This is essentially an experience exercise rather than a conservation exercise. However, when the garden is extensive, it provides close proximity to wild nature and, importantly, when food plants as well as nectar sources are present, then the exercise can have distinct conservation value (Botha and Botha 2006). Also, rare pollinators can be encouraged within urban areas through construction of the scarce resource of nest sites by the making of *condos*—apartment blocks for bees (Shepherd et al. 2003). Furthermore, insects that are reliably present will provide credible additions to the 'experience' of those using insect field guides (after entomologists have established which are the *core resident species*, i.e. those that can be virtually guaranteed to be seen at the right time of year and under the right conditions); (Niba and Samways 2006). Nevertheless, we must always remember one of the maxims of bird field guides: always give descriptions of the rare vagrants to add mystique to the twitching. While rescuing the extinction of experience may not necessarily be conservation at any one moment in time, the investment of introducing children to insects in the urban park or home garden will be an investment in prospective insect conservationists of the future. And, as we saw from the quantitative ecological findings mentioned above, diversity begets stability of ecological integrity and ecosystem health in the long term. In turn, and as Rolston (1994) points out, this valuing of nature provides goodness and a better world.

11.7 Summary

The primary drivers of global environmental change are habitat loss, invasive alien species, climate change, pollution, and over-harvesting. With such major changes currently occurring across so many landscapes, inevitably there is *habitat* loss, as well as habitat *fragmentation* and isolation of populations. These changes can lead to loss of critical resources, and *synergisms* with other adverse factors such as invasive alien organisms and climate change. Additive upon these impacts is the human manipulation of natural resources through use of biotechnology, *classical biological control*, and the harvesting of insect resources. Further impacts include those of contamination from pollution and from pesticides.

Insects are important components of ecosystems, with evidence accumulating that to maintain healthy ecosystems we need to ensure the continuance of *ecological integrity*, such that systems remain vigorous and *resilient* in the face of impacts. This means maintaining intact food webs. To do this requires a *sense of the past*, which gives us a moving backdrop against which we can measure and react to changes, particularly those involving adverse synergisms which risk leading to *discontinuities*. While species with high food web *connectance* are particularly important, every effort should be made to conserve as many populations, species, and thus interactions, as possible (the *precautionary principle*) as a buffer against adverse, perhaps irreversible, changes in the future.

A wide range of research findings is starting to enable us to take an informed *synthetic management approach* to conservation using a set of principles aimed at maintaining high connectance in food webs, hand in hand with sustainable ecological integrity, at both the *ecological* and *evolutionary time scales*. While it is much more ethically sensitive and practical to conserve landscapes in the first place, *restoration* can nevertheless be appropriate for the benefit of insects and other biodiversity. To undertake restoration, the goals need to be clearly established, using *triage* as an adjunct, and being cognizant of whether restoration is really being done, or whether it is more simply *re-greening*, *rehabilitation*, or *ecological landscaping*. All these efforts at management require a watchful eye to see that the state of the environment is indeed improving. To do this, it is important to *monitor* conditions, with insects being an important component in this process.

Scientists are not the sole players in insect conservation, and it is important to engage the wider human population, especially children, whose future is at stake. This is important, as there is an *extinction of experience* (a disconnect between us and nature) taking place. Combining good ethical values with good practical science (based in turn on good theoretical underpinning), will involve *rescuing the extinction of experience*, and will give us all more hope for a continuing, natural tapestry of life, with insects as major players.

Useful software in insect ecology and conservation

Software name, description, website	Freeware	Key references	Chapter reference
BIOTA The Biodiversity Database Manager http://viceroy.eeb.uconn.edu/biota	No		9
CANOCO Canonical community ordination http://www.microcomputerpower.com/	No	Leps and Smilauer (2003)	9.2–9.3
Ecosim Gotelli, N.J. and G.L. Entsminger. 2009. Null models software for ecology. Version 7. Acquired Intelligence Inc. & Kesey-Bear. Jericho, VT05465. http://www.garyentsminger.com/ecosim.htm	Yes	Gotelli and Entsminger (2004)	9.2
EstimateS Biodiversity estimation http://viceroy.eeb.uconn.edu/estimates	Yes	Colwell and Coddington (1994); Gotelli and Colwell (2001); Colwell et al. (2004)	9.4–9.6, ; 9.8
GPOWER General power analysis http://www.psycho.uni-duesseldorf.de/aap/projects/gpower/	Yes	Erdfelder et al. (1996)	3.3, Box 3.2
IndVal Identification of indicator species http://biodiversite.wallonie.be/outils/indval/home.html	Yes	Dufrêne and Legendre (1997)	9.2

PARTITION	Yes	Belkhir and Dawson (2001)	

Assigning individuals to populations based on genotypes at co-dominant marker loci
http://www.genetix.univ-montp2.fr/partition/partition.htm

PATN	No		9.2–9.3

Pattern analysis and multivariate statistics
http://www.patn.com.au/

PERMANOVA	Yes	Anderson (2001); Anderson et al. (2006); McArdle and Anderson (2001)	9.2–9.3

Permutation multivariate analysis of variance
http://www.stat.aukland.ac.nz/~mja/Programs.htm

PRIMER	No	Clark (1993)	9.2–9.3, 9.8

Plymouth routines in multivariate ecological research
http://www.zen87707.zen.co.uk/primer-e/primer6.htm

R-Package	Yes		3.3.5; Box 3.3

Complex multidimensional and spatial analysis procedures
http://www.bio.umontreal.ca/Casgrain/en/labo/R/

SADIE	Yes	Perry (1995)	3.3.5; Box 3.3

Spatial Analysis by Distance Indices
http://www.rothamsted.bbsrc.ac.uk/pie/sadie/SADIE_home_page_1.htm

SAM	Yes	Rangel et al. (2006)	3.3.5; Box 3.3

Statistical tools for spatial analysis
http://www.ecoevol.ufg.br/sam/

SpaceMaker	Yes	Borcard and Legendre (2002); Borcard et al. (2004)	3.3.5

Generates spatial descriptors for use in multiple regression or canonical ordination
http://www.bio.umontreal.ca/casgrain/en/labo/spacemaker.html

SPADE	Yes	Chao et al. (2005)	9.4–9.5; 9.8

Species richness prediction, similarity indices and diversity estimation
http://chao.stat.nthu.edu.tw/softwareCE.html

(Website addresses accurate as at 15 July 2009.)

References

Addison, P. and Samways, M. J. (2006). Surrogate habitats demonstrate the invasion potential of the African pugnacious ant. *Biodiversity and Conservation*, 15, 397–414.

Agosti, D., Majer, J. D., Alonso, L. E., and Schultz, T. R. (eds) (2000). *Ants: Standard Methods for Measuring and Monitoring Biodiversity*. Smithsonian Institution, Washington, DC.

Akcakaya, H. R., Butchart, S. H. M., Mace, G. M., Stuart, S., and Hilton-Taylor, C. (2006). Use and misuse of the IUCN Red List Criteria in projecting climate change impacts on biodiversity. *Global Change Biology*, 12, 2037–2043.

Alonso, L. E. (2000). Ants as indicators of diversity. In: Agosti, D., Majer, J. D., Alonso, L. E., and Schultz, T. R. (eds). *Ants: Standard Methods for Measuring and Monitoring Biodiversity*, pp. 80–88. Smithsonian Institution Press, Washington, DC.

Andersen, A. N. (1995a). Measuring more of biodiversity: genus richness as a surrogate for species richness in Australian ant faunas. *Biological Conservation*, 73, 39–43.

Andersen, A. N. (1995b). A classification of Australian ant communities based on functional groups which parallel plant life-forms in relation to stress and disturbance. *Journal of Biogeography*, 22, 15–29.

Andersen, A. N. (1999). My bioindicator or yours? Making the selection. *Journal of Insect Conservation*, 3, 61–64.

Andersen, A. N., Hoffmann, B. D., Müller, W. J., and Griffiths, A. D. (2002). Using ants as bioindicators in land management: simplifying assessment of ant community responses. *Journal of Applied Ecology*, 39, 8–17.

Anderson, M. J. (2001). A new method for non-parametric multivariate analysis of variance. *Austral Ecology*, 26, 32–46.

Anderson, M. J., Ellingsen, K. E. and Ardle, B. H. (2006). Multivariate dispersion as a measure of beta diversity. *Ecology Letters*, 9, 683–693.

Ando, N., Shimoyama I. and Kanzaki, R. (2002). A dual-channel FM transmitter for acquisition of flight muscle activities from the freely flying hawkmoth, *Agrius convolvuli*. *Journal of Neuroscience Methods*, 115, 181–187.

Andow, D. A. and Hilbeck, A. (2004). Science-based risk assessment for non-target effects of transgenic crops. *BioScience*, 54, 637–649.

Araujo, M. B., Humphries, C. J., Densham, P. J. *et al.* (2001). Would environmental diversity be a good surrogate for species diversity? *Ecography*, 24, 103–110.

Arnold, R. A. (1983). Ecological studies of six endangered butterflies (Lepidoptera: Lycaenidae): island biogeography, patch dynamics and the design of habitat preserves. *University of California Publications in Entomology*, 99, 1–161.

Asher, J., Warren, M., Fox, R., Harding, P., Jeffcoate, G., and Jeffcoate, S. (2001). *The Millenium Atlas of Butterflies in Britain and Ireland*. Oxford University Press, Oxford.

Baillie, J. E. M., Hilton-Taylor, C., and Stuart, S. N. (eds) (2004). *2004 IUCN Red List of Threatened Species. A Global Assessment.* IUCN, Gland, Switzerland.

Bale, J. S., Masters, G. J., Hodkinson, I. D. et al. (2002). Herbivory in global climate change research: direct effects of rising temperature on insect herbivores. *Global Change Biology*, 8, 1–16.

Balvanera, P., Kremen, C. and Martinez-Ramos, M. (2005). Applying community structure analysis to ecosystem function: examples from pollination and carbon storage. *Ecological Applications*, 15, 360–375.

Barnard, P. C. (ed.) (2000). *Identifying British Insects and Arachnids.* Cambridge University Press, Cambridge.

Bartlett, A. C. (1985). Guidelines for genetic diversity in laboratory colony establishment and maintenance. In: Singh, P. and Moore, R. F. (eds) *Handbook of Insect Rearing*, Vol. 1. pp. 7–17. Elsevier, Amsterdam.

Basset, Y., Mavoungou, J. F., Mikissa, J. B., Missa, O., Miller, S. E. Kitching, R. L., and Alonso, A. (2004). Discriminatory power of different arthropod data sets for the biological monitoring of anthropogenic disturbance in tropical forests. *Biodiversity and Conservation*, 13, 709–732.

Basset, Y., Novotny, V., Miller, S. E., and Pyle, R. (2000). Quantifying biodiversity: experience with parataxonomists and digital photography in Papua New Guinea and Guyana. *BioScience,* 50, 899–908.

Beaudoin-Ollivier, L., Bonaccorso, F., Aloysius, M., and Kasiki, M. (2003). Flight movement of *Scapanes australis australis* (Boisduval) (Coleoptera: Scarabaeidae: Dynastinae) in Papua New Guinea: a radiotelemetry study. *Australian Journal of Entomology*, 42, 367–372.

Begon, M. (1979). *Investigating Animal Abundance: Capture-recapture for Biologists.* Edward Arnold, London.

Belkhir, K. and Dawson, K. J. (2001). A Baysian approach to the identification of panmictic populations and the assignment of individuals. *Genetical Research*, 78, 59–77.

Bell, G. (2003). The interpretation of biological surveys. *Proceedings of the Royal Society of London, Series B*, 270, 2531–2542.

Betrus, C. J., Fleishmann, E. and Blair, R. B. (2005). Cross-taxonomic potential and spatial transferability of an umbrella species index. *Journal of Environmental Management*, 74, 79–87.

Bickel, T. O., Brühl, C. A., Gadau, J. R., Hölldobler, B., and Linsenmair, K. E. (2006). Influence of habitat fragmentation on the genetic variability in leaf litter ant populations in tropical rainforests of Sabah, Borneo. *Biodiversity and Conservation*, 15, 143–161.

Birch, L. C., Wheatley, R., Anyango, B. *et al.* (2004). Biodiversity and non-target impacts: a case study of Bt maize in Kenya. In: Hilbeck, A. and Andow, D. A. (eds) *Environmental Risk Assessment of Genetically Modified Organisms: A Case Study of Bt Maize in Kenya*, Vol. 1, pp. 117–186. CABI Publishing, Wallingford.

Birkinshaw, N. and Thomas, C. D. (1999). Torch-light transect surveys for moths. *Journal of Insect Conservation*, 3, 15–24.

Bisevac, L. and Majer, J. (2002). Cost effectiveness and data-yield of biodiversity surveys. *Journal of the Royal Society of Western Australia*, 85, 129–132.

Boivin, G., Kölliker-Ott, U. M., Bale, J. S., and Bigler, F. (2006). Assessing the establishment potential of inundative biological control agents. In: Bigler, F., Babendrier, D., and Kuhlmann, U. (eds) *Environmental Impacts of Invertebrates for Biological Control of Arthropods: Methods and Risk Assessment*, pp. 98–113. CABI, Wallingford.

Bonada, N., Prat, N., Resh, V. H., and Statzner, B. (2006). Developments in aquatic insect biomonitoring: A comparative analysis of recent approaches. *Annual Review of Entomology*, 51, 495–523.

Bonan, G. B. (2002). *Ecological Climatology. Concepts and Applications*. Cambridge University Press, Cambridge.

Bond, E. M. and Chase, J. M. (2002). Biodiversity and ecosystem functioning at local and regional scales. *Ecology Letters*, 5, 467–470.

Borcard, D. and Legendre, P. (2002). All-scale spatial analysis of ecological data by means of principal coordinates of neighbour matrices. *Ecological Modelling*, 153, 51–68.

Borcard, D., Legendre, P., Avois-Jacquet, C. and Tuomisto, H. (2004). Dissecting the spatial structure of ecological data at multiple scales. *Ecology*, 85, 1826–1832.

Borgelt, A. and New, T. R. (2006). Pitfall trapping for ants (Hymenoptera, Formicidae) in mesic Australia: what is the best trapping period? *Journal of Insect Conservation*, 10, 75–77.

Botha, C. and Botha, J. (2006). *Bring Butterflies Back to Your Garden*. Botanical Society of South Africa, Mayville.

Brakefield, P. M. (1991). Genetics and the conservation of invertebrates. In: Spellerberg, I. F., Goldsmith, F. B. and Morris, M. G. (eds) *The Scientific Management of Temperate Communities for Conservation*, pp. 45–79. Blackwell, Oxford.

Brakefield, P. M. and Saccheri, I. J. (1994). Guidelines in conservation genetics and the use of population cage experiments with butterflies to investigate the effects of genetic drift and inbreeding. In: Loeschcke,V., Tomiuk, J. and Jain, S. K.(eds) *Conservation Genetics,* pp. 165–179. Birkhauser Verlag, Basel.

Bried, J. T., Herman, B. D. and Ervin, G. N. (2007). Umbrella potential of plants and dragonflies for wetland conservation: a quantitative case study using the umbrella index. *Journal of Applied Ecology*, 44, 833–842.

Brook, B. W., O'Grady. J. J., Chapman, A. P., Burgman, M. A., Akcakaya, R., and Frankham, R. (2000). Predictive accuracy of population viability analysis in conservation biology. *Nature*, 404, 385–387.

Brooks, S. J. (1993). Guidelines for invertebrate site surveys. *British Wildlife*, 283–286.

Brooks, T. M., da Fonseca, G. A. B., and Rodrigues, A. S. L. (2004). Protected areas and species. *Conservation Biology*, 18, 616–618.

Brose, U. and Martinez, N. D. (2004). Estimating the richness of species with variable mobility. *Oikos*, 105, 292–300.

Brose, U., Martinez, N. D., and Williams, R. J. (2003). Estimating species richness: sensitivity to sample coverage and insensitivity to spatial patterns. *Ecology*, 84, 2364–2377.

Buda, V, Mäeorg, U., Karalius, V. *et al.* (1993). C18 dienes as attractants for eighteen clearwing (Sesiidae), tineid (Tineidae) and choreutid (Choreutidae) moth species. *Journal of Chemical Ecology*, 19, 799–813.

Butchart, S. H. M., Stattersfield, A. J., Bailie, J. *et al.* (2005). Using Red List Indices to measure progress toward the 2010 target and beyond. *Philosophical Transactions of the Royal Society of London, Series B*, 360, 255–268.

Butterfly Conservation (1995). Butterfly Conservation Lepidoptera restoration: policy, code of conduct, and guidelines for action. *Butterfly Conservation News*, 60, 20–21.

Campbell, J. W. and Hanula, J. L. (2007).Efficiency of Malaise and coloured pan traps for collecting floral visiting insects from three forested ecosystems. *Journal of Insect Conservation*, 11, 399–408.

Cappuccino, N. and Price, P. W. (eds) (1995).*Population Dynamics. New Approaches and Synthesis*. Academic Press, San Diego, CA.

Carter, M. R. and Gregorich, E. G. (2007). *Soil Sampling and Methods of Analysis*, 2nd edn. CRC Press, Boca Raton, FL.

Caughley, G. (1994). Directions in conservation biology. *Journal of Animal Ecology*, 63, 215–244.

CBD (2002). Cartagena Protocol on Biosafety: Status of ratification and entry into force. Convention on Biological Diversity. http://www.biodiv.org/biosafety/signinglist.asp

Chao, A. and Shen, T.-J. (2003–05). *Program SPADE (Species Prediction and Diversity Estimation). Program and Users Guide*. http://chao.stat.nthu.edu.tw.

Chao, A., Chazdon, R. L., Colwell, R. K. and Shen, T.-J. (2005). A new statistical approach for assessing similarity of species composition with incidence and abundance data. *Ecology Letters*, 8, 148–159.

Cheesman, O. D. and Key, R. S. (2007). The extinction of experience: a threat to insect conservation? In: Stewart, A. J. A., New, T. R., and Lewis, O. T. (eds) *Insect Conservation Biology*, pp. 322–348. CABI, Wallingford.

Chesmore, E. D. (2004). Automatic bioacoustic identification of species. *Anais da Academia Brasiliera de Ciências*, 76, 435–440.

Chiesura, A. (2004). The role of urban parks for the sustainable city. *Landscape and Urban Planning*, 68, 129–138.

Chown, S. L. and Nicolson, S. W. (2004). *Insect Physiological Ecology*. Oxford University Press, Oxford.

Chown, S. L., S. Slabber, M. A. McGeoch, C. Janion, and H. P. Leinaas. 2007. Phenotypic plasticity mediates climate change responses among invasive and indigenous arthropods. Proceedings of the Royal Society B 274:2531-2537.

Cilgi, T. and Jepson, P. (1995). Pesticide spray drift into field boundaries and hedgerows: toxicity to non-target Lepidoptera. *Journal of Environment and Pollution*, 87, 1–9.

Clark, T. E. and Samways, M. J. (1997). Sampling arthropod diversity for urban ecological landscaping in a species-rich southern hemisphere botanic garden. *Journal of Insect Conservation*, 1, 221–234.

Clarke, G. M. (2000). Inferring demography from genetics: a case study of the endangered golden sun moth, *Synemon plana*. In Young, A. G. and Clarke, G. M. (eds) (2000). *Genetics, Demography and Viability of Fragmented Populations*, pp. 313–225. Cambridge University Press, Cambridge.

Clarke, G. M. and O'Dwyer, C. (2000). Genetic variability and population structure of the endangered golden sun moth, *Synemon plana*. *Biological Conservation*, 92, 371–381.

Clarke, K. R. and Warwick, R. M. (1998). A taxonomic distinctness index and its statistical properties. *Journal of Applied Ecology*, 35, 523–531.

Colwell, R. K. (2004). *EstimateS: Statistical estimation of Species Richness and Shared Species from Samples. Page User's Guide and Application*. http://purl.oclc.org/estimates.

Colwell, R. K. and Coddington, J. A. (1994). Estimating terrestrial biodiversity through extrapolation. *Philosophical Transactions of the Royal Society of London, Series B*, 345, 101–118.

Colwell, R. K., Mao, C. X., and Chang, J. (2004). Interpolating, extrapolating, and comparing incidence-based species accumulation curves. *Ecology*, 85, 2717–2727.

Cooke, S. J., Hinch, S. G., Wikelski, M. *et al.* (2004). Biotelemetry: a mechanistic approach to ecology. *Trends in Ecology and Evolution*, 19, 334–339.

Coope, G. R. (1995). Insect faunas in ice age environments: why so little extinction? In: Lawton, J. H. and May, R. M. (eds) *Extinction Rates*, pp. 55–74. Oxford University Press, Oxford.

Cooter, J. and Barclay, M. V. L. (eds) (2006). *A Coleopterist's Handbook*, 4th edn. Amateur Entomologists' Society, Orpington.

Corbet, P. S. (1999). *Dragonflies: Behaviour and Ecology of Odonata*. Harley Books, Colchester.

Costanza, R. and Mageau, M. (2000). What is a healthy ecosystem? *Aquatic Ecology*, 33, 105–115.

Crandall, K. A., Bininda-Emonds, O. R. P., Mace, G. M., and Wayne, R. K. (2000). Considering evolutionary processes in conservation biology. *Trends in Ecology and Evolution*, 15, 290–295.

Cranston, P. and Hillman, T. (1992). Rapid assessment of biodiversity using 'Biological Diversity Technicians'. *Australian Biologist,* 5, 144–154.

Cranston, P. S. and Trueman, J. W. H. (1997). Indicator taxa in invertebrate biodiversity assessment. *Memoirs of the Museum of Victoria*, 56, 267–274.

Crawley, M. J. (1993). *GLIM for Ecologists*. Blackwell Science, Oxford.

Crist, T. O., Veech, J. A., Gering, J. C., and Summerville, K. S. (2003). Partitioning species diversity across landscapes and regions: a hierarchical analysis of alpha, beta, and gamma diversity. *American Naturalist*, 162, 734–743.

Dallas, H. F. and Day, J. A. (1993). *The Effect of Water Quality Variables on Riverine Ecosystems: a Review*. Technical Report Series No. TT 61/93. Water Research Commission, South Africa.

Danks, H. V. (1996). *How to Assess Insect Biodiversity Without Wasting Your Time*. Biological Surveys of Canada (Terrestrial Arthropods), Document Series No. 5, Ottawa.

Danks, H. V., Wiggins, G. B., and Rosenberg, D. M. (1987). Ecological collections and long-term monitoring. *Bulletin of the Entomological Society of Canada*, 19, 16–18.

Davey, J. T. (1956). A method of marking isolated adult locusts in large numbers as an aid to the study of their seasonal migrations. *Bulletin of Entomological Research*, 46, 797–802.

Davies, Z. G., Wilson, R. J., Coles, S. and Thomas, C. D. (2006). Changing habitat associations of a thermally constrained species, the silver-spotted skipper butterfly, in response to climate warming. *Journal of Applied Ecology*, 75, 247–256.

Davis, J. C. 2002. *Statistics and Data Analysis in Geology*, 3rd edn. Wiley, New York.

Debuse, V. J., King, J. and House, A. P. N. (2007). Effect of fragmentation, habitat loss and within-patch characteristics on ant assemblages in semi-arid Australia. *Landscape Ecology*, 22, 731–745.

Dempster, J. P. (1989). Insect introductions: natural dispersal and population persistence in insects. *The Entomologist*, 108, 5–13.

Dennis, R. L. H., Shreeve, T. G., Isaac, N. J. B. *et al.* (2006). The effects of visual apparency on bias in butterfly recording and monitoring. *Biological Conservation*, 128, 486–492.

Dennis, R. L. H., Shreeve, T. G., and Sheppard, D. A. (2007). Species conservation and landscape management: a habitat perspective. In: Stewart, A. J. A., New, T. R. and Lewis, O. T. (eds) *Insect Conservation Biology*, pp.92–126. CABI, Wallingford.

DeWeerdt, S. (2002). What really is an evolutionarily significant unit? The debate over integrating genetics and ecology in conservation biology. *Conservation Biology in Practice*, 3, 10–17.

Dial, R. and Tobin, S. C. (1994).Description of arborist methods for forest canopy access and movement. *Selbyana*, 15, 24–37.

Diamond, J. (1986). Overview: laboratory experiments, field experiments and natural experiments. In: Diamond, J. and Case, T. J. (eds) *Community Ecology*, pp. 3–22. Harper & Row, New York.

Diaz, S. and Cabido, M. (2001). Vive la difference: plant functional diversity matters to ecosystem processes. *Trends in Ecology and Evolution*, 16, 646–655.

Diaz, S., Fargione, J., Stuart Chapin III, F., and Tilman, D. (2006). Biodiversity loss threatens human wellbeing. *PLoS Biology*, 4, 1300–1305.

Dijkstra, K.-D., Samways, M. J., and Simaika, J. P. (2007). Two new relict *Syncordulia* species found during museum and field studies of threatened dragonflies in the Cape Floristic Region (Odonata: Corduliidae). *Zootaxa*, 1467, 19–34.

Disney, R. H. L. (1986). Assessments using invertebrates, posing the problem. In Usher, M. B. (ed.) *Wildlife Conservation Evaluation*, pp. 271–193. Chapman & Hall, London.

Diwakar, S., Jain, M., and Balakrishnan, R. (2007).Psychoacoustic sampling as a reliable, non-invasive method to monitor orthopteran species diversity in tropical forests. *Biodiversity and Conservation*, 16, 4081–4093.

Dobson, A. (2005). Monitoring global rates of biodiversity change: challenges that arise in meeting the Convention on Biological Diversity (CBD) 2010 goals. *Philosophical Transactions of the Royal Society of London, Series B*, 360, 229–241.

Dover, J. W. (1991). The conservation of insects on arable farmland. In: Collins, N. M. and Thomas, J. A. (eds) *The Conservation of Insects and their Habitats*, pp. 293–318. Academic Press, London.

Dover, J. W. and Rowlinson, B. (2005). The western jewel butterfly (*Hypochrysops halyaetus*): factors affecting adult butterfly distribution within native *Banksia* bushland in an urban setting. *Biological Conservation* 122, 599–609.

Duelli, P. and Obrist, M. K. (2003). Biodiversity indicators: the choice of values and measures. *Agriculture, Ecosystems and Environment*, 98, 87–98.

Duffey, E. (1977). The re-establishment of the large copper butterfly *Lycaena dispar batava* Obthr. at Woodwalton Fen National Nature Reserve, Cambridgeshire, England, 1969–1973. *Biological Conservation*, 12, 145–158.

Duffy, J. E. (2003). Biodiversity loss, trophic skew and ecosystem functioning. *Ecology Letters*, 6, 680–687.

Dufrêne, M. and Legendre, P. (1997). Species assemblages and indicator species: the need for a flexible asymmetrical approach. *Ecological Monographs*, 67, 345–366.

Dunne, A., Saleska, S. R., Fischer, M. L. and Harte, J. (2004). Integrating experimental and gradient methods in ecological climate change research. *Ecology*, 85, 904–917.

Dunne, J. A., Williams, R. J. and Martinez, N. D. (2002). Network structure and biodiversity loss in food webs: robustness increases with connectance. *Ecology Letters*, 5, 558–567.

Ehrlich, P. R. and Davidson, S. E. (1960). Techniques for capture-recapture studies of Lepidoptera populations. *Journal of the Lepidopterists' Society*, 14, 227–229.

Ehrlich, P. R., Hanski, I. and Boggs, C. L. (2004). What have we learned? In: Ehrlich, P. R. and Hanski, I. (eds.) *On the Wings of Checkerspots. A Model System for Population Biology*, pp. 288–300. Oxford University Press, Oxford.

Engen, S., Lande, R., Walla, T. and DeVries, P. J. (2002). Analyzing spatial structure of communities using the two-dimensional Poisson lognormal species abundance model. *American Naturalist*, 160, 60–73.

Erdfelder, E., Faul, F. and Buchner, A. (1996). GPOWER: a general power analysis program. *Behavior Research Methods, Instruments and Computers*, 28, 1–11.

Erwin, T. L. (1982).Tropical forests: their richness in Coleoptera and other Arthropod species. *Coleopterists' Bulletin*, 36, 74–75.

Fahrig, L. (2003). Effects of habitat fragmentation on biodiversity. *Annual Review of Ecology and Systematics*, 34, 487–515.

Faith, D. P. (1992). Conservation evaluation and phylogenetic diversity. *Biological Conservation*, 61, 1–10.

Faith, D. P. (1994). Phylogenetic pattern and the quantification of organismal diversity. *Philosophical Transactions of the Royal Society of London, Series B*, 345, 45–58.

Farina, A. (2000). *Principles and Methods in Landscape Ecology*. Kluwer, Dordrecht.

Ferreira, L. B. and Tidon, R. (2005). Colonizing potential of Drosophilidae (Insecta, Diptera) in environments with different grades of urbanization. *Biodiversity and Conservation*, 14, 1809–1821.

Finch, J. M., Samways, M. J., Hill, T. R., Piper, S. E., and Taylor, S. (2006). Application of predictive distribution modelling to invertebrates: Odonata in South Africa. *Biodiversity and Conservaion*, 15, 4239–4251.

Fleishman, E., Blair, R. B., and Murphy, D. D. (2001). Empirical validation of a method for umbrella species selection. *Ecological Applications*, 11, 1489–1501.

Fleishman, E., Murphy, D. D., and Brussard, P. F. (2000). A new method for selection of umbrella species for conservation planning. *Ecological Applications*, 10, 569–579.

Fleishman, E., Thomson, J. R., Mac Nally, R., Murphy, D. D., and Fay, J. P. (2005). Using indicator species to predict species richness of multiple taxonomic groups. *Conservation Biology*, 19, 1125–1137.

Fontaine, C, Dajoz, I., Meriguet, J., and Loreau, M. (2006). Functional diversity of plant-pollinator interaction webs enhances the persistence of plant communities. *PLoS Biology*, 4, 129–135.

Forman, R. T. T. (1995). *Land Mosaics*. Cambridge University Press, Cambridge.

Fortin, M.-J. and Dale, M. R. T. (2005). *Spatial Analysis: A Guide for Ecologists*. Cambridge University Press, Cambridge.

Foster, S. E. and Soluk, D. A. (2006). Protecting more than the wetland: the importance of biased sex ratios and habitat segregation for conservation of Hine's emerald dragonfly. *Biological Conservation*, 127, 158–166.

Fox, R., Conrad, K. F., Parsons, M. S., Warren, M. S. and Woiwod, I. P. (2006). *The State of Britain's Larger Moths*. Butterfly Conservation and Rothamsted Research, Wareham.

Fox, W. (1993). Why care about the world around us? *Resurgence*, 161, 10–12.

Frankham, R., Ballou, J. D., and Briscoe, D. A. (2002). *Introduction to Conservation Genetics*. Cambridge University Press, Cambridge.

Fraser, S. E. M., Dytham, C., and Mayhew, P. J. (2008). The effectiveness and optimal use of Malaise traps for monitoring parasitoid wasps. *Insect Conservation and Diversity*, 1, 22–31.

Freckleton, R. P. (2000). Phylogenetic tests of ecological and evolutionary hypotheses: checking for the phylogenetic independence. *Functional Ecology*, 14, 129–134.

Freitag, H. (2004). Adaptations of an emergence trap for use in tropical streams. *International Review of Hydrobiology*, 89, 363–374.

Fry, G. (1994). The role of field margins in the landscape. *British Crop Protection Memoirs*, 58, 31–40.

Fry, G., Robson, W. and Banham, A. (1992). *Corridors and Barriers to Butterfly Movement in Contrasting Landscapes*. NINA Research Report, Norwegian Institute for Nature Preservation, Trondheim.

Fry, R. and Waring, P. (1996). *A Guide to Moth Traps and their Use*. Volume 24, Amateur Entomologists' Society Publications, Manningtree.

Gall, L. F. (1984). The effects of capturing and marking on subsequent activity in *Boloria acrocnema* (Lepidoptera: Nymphalidae), with a comparison of different numerical models that estimate population size. *Biological Conservation*, 28, 139–154.

Gardiner, T., Hill, J., and Chesmore, D. (2005). Review of the methods frequently used to estimate the abundance of Orthoptera in grassland ecosystems. *Journal of Insect Conservation*, 9, 151–173.

Gaston, K. J. (1994). *Rarity*. Chapman & Hall, London.

Gaston, K. J. (1996). *Biodiversity: A Biology of Numbers and Difference*. Blackwell, Oxford.

Gaston, K. J. (2003). *The Structure and Dynamics of Geographic Ranges*. Oxford University Press, Oxford.

Gaston, K. J. and Blackburn, T. M. (1996). Range size-body size relationships: evidence of scale dependence. *Oikos*, 75, 479–485.

Gaston, K. J. and Blackburn, T. M. (2000). *Pattern and Process in Macroecology*. Blackwell Science, Oxford.

Gaston, K. J. and Spicer, J. I. (1998). *Biodiversity: an Introduction*. Blackwell, Oxford.

Gerber, A. and Gabriel, M. J. M. (2002). *Aquatic Invertebrates of South African Rivers: Field Guide*. Institute for Water Quality Studies, Department of Water Affairs and Forestry, Pretoria.

Gergel, S. E. and Turner, M. G. (2002). *Learning Landscape Ecology: A Practical Guide to Concepts and Techniques*. Springer, New York.

Gering, J. C., Crist, T. O. and Veech, J. A. (2003). Additive partitioning of species diversity across multiple spatial scales: implications for regional conservation of diversity. *Conservation Biology*, 17, 488–499.

Gibson, L and New, T. R. (2007). Problems in studying populations of the golden sun-moth, *Synemon plana* (Lepidoptera: Castniidae), in south eastern Australia. *Journal of Insect Conservation*, 11, 309–313.

Gillman, M. P. and Hails, R. S. (1997). *An Introduction to Ecological Modelling: Putting Practice into Theory*. Blackwell Science, Oxford.

Goldstein, P. Z. (2004). Systematic collection data in North American invertebrate conservation and monitoring programmes. *Journal of Applied Ecology*, 41, 175–180.

Gotelli, N. J. (2000). Null model analysis of species co-occurrence patterns. *Ecology*, 81, 2606–2621.

Gotelli, N. J., and Colwell, R. K. (2001). Quantifying biodiversity: procedures and pitfalls in the measurement and comparison of species richness. *Ecology Letters*, 4, 379–391.

Gotelli, N. J. and Entsminger, G. L. (2001). *Ecosim: Null Models Software for Ecology*. Acquired Intelligence Inc. and Kessey Bear, Jericho, VT. http://homepages.together.net/~gentsmin/ecosim.htm

Graham, C. H., Moritz, C., and Williams, S. E. (2006). Habitat history improves prediction of biodiversity in rainforest fauna. *Proceedings of the National Academy of Sciences of the USA*, 103, 632–636.

Clarke, G. and Spier, F. (2003). *A Review of the Conservation Status of Selected Non-marine Invertebrates*. Environment Australia/Natural Heritage Trust, Canberra, ACT.

Green, R. H. (1979). *Sampling Design and Statistical Methods for Environmental Biologists*. Wiley, Chichester.

Greenslade, P. (2007). The potential of Collembola to act as indicators of landscape stress in Australia. *Australian Journal of Experimental Agriculture*, 47, 424–434.

Greenwood, S. R. (1987). The role of insects in tropical forest food webs. *Ambio*, 16, 267–270.

Griffiths, G. J. K., Alexander, C. J., Birt, A. *et al.* (2005). A method for rapidly mass laser-marking individually coded ground beetles (Coleoptera: Carabidae) in the field. *Ecological Entomology*, 30, 391–396.

Grimaldi, D. and Engel, M. S. (2005). *Evolution of the Insects*. Cambridge University Press, Cambridge.

Gurr, G. M., Wratten, S. D. and Altieri, M. A. (eds) (2004). *Ecological Engineering for Pest Management: Advances in Habitat Manipulation for Arthropods*. CSIRO, Collingwood, VIC.

Hagler, J. R. and Jackson, C. G. (2001). Methods for marking insects: current techniques and future prospects. *Annual Review of Entomology*, 46, 511–543.

Haider, S. and Jax, K. (2007). The application of environmental ethics in biological conservation: a case study from the southernmost tip of the Americas. *Biodiversity and Conservation*, 16, 2559–2573.

Hammond, P. M. (1994). Practical approaches to the estimation of the extent of biodiversity in speciose groups. *Philosophical Transactions of the Royal Society of London, Series B*, 345, 119–136.

Hanski, I. (1999). *Metapopulation Ecology*. Oxford University Press, Oxford.

Hanski, I. and Pöyry, J. (2007). Insect populations in fragmented habitats. In: Stewart, A.J.A., New, T. R., and Lewis, O. T. (eds) *Insect Conservation Biology*, pp. 175–202. CABI, Wallingford.

Hardersen, S. (2007).Telemetry of Anisoptera after emergence—first results (Odonata). *International Journal of Odonatology*, 10, 189–202.

Harding, P. T., Asher, J., and Yates, T. J. (1995). Butterfly monitoring. 1—recording the changes. In: Pullin, A. S. (ed.) *Ecology and Conservation of Butterflies,* Chapman & Hall, London.

Harrison, S. and Hastings, A. (1996). Genetic and evolutionary consequences of metapopulation structure. *Trends in Ecology and Evolution*, 11, 180–183.

Harvey, P. H. and Pagel, M. D. (1991). *The Comparative Method in Evolutionary Biology*. Oxford University Press, Oxford.

Hassall, C., Thompson, D. J., French, G. C., and Harvey, I. F. (2007). Historical changes in the phenology of British Odonata are related to climate. *Global Change Biology*, 13, 1–9.

Hauer, F. R. and Lamberti, G. A. (eds) (2007). *Methods in Stream Ecology*. Academic Press, London.

Hawking, J. H. and New, T. R. (1999). The distribution patterns of dragonflies (Insecta: Odonata) along the Kiewa River, Australia, and their relevance in conservation assessment. *Hydrobiologia*, 392, 249–260.

Hebert, P. D. N., Penton, E. H., Burns, J. M., Janzen, D. H., and Hallwachs, W. (2004). Ten species in one: DNA barcoding reveals cryptic species in the neotropical skipper butterfly *Astraptes fulgerator*. *Proceedings of the National Academy of Sciences of the USA,* 101, 14812–14817.

Hedin, J. and Ranius, T. (2002). Using radio telemetry to study dispersal of the beetle *Osmoderma eremita*, an inhabitant of tree hollows. *Computers and Electronics in Agriculture*, 35, 171–180.

Hellawell, J. M. (1991). Development of a rationale for monitoring. In: Goldsmith, B. (ed.) *Monitoring for Conservation and Ecology*. pp. 1–22. Chapman & Hall, London.

Hellmann, J. J. and Fowler, G. W. (1999). Bias, precision, and accuracy of four measures of species richness. *Ecological Applications*, 9, 824–834.

Hess, G. R. and Fischer, R. A. (2001). Communicating clearly about corridors. *Landscape and Urban Planning*, 55, 195–208.

Hewitt, G. (2003). Ice ages, species distributions, and evolution. In: Rothschild, L. J. and Lister, A. M. (eds) *Evolution on Planet Earth*, pp. 339–361 Academic Press, Amsterdam.

Heywood, V. H. (1994). The measurement of biodiversity and the politics of implementation. In: Forey, P. L., Humphries, C. J., and Vane-Wright, R. I. (eds) *Systematics and Conservation Evaluation*, pp. 15–22. Clarendon Press, Oxford.

Hijmans, R. J. and Graham, C. H. (2006). The ability of climate envelope models to predict the effect of climate change on species distributions. *Global Change Biology*, 12, 2272–2281.

Hilbeck, A. andAndow, D. A. (2004). *Environmental Risk Assessment of Genetically Modified Organisms: A case study of Bt maize in Kenya*. CABI Publishing, Wallingford.

Hilbeck, A., Andow, D. A., and Fontes, E. M. G. (2006). *Methodologies for Assessing BT Cotton in Brazil*. CABI Publishing, Wallingford.

Hill, J. K., Thomas, C. D., Fox, R. et al. (2002). Responses of butterflies to twentieth century climate warming: implication for future ranges. *Proceedings of the Royal Society of London, Series B*, 269, 2163–2171.

Hill, P. and Twist, C. (1998). *Butterflies and Dragonflies: A Site Guide,* 2nd edn. Arlequin, Chelmsford.

Hill, R. and Sendashonga, C. (2006). Conservation biology, genetically modified organisms, and the Biosafety Protocol. *Conservation Biology*, 20, 1620–1625.

Hilty, J. and Merenlender, A. (2000). Faunal indicator taxa selection for monitoring ecosystem health. *Biological Conservation*, 92, 185–197.

Hilty, J. A., Lidicker, W. Z. Jr, Merenlender, A. M. (2006). *Corridor Ecology: The Science and Practice of Linking Landscapes for Biodiversity Conservation*. Island Press, Washington, DC.

Hirao, T., Murakami, M. and Kashizaki, A. (2008). Effects of mobility on daily attraction to light traps: comparison between lepidopteran and coleopteran communities. *Insect Conservation and Diversity*, 1, 32–39.

Holloway, G. J., Griffiths, G. H., and Richardson, P. (2003). Conservation strategy maps: a tool to facilitate biodiversity action planning illustrated using the heath fritillary butterfly. *Journal of Applied Ecology*, 40, 413–421.

Honan, P. (2008). Notes on the biology, captive management and conservation status of the Lord Howe Island stick insect (*Dryococelus australis*) (Phasmatodea). *Journal of Insect Conservation*, 12, 399–413.

Hooper, D. U. and Vitousek, P. M. (1998). Effects of plant composition and diversity on nutrient cycling. *Ecological Monographs*, 68, 121–149.

Hopkins, J. J., Allison, H. M., Walmsley, C. A., Gaywood, M., and Thurgate, G. (2007). *Conserving Biodiversity in a Changing Climate: Guidance on a Building Capacity to Adapt*. Department for Environment, Food and Rural Affairs, London.

Horne, P. A. (2007). Carabids as potential indicators of sustainable farming systems. *Australian Journal of Experimental Agriculture*, 47, 455–459.

Hortal, J., Borges, P. A. V., and Gaspar, C. (2006). Evaluating the performance of species richness estimators: sensitivity to sample grain size. *Journal of Animal Ecology*, 75, 274–287.

Howarth, F. G. (1991). Environmental impacts of classical biological control. *Annual Reviews of Entomology*, 36, 485–509.

Hudson, P. J., Dobson, A. P., and Lafferty, K. D. (2006). Is a healthy ecosystem one that is rich in parasites? *Trends in Ecology and Evolution,* 21, 381–385.

Hughes, L. (2000). Biological consequences of global warming: is the signal already apparent? *Trends in Ecology and Evolution*, 15, 56–61.

Human, K. G., Weiss, S., Weiss, A., Sandler, B., and Gordon, D. M. (1998). Effects of abiotic factors on the distribution and activity of the invasive Argentine ant (Hymenoptera : Formicidae). *Environmental Entomology*, 27, 822–833.

Hung, Y. P. and Prestwich, K. N. (2004). Is significant acoustic energy found in the audible and ultrasonic harmonics in cricket calling songs. *Journal of Orthoptera Research*, 13, 231–238.

Hunter, M. L. Jr (1996). *Fundamentals of Conservation Biology.* Blackwell Science, Cambridge, MA.

Hunter, M. L. Jr (2005). A mesofilter complement to coarse and fine filters. *Conservation Biology*, 19, 1025–1029.

Hurlbert, S. H. (1984). Pseudoreplication and the design of ecological field experiments. *Ecological Monographs*, 54, 187–211.

Hurlbert, S. H. (1990). Spatial distribution of the montane unicorn. *Oikos*, 58, 257–271.

Ingham, D. S. and Samways, M. J. (1996). Application of fragmentation and variegation models to epigaeic invertebrates in South Africa. *Conservation Biology*, 10, 1353–1358.

IUCN (1995). *IUCN/SSC Guidelines for Re-introductions*. International Union for the Conservation of Nature, Gland.

IUCN (2001). *IUCN Red List Categories and Criteria, Version 3.1*. International Union for the Conservation of Nature, Gland.

IUCN (2003). *Guidelines for Application of IUCN Red List Criteria at Regional Levels, Version 3.0*. International Union for the Conservation of Nature, Gland.

Ives, A. R. (2007). Diversity and stability in ecological communties. In: May, R. M. and McLean, A. (eds). *Theoretical Ecology: Principles and Applications*, 3rd edn. Oxford University Press, Oxford.

Janzen, D. H. (1997). Wildland biodiversity management in the tropics. In: Reaka-Kudla, M. L., Wilson, D. E., and Wilson, E. O. (eds). *Biodiversity II*, pp. 411–431. Joseph Henry Press, Washington, DC.

Janzen, D. H., Hajibabaei, M., Burns, J. M., Hallwachs, W., Remigio, E. and Hebert, P. D. N. (2005). Wedding biodiversity inventory of a large and complex Lepidoptera fauna with DNA barcoding. *Philosophical Transactions of the Royal Society of London, Series B*, 360, 1835–1845.

JCCBI (1986). Joint Committee for the Conservation of British Insects. Insect re-establishment—a code of conservation practice. *Antenna*, 10, 13–18.

Jetz, W., Sekercioglu, C. H., and Watson, J. E. M. (2008). Ecological correlates and conservation implications of overestimating species geographic ranges. *Conservation Biology*, 22, 110–119.

Johansson, T., Gibb, H., Hilszański, J. *et al.* (2006). Conservation-oriented manipulations of coarse woody debris affect its value as habitat for spruce-infesting bark and ambrosia beetles (Coleoptera: Scolytidae) in northern Sweden. *Canadian Journal of Forestry Research*, 36, 174–185.

Johnson, N. F. (2007). Biodiversity informatics. *Annual Review of Entomology*, 52, 421–438.

Jones, D. T., Verkerk, R. H. J., and Eggleton, P. (2005). Methods for sampling termites. In: Leather, S. (ed.) *Insect Sampling in Forest Ecosystems*, pp. 221–253. Blackwell, Oxford

Kasperi, M. and Majer, J. D. (2000). Using ants to monitor environmental change. In: Agosti, D., Majer, J. D., Alonso, L. E., and Schultz, T R. (eds) *Ants: Standard Methods for Measuring and Monitoring Biodiversity*, pp. 89–98. Smithsonian Institution Press, Washington, DC.

Kelly, J. A. and Samways, M. J. (2003). Diversity and conservation of forest-floor arthropods on a small Seychelles island. *Biodiversity and Conservation*, 12, 1793–1813.

Kennedy, A. D. (1995). Antarctic terrestrial ecosystem response to global environmental change. *Annual Review of Ecology and Systematics*, 26, 683–704.

Kent, M. and Coker, P. (1992). *Vegetation Description and Analysis: A Practical Approach*. Wiley, Chichester.

King, E. G., Hopper, K. R., and Powell, J. E. (1985). Analysis for systems for biological control of crop arthropod pests in the United States by augmentation of predators and parasites. In: Hoy, M. A. and Herzog, R. C. (eds) *Biological Control in Agricultural IPM Systems*, pp. 210–227. Academic Press, New York.

Kinzig, A. P., Pacala, S. W., and D. Tilman, D. (2001). *The Functional Consequences of Biodiversity—Empirical Progress and Theoretical Extensions*. Princeton University Press, Princeton, NJ.

Kirk, W. D. J. (1984). Ecologically selective coloured traps. *Ecological Entomology*, 9, 35–41.

Klein, A. M., Steffan-Dewenter, I., Buchori, D., and Tscharntke, T. (2002). Effects of land-use intensity in tropical agroforestry systems on coffee flower-visiting and trap-nesting bees and wasps. *Conservation Biology*, 16, 1003–1014.

Knight, T. M., McCoy, M. W., Chase, J. M., McCoy, K. A., and Holt, R. D. (2005). Trophic cascades across ecosystems. *Nature*, 437, 880–883.

Koleff, P., Gaston, K. J., and Lennon, J. J. (2003). Measuring beta diversity for presence-absence data. *Journal of Animal Ecology*, 72, 367–382.

Kotze, D. J., Niemelä, J., O'Hara, R. B., and Turin, H. (2003). Testing abundance-range size relationships in European carabid beetles (Coleoptera, Carabidae). *Ecography*, 26, 553–566.

Krebs, C. J. (1989). *Ecological Methodology*. HarperCollins, New York.

Krell, F.-T. (2004). Parataxonomy vs taxonomy in biodiversity studies—pitfalls and applicability of 'morphospecies' sorting. *Biodiversity and Conservation,* 13, 795–812.

Kremen, C., Colwell, R. K., Erwin, T. L., Murphy, D. D., Noss, R. F., and Sanjayan, M. A. (1993). Terrestrial arthropod assemblages: their use in conservation planning. *Conservation Biology*, 11, 849–856.

Kristensen, T. N., Hoffmann, A. A., Overgaard, J., Sørensen, J. G., Hallas, R., and Loeschcke, V. (2008). Costs and benefits of cold acclimation in field-released *Drosophila*. *Proceedings of the National Academy of Sciences of the USA*, 105, 216–221.

Kuchlein, J. H. and Ellis, W. N. (1997). Climate-induced changes in the microlepidoptera fauna of the Netherlands and implications for nature conservation. *Journal of Insect Conservation*, 1, 73–80.

Kutsch, W. (2002). Transmission of muscle potentials during free flight of locusts. *Computers and Electronics in Agriculture*, 35, 181–199.

Labandeira, C. C. and Sepkoski, J. J. Jr (1993). Insect diversity and the fossil record. *Science,* 261, 310–315.

Labandeira, C. C., Johnson, K. R., and Wilf, P. (2002). Impact of the terminal Cretaceous event on plant-insect associations. *Proceedings of the National Academy of Sciences of the USA*, 99, 2061–2066.

Lande, R. (1996). Statistics and partitioning of species diversity, and similarity among multiple communities. *Oikos*, 76, 5–13.

Larsson, M. C., Hedin, J., Svensson, G. P., Tolasch, T., and Francke, W. (2003). Characteristic odor of *Osmoderma eremita* identified as a male-released pheromone. *Journal of Chemical Ecology*, 29, 575–587.

Larsson, M. C., Svensson, G. P., and Ryrholm, N. (2008). Pheromones and biodiversity—monitoring change in a changing world. Meeting abstract: International Congress of Entomology, Durban, South Africa.

Lawler, J. L., White, D., Sifneos, J. C., and Master, L. L. (2003). Rare species and the use of indicator groups for conservation planning. *Conservation Biology*, 17, 875–882.

Lawrence, J. M. and Samways, M. J. (2002). Influence of hilltop vegetation type on an African butterfly assemblage and its conservation. *Biodiversity and Conservation*, 11, 1163–1171.

Le Roux, P. C. and McGeoch, M. A. (2008). Changes in climatic extremes, variability and signature on sub-Antarctic Marion Island. *Climatic Change*, 86, 309–329

Le Roux, P. C., McGeoch, M. A., Nyakatya, M. J., and Chown, S. L. (2005). Effects of a short-term climate change experiment on a sub-Antarctic keystone plant species. *Global Change Biology*, 11, 1628–1639.

Leamy, L. J. and Klingenberg, C. P. (2005). The genetics and evolution of fluctuating asymmetry. *Annual Review of Ecology Evolution and Systematics*, 36, 1–21.

Leather, S. (ed.) (2005). *Insect Sampling in Forest Ecosystems*. Blackwell, Oxford.

Lee, W. L., Bell, B. M. and Sutton, J. F. (1982). *Guidelines for Acquisition and Management of Biological Specimens*. Association of Systematics Collections, Lawrence, KS.

Legendre, L., and Legendre, P. (1998). *Numerical Ecology*. Elsevier, Amsterdam.

Legendre, P. (1993). Spatial autocorrelation: trouble or new paradigm? *Ecology*, 74, 1659–1673.

Legge, J. T., Roush, R., DeSalle, R., Vogler, A. P., and May, B. (1996). Genetic criteria for establishing evolutionarily significant units in Cryan's buckmoth. *Conservation Biology*, 10, 85–98.

Lemelin, R. H. (2007). Finding beauty in the dragon: the role of dragonflies in recreation and tourism. *Journal of Ecotourism*, 6, 139–145.

Lennon, J. J., Koleff, P., Greenwood, J. J. D., and Gaston, K. J. (2001). The geographical structure of British bird distributions: diversity, spatial turnover and scale. *Journal of Animal Ecology*, 70, 966–979.

Leps, J. and Smilauer, P. (2003). *Multivariate Analysis of Ecological Data using CANOCO*. Cambridge University Press, Cambridge.

Lewis, O. T. and Basset, Y. (2007). Insect conservation in tropical forests. In: Stewart, A. J. A., New, T. R. and Lewis, O. T. (eds) *Insect Conservation Biology*, pp. 34–56. CABI, Wallingford.

Lewis, O. T. and Thomas, C. D. (2001). Adaptations to captivity in the butterfly *Pieris brassicae* (L.) and the implications for *ex situ* conservation. *Journal of Insect Conservation*, 5, 55–63.

Lewis, O. T., Thomas, C. D., Hill, J. K. *et al.* (1997). Three ways of assessing metapopulation structure in the butterfly *Plebejus argus*. *Ecological Entomology*, 22, 283–293.

Lindborg, R. and Eriksson, O. (2004). Historical landscape connectivity affects present plant species diversity. *Ecology*, 85, 1840–1845.

Lindenmayer, D. B. and Fischer, J. (2006). *Habitat Fragmentation and Landscape Change: An Ecological and Conservation Synthesis*. Island Press, Washington, DC.

Lockwood, J.A. (2001). The ethics of 'classical' biological control and the value of place. In: Lockwood, J. A., Howarth, F. G. and Purcell, M. F. (eds) *Balancing Nature: Assessing the Impact of Importing Non-native Biological Control Agents (an International Perspective)*, pp. 100–119. Entomological Society of America, Lanham, MA.

Lorch, P. D., Sword, G. A., Gwynne, D. T., and Anderson, G. L. (2005). Radiotelemetry reveals differences in individual patterns between outbreak and non-outbreak Mormon cricket populations. *Ecological Entomology*, 30, 548–555

Loreau, M., Naeem, S., and Inchausti, P. (eds) (2002). *Biodiversity and Ecosystem Functioning: Synthesis and Perspectives*. Oxford University Press, Oxford.

Losey, J. E. and Vaughan, M. (2006). The economic value of ecological services provided by insects. *BioScience,* 56, 311–323.

Lovett, G. M., Burns, D. A., Driscoll, C. T. *et al.* (2007). Who needs environmental monitoring? *Frontiers in Ecology and the Environment*, 5, 253–260.

Lowman, M. D., Kitching, R. L., and Carruthers, G. (1996). Arthropod sampling in Australian sub-tropical rain forests: how accurate are some of the more common techniques? *Selbyana*, 17, 36–42.

Lu, S.-S. and Samways, M. J. (2001). Life history of the threatened Karkloof blue butterfly *Orachrysops ariadne* (Lepidoptera: Lycaenidae). *African Entomology*, 9, 137–151.

Ludwig, J. A. and Reynolds, J. F. (1988). *Statistical Ecology: A Primer on Methods and Computing*. Wiley, New York.

Lyons, K. G. and Schwartz, M. W. (2001). Rare species loss alters ecosystem function—invasion resistance. *Ecology Letters*, 4, 358–365.

Lyons, K. G., Brigham, C. A., Traut, B. H., and Schwartz, M. W. (2005). Rare species and ecosystem functioning. *Conservation Biology*, 19, 1019–1024.

Macadam, C. (2006). 'Voucher' photographs with digital cameras. *Antenna*, 30, 131–133.

MacArthur, R. A. and Wilson, E. O. (1967). *The Theory of Island Biogeography*. Princeton University Press, Princeton, NJ.

MacNally, R. and Fleishman, E. (2002). Using "indicator" species to model species richness: Model development and predictions. *Ecological Applications*, 12, 79–92.

MacNally, R. and Fleishman, E. (2004). A successful predictive model of species richness based on indicator species. *Conservation Biology*, 18, 646–654.

Mader, H. J. (1984). Animal habitat isolation by roads and agricultural fields. *Biological Conservation*, 29, 81–96.

Maes, D. and Van Dyck, H. (2005). Habitat quality and biodiversity indicator performances of a threatened butterfly versus a multispecies group for wet heathlands in Belgium. *Biological Conservation*, 123, 177–187.

Magurran, A. E. (2004). *Measuring Biological Diversity*, 2nd edn. Blackwell Publishing, Oxford.

Majer, J. D. (1983). Ants: bioindicators of minesite rehabilitation land use and land conservation. *Environmental Management*, 7, 375–383.

Majer, J. D., Shattuck, S. O., Andersen, A. N., and Beattie, A. J. (2004). Australian ant research: fabulous fauna, functional groups, pharmaceuticals and the Fatherhood. *Australian Journal of Entomology*, 43, 235–247.

Malcolm, J. R., Markham, A., Neilson, R. P., and Garaci, M. (2002). Estimated migration rates under scenarios of global climate change. *Journal of Biogeography*, 29, 835–849.

Manly, B. F. J. (1997). *Randomization, Bootstrap and Monte Carlo Methods in Biology*, 2nd edn. Chapman & Hall, London.

Markow, T. A. (1995). Evolutionary ecology and developmental instability. *Annual Review of Entomology*, 40, 105–120.

Martin, J. E. H. (1977). *The Insects and Arachnids of Canada. Part 1. Collecting, Preparing and Preserving Insects, Mites and Spiders*. Agriculture Canada, Québec.

Mavárez, J., Salazar, C. A., Bermigham, E., Salcedo, C., and Jiggins, C. D. (2006). Speciation by hybridization in *Heliconius* butterflies. *Nature,* 441, 868–871.

Mawdsley, N. A. and Stork, N. E. (1995). Species extinctions in insects: ecological and biogeographical considerations. In: Harrington, R. and Stork, N. E. (eds) *Insects in a Changing Environment*, , pp. 321–369. Academic Press, London.

May, R. M. and McLean, A. R. (eds) (2007). *Theoretical Ecology: Principles and Applications*. Oxford University Press, Oxford.

Mayhew, P. J. (2007). Why are there so many insect species? Perspectives from fossils and phylogenies. *Biological Reviews,* 82, 425–454.

McArdle, B. H. (1990). When are rare species not there? *Oikos*, 57, 276–277.

McArdle, B. H. and Anderson, M. J. (2001). Fitting multivariate models to community data: a comment on distance-based redundancy analysis. *Ecology*, 82, 290–297.

McCann, K. (2007). Protecting biostructure. *Nature*, 446, 29.

McGeoch, M. A. (1998). The selection, testing and application of terrestrial insects as bioindicators. *Biological Reviews*, 73, 181–201.

McGeoch, M. A. (2002).Bioindicators. In: El-Shaarawi, A. H. and Piergorsch, W. W. (eds) *Encyclopedia of Environmetrics*, p. 186–189. Wiley, Chichester

McGeoch, M. A. (2007). Insects and bioindication: theory and progress. In: Stewart, A. J. A., New, T. R., and Lewis, O. T. (eds). *Insect Conservation Biology*, pp. 144–174. CABI, Wallingford.

McGeoch, M. A. and Chown, S. L. (1998). Scaling up the value of bioindicators. *Trends in Ecology and Evolution*, 13, 46–47.

McGeoch, M. A. and Gaston, K. J. (2000). Edge effects on the prevalence and mortality factors of *Phytomyza ilicis* (Diptera, Agromyzidae) in a suburban woodland. *Ecology Letters,* 3, 23–29.

McGeoch, M. A. and Gaston, K. J. (2002). Occupancy frequency distributions: patterns, artefacts and mechanisms. *Biological Reviews*, 77, 311–331.

McGeoch, M. A. and Price, P. W. (2004). Spatial abundance structures in an assemblage of gall-forming sawflies. *Journal of Animal Ecology*, 73, 506–516.

McGeoch, M. A., Van Rensburg, B. J., and Botes, A. (2002). The verification and application of bioindicators: a case study of dung beetles in a savanna ecosystem. *Journal of Applied Ecology*, 39, 661–672.

McGeoch, M. A., Le Roux, P. C., Hugo, A. E., and Chown, S. L. (2006). Species and community responses to short-term climate manipulation: microarthropods in the sub-Antarctic. *Austral Ecology*, 31, 719–731.

McGeoch, M. A., Schroeder, M., Ekbom, B., and Larsson, S. (2007). Saproxylic beetle diversity in a managed boreal forest: importance of stand characteristics and forestry conservation measures. *Diversity and Distributions*, 13, 418–429.

McKinney, M. L. (1999). High rates of extinction and threat in poorly studied taxa. *Conservation Biology,* 13, 1273–1281.

Meads, M. J. (1995). Translocation of New Zealand's endangered insects as a tool for conservation. In: Serena, M. (ed.) *Reintroduction Biology of Australian and New Zealand Fauna*, pp. 53–56. Surrey Beatty, Chipping Norton, NSW.

Memmott, J. (2000). Food webs as a tool for studying nontarget effects in biological control. In: Follett, P. A. and Duan, J. J. (eds) *Nontarget Effects of Biological Control*, pp.147–163. Kluwer, Boston, MA.

Memmott, J., Gibson, R., Carvalheiro, L. G., Henson, K., Heleno, R. H., Mikel, M. L., and Pearce, S. (2007). The conservation of ecological interactions. In: Stewart, A. J. A., New, T. R., and Lewis, O. T. (eds) *Insect Conservation Biology*, pp. 226–244. CABI, Wallingford.

Menéndez, R. (2007). How are insects responding to global warming? *Tijdschrift voor Entomologie*, 150, 355–365.

Michaels, K. F. (2007). Using staphylinid and tenebrionid beetles as indicators of sustainable landscape management in Australia: a review. *Australian Journal of Experimental Agriculture*, 47, 435–449.

Millennium Ecosystem Assessment (2005). *Ecosystems and Human Well-being: Biodiversity Synthesis*. World Resources Institute, Washington, DC.

Miller, J. R. (2005). Biodiversity conservation and the extinction of experience. *Trends in Ecology and Conservation*, 20, 430–434.

Miller, J. C., Janzen, D. H., and Hallwachs, W. (2006). *100 Caterpillars. Portraits from the Tropical Forests of Cost Rica*. Belknap Press of Harvard University Press, Cambridge, MA.

Mitchell, A., Secoy, K. and Jackson, T. (eds) (2002).*The Global Canopy Handbook*. GCP, Oxford.

Moffet, M. and Lowman, M. D. (1995).Canopy access techniques. In: Lowman, M. D. and Nadkarni, N. (eds) *Forest Canopies*, pp. 3–26. Academic Press, San Diego, CA.

Moller, H. (1996). Lessons for invasion theory from social insects. *Biological Conservation*, 78, 125–142.

Moore, R. F., Odell, T. M., and Calkins, C. O. (1985). Quality assessment in laboratory-reared insects. In: Singh, P. and Moore, R. F. (eds) *Handbook of Insect Rearing*, Vol. 1, pp. 107–135. Elsevier, Amsterdam.

Moritz, C. (1994). Defining 'evolutionarily significant units' for conservation. *Trends in Ecology and Evolution*, 9, 373-375.

Morris, R. F. (1960). Sampling insect populations. *Annual Review of Entomology*, 5, 243–264.

Morton, A. C. (1981). Raising butterflies on artificial diets. *Journal of Research on Lepidoptera*, 18, 221–227.

Muirhead-Thomson, R. C. (1991). *Trap Responses of Flying Insects*. Academic Press, London.

Murphy, D. D. (1989). Are we studying our endangered butterflies to death? *Journal of Research on Lepidoptera*, 26, 236–239.

Naef-Daenzer, B., Früh, D., Stalder, M., Wetli, P., and Weise, E. (2005). Miniaturization (0.2 g) and evaluation of attachment techniques of telemetry transmitters. *Journal of Experimental Biology*, 208, 4063–4068.

Negro, M., Casale, A., Migliore, L., Palestrini, C., and Rolando, A. (2008). Habitat use and movement patterns in the endangered ground beetle species, *Carabus olympiae* (Coleoptera: Carabidae). *European Journal of Entomology*, 105, 105–112.

Nekola, J. C. and Kraft, C. E. (2002). Spatial constraint of peatland butterfly occurrences within a heterogeneous landscape. *Oecologia*, 130, 53–61.

Neville, P. J. and New, T. R. (1999). Ant genus to species ratios: a practical trial for surrogacy in Victorian forests. *Transactions of the Royal Zoological Society of New South Wales*, 133–137.

Neville, P. J. and Yen, A. L. (2007). Standardizing terrestrial invertebrate biomonitoring techniques across natural and agricultural systems. *Australian Journal of Experimental Agriculture*, 47, 384–391.

New, T. R. (1997a). *Butterfly Conservation*, 2nd edn. Oxford University Press, Melbourne, VIC.

New, T. R. (1997b). Exploitation and conservation of butterflies in the Indo-Pacific region. In: Bolton, M. (ed.) *Conservation and the Use of Wildlife Resources*, pp. 97–109. Chapman & Hall, London.

New, T. R. (1998). *Invertebrate Surveys for Conservation*. Oxford University Press, Oxford.

New, T. R. (1999). Entomology and nature conservation. *European Journal of Entomology* 96, 11–17.

New, T. R. (2005). *Invertebrate Conservation and Agricultural Ecosystems*. Cambridge University Press, Cambridge.

New, T. R. (2007). Are predatory arthropods useful indicators in Australian agroecosystems? *Australian Journal of Experimental Agriculture*, 47, 450–454.

New, T. R. and Britton, D. R. (1997). Refining a conservation plan for an endangered lycaenid butterfly, *Acrodipsas myrmecophila*, in Victoria, Australia. *Journal of Insect Conservation*, 1, 65–72.

New, T. R. and Collins, N. M. (1991). *Swallowtail Butterflies: an Action Plan for their Conservation*. IUCN, Gland.

Niba, A. S. and Samways, M. J. (2006). Development of the concept of 'core resident species' for quality assurance of an insect reserve. *Biodiversity and Conservation*, 15, 4181–4196.

Niemelä, J., Kotze, J., Ashworth, A. *et al.* (2000). The search for common anthropogenic impacts on biodiversity: a global network. *Journal of Insect Conservation*, 4, 3–9.

Nollet, L. M. L. (2007). *Handbook of Water Analysis*. CRC Press, Boca Raton, FL.

O'Neal, M. E., Landis, D. A., Rothwell, E., Kempel, L., and Reinhard, D. (2004). Tracking insects with harmonic radar: a case study. *American Entomologist*, 50, 212–218.

Oates, M. R. and Warren, M. S. (1990). *A Review of Butterfly Introductions in Britain and Ireland*. Joint Committee for the Conservation of British Insects and Worldwide Fund for Nature, Godalming, Surrey.

O'Hara, R. B. (2005). Species richness estimators: how many species can dance on the head of a pin? *Journal of Animal Ecology*, 74, 375–386.

Økland, B. (1996). A comparison of three methods of trapping saproxylic beetles. *European Journal of Entomology*, 93, 195–209.

Oldroyd, H. (1968). *Collecting, Preserving and Studying Insects*. Hutchinson, London.

Oliver, I. and Beattie, A. J. (1996). Invertebrate morphospecies as surrogates for species: a case study. *Conservation Biology*, 10, 99–109.

Oliver, I., Pik, A., Britton, D., Dangerfield, J. M., Colwell, R. K., and Beattie, A. J. (2000). Virtual biodiversity assessment systems: the application of bioinformatics technologies to the accelerated accumulation of biodiversity information. *BioScience,* 50, 441–450.

Orme, C. D. L., Davies, R. G., Burgess, M. *et al.* (2005). Global hotspots of species richness are not congruent with endemism or threat. *Nature*, 436, 1016–1019.

Osborne, J. L., Loxdale, H. D., and Woiwod, I. P. (2002). Monitoring insect dispersal: methods and approaches. In: Bullock, J. M., Kenwood, R E., and Hails, R. S. (eds). *Dispersal Ecology*, pp. 24–48. Cambridge University Press, Cambridge.

Ott, J. (1996). Zeigt die Ausbreitung der Feuerlibelle *Crocothemis erythraea* Brullé in Deutschland eine Klimaveränderung an? *Naturschutz und Landschaftsplanung*, 2/96: 53–61.

Ott, J. (2001). Expansion of Mediterranean Odonata in Germany and Europe—consequences of climatic changes. In: Walter, G.-R., Burga C. A., and Edwards P. J. (eds.) *'Fingerprints' of Climate Change—Adapted Behaviour and Shifting Species Ranges*, pp. 89–111. Kluwer, New York.

Ott, J. (2007). The expansion of *Crocothemis erythraea* (Brullé, 1832) in Germany—an indicator of climatic changes. In: Tyagi, B. K. (ed.) *Odonata: Biology of Dragonflies*, pp: 201–222, Scientific Publications, Jodhpur.

Ott, J. (2009). Effects of climatic changes on dragonflies—results and recent observations in Europe. In: Ott, J. (ed.) *Monitoring Climate Changes Using Dragonflies*, Pensoft, Sofia. In press.

Ott, J., Schorr, M., Trockur, B., and Lingenfelder, U. (2007). *Species Protection Programme for the Orange-spotted Emerald (*Oxygastra curtisii*, Insecta: Odonata) in Germany—the Example of the River Our Population*. Invertebrate Ecology and Conservation Series, No. 3. Pensoft, Sofia.

Ozanne, C. M. P. (2005). Techniques and methods for sampling canopy insects. In: Leather, S. (ed.) *Insect Sampling in Forest Ecosystems*, pp. 146–167. Blackwell, Oxford.

Parmesan, C., Ryrholm, N., Stefanescy, C. *et al.* (1999). Poleward shifts in geographical ranges of butterfly species associated with regional warming. *Nature*, 399, 579–583.

Parsons, M. (1978). *Farming Manual: Insect Farming and Trading Agency*. Division of Wildlife, Papua New Guinea.

Parsons, M. (1999). *The Butterflies of Papua New Guinea*. Academic Press, London.

Parsons, M. J. (1992). The butterfly farming and trading industry in the Indo-Australian region and its role in tropical forest conservation. *Tropical Lepidoptera* 3 (supplement 1), 1–31.

Pearce-Kelly, P., Morgan, R., Honan, P. *et al.* (2007). The conservation value of insect breeding programmes: rationale, evaluation tools and example programme case studies. In: Stewart, A. J. A., New, T. R. and Lewis, O. T. (eds) *Insect Conservation Biology*, pp. 57–75. CABI, Wallingford.

Pearson, D. L. (1994). Selecting indicator taxa for the quantitative assessment of biodiversity. *Philosophical Transactions of the Royal Society of London, Series B*, 345, 75–79.

Pearson, D. L. and Cassola, F. (1992). World-wide species richness patterns of tiger beetles (Coleoptera: Cicindelidae): indicator taxon for biodiversity and conservation studies. *Conservation Biology*, 6, 376–390.

Pearson, R. G., Thuiller, W., Araujo, M. B. *et al.* (2006). Model-based uncertainty in species range prediction. *Journal of Biogeography*, 33, 1704–1711.

Perry, J. N. (1995). Spatial analysis by distance indices. *Journal of Animal Ecology*, 64, 303–314.

Perry, J. N. (1998). Measures of spatial pattern for counts. *Ecology*, 79, 1008–1017.

Petchey, O. L. (2004). On the statistical significance of functional diversity effects. *Functional Ecology*, 18, 297–303.

Petchey, O. L. and Gaston, K. J. (2006). Functional diversity: back to basics and looking forward. *Ecology Letters*, 9, 741–758.

Petchey, O. L., Hector, A., and Gaston, K. J. (2004). How do different measures of functional diversity perform? *Ecology*, 85, 847–857.

Peterson, G., Allen, C. R. and Holling, C. S. (1998). Ecological resilience, biodiversity, and scale. *Ecosystems*, 1, 6–18.

Pollard, E. and Yates, T. J. (1993). *Monitoring Butterflies for Ecology and Conservation*. Chapman & Hall, London.

Ponel, P., Orgeas, J., Samways, M. J. *et al.* (2003). 110 000 years of Quaternary beetle diversity change. *Biodiversity and Conservation*, 12, 2077–2089.

Price, P. W. (1991a). The plant vigor hypothesis and herbivore attack. *Oikos*, 62, 244–251.

Price, P. W. (1991b). Darwinian methodology and the theory of insect herbivore population dynamics. *Annals of the Entomological Society of America*, 48, 464–473.

Price, P. W. (1997). *Insect Ecology*, 3rd edn. Wiley, New York.

Priddel, D., Carlile, N., Humphrey, M., Fellenberg, S., and Hiscox, D. (2003). Rediscovery of the 'extinct' Lord Howe Island stick insect (*Dryococelus australis* (Montrouzier)) (Phasmatodea) and recommendations for its conservation. *Biodiversity and Conservation*, 12, 1391–1403.

Priesner, E., Ryrholm, N. and Dobler, G. (1989). Der Glasflugler *Synanthedon polaris* (Stgr.) in den schweizer Hochalpen, nach gewiesen mit Sexual pheromonen [*Synanthedon polaris* found in the high Alps of Switzerland, by using pheromones]. *Nachrichtenblatt der bayerichen Entomologen*, 38, 89–97.

Primack, R., Kobori, H., and Mori, S. (2005). Dragonfly pond restoration promotes conservation awareness in Japan. *Conservation Biology*, 14, 1553–1554.

Pryke, S. R. and Samways, M. J. (2001). Width of grassland linkages for the conservation of butterflies in South African afforested areas. *Biological Conservation*, 101, 85–96.

Pullin, A. S., McLean, I. G., and Webb, M. R. (1995). Ecology and conservation of *Lycaena dispar*: British and European perspectives. In: Pullin, A. S. (ed.) *Ecology and Conservation of Butterflies*, pp. 150–164. Chapman & Hall, London.

Purse, B. V., Hopkins, G. W., Day, K. J., and Thompson, D. J. (2003). Dispersal characteristics and management of a rare damselfly. *Journal of Applied Ecology*, 40, 716–728.

Pyle, R. M. (1976). Conservation of Lepidoptera in the United States. *Biological Conservation*, 9, 55–75.

Quinn, G. P. and Keough, M. J. (2002). *Experimental Design and Data Analysis for Biologists*. Cambridge University Press, Cambridge.

Rangel, T. F. L. V. B., Diniz-Filho, J. A. F., and Bini, L. M. (2006). Towards an integrated computational tool for spatial analysis in macroecology and biogeography. *Global Ecology and Biogeography*, 15, 321–327.

Remsburg, A. J. Olson, A. C., and Samways, M. J. (2008). Shade alone reduces adult dragonfly (Odonata: Libellulidae) abundance. *Journal of Insect Behaviour*, 21, 460–468.

Resh, V. H. and Jackson, J. K. (1993). Rapid assessment approaches to biomonitoring using benthic macroinvertebrates. In: Rosenberg, D. M. and Resh, V. H. (eds) *Freshwater Biomonitoring and Benthic Macroinvertebrates*, pp. 195–233. Chapman & Hall, New York.

Reusch, T. B. H., Ehlers, A., Hämmerli, A., and Worm, B. (2005). Ecosystem recovery after climatic extremes enhanced by genotypic diversity. *Proceedings of the National Academy of Sciences of the USA*, 102, 2826–2831.

Revenga, C., Campbell, I., Abell, R., de Villiers, P., and Bryer, M. (2005). Prospects for monitoring freshwater ecosystems towards the 2010 targets. *Philosophical Transactions of the Royal Society of London, Series B*, 360, 397–413.

Reyers, B., Wessels, K. J., and van Jaarsveld, A. S. (2002). An assessment of biodiversity surrogacy options in the Limpopo Province of South Africa. *African Zoology*, 37, 185–195.

Rice, T. M. (2008). A review of methods for maintaining odonate larvae in the laboratory, with a description of a new technique. *Odonatologica*, 37, 41–54.

Richards, O. W. and Waloff, N. (1954). Studies on the biology and population dynamics of British grasshoppers. *Anti-locust Bulletin*, 17, 1–182.

Riecken, U. and Raths, U. (1996). Use of radio telemetry for studying dispersal and habitat use of *Carabus coriaceus* L. *Annales Zoologici Fennici*, 33, 109–116.

Riede, K. (1998). Acoustic monitoring of Orthoptera and its potential for conservation. *Journal of Insect Conservation*, 2, 217–223.

Riede, K., Nischk, F., Dietrich, C., Thiel, C., and Schwenker, F. (2006). Automated annotation of Orthoptera songs: first results from analysing DORSA sound repository. *Journal of Orthoptera Research*, 15, 105–113.

Ries, L., Fletcher, R. J., Battin, J., and Sisk, T. D. (2004). Ecological responses to habitat edges: mechanisms, models and variability explained. *Annual Review of Ecology and Systematics*, 35, 491–522.

Roberge, J.-M., and Angelstam, P. (2004). Usefulness of the umbrella species concept as a conservation tool. *Conservation Biology*, 18, 76–85.

Rohr, J. R., Mahan, C. G., and Kim, K. C. (2007). Developing a monitoring program for invertebrates: guidelines and a case study. *Conservation Biology*, 21, 422–433.

Rolston, H. III (1994). *Conserving Natural Value*. Columbia University Press, New York.

Rondinini, C., Wilson, K. A., Boitani, L., Grantham, H., and Possingham, H. P. (2006). Tradeoffs of different types of species occurrence data for use in systematic conservation planning. *Ecology Letters*, 9, 1136–1145.

Rosenberg, D. M. and Resh, V. H. (eds) (2003). *Freshwater Biomonitoring and Benthic Macroinvertebrates*. Chapman & Hall, London.

Roura-Pascual, N., Suarez, A. V., Gomez, C. *et al.* (2004). Geographical potential of Argentine ants (*Linepithema humile* Mayr) in the face of global climate change. *Proceedings of the Royal Society of London, Series B*, 271, 2527–2534.

Roy, D. B. and Sparks, T. H. (2000). Phenology of British butterflies and climate change. *Global Change Biology*, 6, 407–416.

Saccheri, I., Kuussaari, M., Kankare, M., Vikman, P., Fortelius, W. and Hanski, I. (1998). Inbreeding and extinction in a butterfly metapopulation. *Nature*, 392, 491–494.

Samways, M. J. (1979). Immigration, population growth and mortality of insects and mites on cassava (*Manihot esculenta*) in Brazil. *Bulletin of Entomological Research*, 69, 491–505.

Samways, M. J. (1985). Relationship between red scale (*Aonidiella aurantii* Maskell) (Hemiptera: Diaspididae), and its natural enemies in the upper and lower parts of citrus trees in South Africa. *Bulletin of Entomological Research*, 75, 379–393.

Samways, M. J. (1987). Weather and monitoring the abundance of the adult citrus psylla, *Trioza erytreae* (Del Guercio) (Hemiptera: Triozidae). *Journal of Applied Entomology*, 103, 502–508.

Samways, M. J. (1988b). Comparative monitoring of red scale *Aonidiella aurantii* (Mask.) (Hom., Diaspididae) and its *Aphytis* spp. (Hym., Aphelinidae) parasitoids. *Journal of Applied Entomology*,105, 483–489.

Samways, M. J. (1990b). Species temporal variability: epigaeic ant assemblages and management for abundance and scarcity. *Oecologia*, 84, 482–490.

Samways, M. J. (1996). Insects on the brink of a major discontinuity. *Biodiversity and Conservation*, 5, 1047–1058.

Samways, M. J. (1997a) Conservation biology of Orthoptera. In: Gangwere, S. K., Muralirangan, M. C., and Muralirangan, M. (eds) *The Bionomics of Grasshoppers, Katydids and their Kin*, pp. 481–496. CABI, Wallingford.

Samways, M. J. (1997b). Classical biological control and biodiversity conservation: what risks are we prepared to accept? *Biodiversity and Conservation*, 6, 1309–1316.

Samways, M. J. (2000). A conceptual model of restoration triage based on experiences from three remote oceanic islands. *Biodiversity and Conservation*, 9, 1073–1083.

Samways, M. J. (2002). A strategy for national red listing invertebrates based on experiences with Odonata in South Africa. *African Entomology*, 10, 43–52.

Samways, M. J. (2003). Threats to the tropical island dragonfly fauna (Odonata) of Mayotte, Comoro Archipelago. *Biodiversity and Conservation*, 12, 1785–1792.

Samways, M. J. (2005). *Insect Diversity Conservation*. Cambridge University Press, Cambridge.

Samways, M. J. (2006a). Insect extinctions, and insect survival. *Conservation Biology*, 20, 245–246.

Samways, M. J. (2006b). National Red List of South African dragonflies (Odonata). *Odonatologica*, 35, 341–368.

Samways, M. J. (2007a). Implementing ecological networks for conserving insect and other biodiversity. In: Stewart, A. J. A., New, T. R. and Lewis, O. T. (eds) *Insect Conservation Biology*, pp. 127–143. CABI, Wallingford.

Samways, M. J. (2007b). Insect conservation: a synthetic management approach. *Annual Review of Entomology*, 52, 465–487.

Samways, M. J. (2007c). Rescuing the extinction of experience. *Biodiversity and Conservation*, 16, 1995–1997.

Samways, M. J. (2008). *Dragonflies and Damselflies of South Africa*. Pensoft, Sofia.

Samways, M. J. (2009). Extreme weather and climate change impacts on South African dragonflies. In: Ott, J. (ed.) *Monitoring Climate Change Using Dragonflies*. Pensoft, Sofia. In press.

Samways, M. J. and Grant, P. B. C. (2006). Honing Red List assessments of lesser-known taxa in biodiversity hotspots. *Biodiversity and Conservation*, 16, 2575–2586.

Samways, M. J. and Kreutzinger, K. (2001). Vegetation, ungulate and grasshopper interactions inside vs. outside an African savanna game park. *Biodiversity and Conservation*, 10, 1963–1981.

Samways, M. J. and Lu, S.-S. (2007). Key traits in a threatened butterfly and its common sibling: implications for conservation. *Biodiversity and Conservation*, 16, 4095–4107.

Samways, M. J. and Taylor, S. (2004). Impacts of invasive alien plants on red-listed South African dragonflies (Odonata). *South African Journal of Science*, 100, 78–80.

Samways, M. J., Osborn, R., and Van Heerden, I. (1996). Distribution of benthic invertebrates at different depths in a shallow reservoir in the KwaZulu-Natal Midlands. *Koedoe*, 39, 69–76.

Samways, M. J., Osborn, R., Hastings, H., and Hattingh, V. (1999). Global climate change and accuracy of prediction of species' geographical ranges: establishment success of introduced ladybirds (Coccinellidae, *Chilocorus* spp.) worldwide. *Journal of Biogeography*, 26, 795–812.

Samways, M. J., Taylor, S., and Tarboton, W. (2005). Extinction reprieve following alien removal. *Conservation Biology*, 19, 1329–1330.

Sands, D. P. A. and New, T. R. (2002). *The Action Plan for Australian Butterflies* Environment Australia, Canberra, ACT.

Sato, M., Kohmatsu, Y., Yuma, M., and Tsubaki, Y. (2008). Population genetic differentiation in three sympatric damselfly species in a highly fragmented urban landscape (Zygoptera: Coenagrionidae). *Odonatologica*, 37, 131–144.

Scheffer, M. and Carpenter, S. R. (2003). Catastrophic regime shifts in ecosystems: linking theory to observation. *Trends in Ecology and Evolution*, 18, 648–656.

Scheiner, S. M., and Gurevitch, J. (2001). *Design and Analysis of Ecological Experiments*. Oxford University Press, Oxford.

Schmitt, T., Cizek, O., and Konvicka, M. (2005). Genetics of a butterfly relocation: large, small and introduced populations of the mountain endemic *Erebia epiphron silesiana*. *Biological Conservation*, 123, 11–18.

Schrader-Frechette, K. S. and McCoy, E. D. (1992). Statistics, costs and rationality in ecological inference. *Trends in Ecology and Evolution*, 3, 96–99.

Schulze, R. E. (ed.) (2006). *South African Atlas of Climatology and Agrohydrology*. WRC Report 1489/1/06, Water Research Commission, Pretoria.

Schwenker, F., Dietrich, C., Kestler, H. A., Riede, K., and Palm, G. (2003). Radial basis function neural networks and temporal fusion for the classification of bioacoustic time series. *Neurocomputing*, 51, 265–275.

Scudder, G. G. E. (1996). *Terrestrial and Freshwater Invertebrates of British Columbia: Priorities for Inventory and Descriptive Research*. Research Branch, B. C. Ministry of Forests and Wildlife Branch, Vancouver.

Shattuck, S. O. (1999). *Australian Ants. Their Biology and Identification*. CSIRO Publishing, Collingwood, VIC.

Shepherd, M., Buchmann, S. L., Vaughan, M., and Black, S. H. (2003). *Pollinator Conservation Handbook*. Xerces Society, Portland, OR.

Sherley, G. H. (1998a). Translocating a threatened New Zealand giant orthopteran, *Deinacrida* sp. (Stenopelmatidae): some lessons. *Journal of Insect Conservation*, 2, 195–199.

Sherley, G. H. (1998b). *Threatened Weta Recovery Plan*. Department of Conservation, Wellington.

Shields, O. (1967). Hilltopping. *Journal of Research on Lepidoptera*, 6, 69–178.

Silberbauer, L. X. and Britton, D. R. (1999). Holiday houses or habitat: conservation of the Brenton blue butterfly, *Orachrysops niobe* (Trimen) (Lepidoptera: Lycaenidae) in Knysna, South Africa. *Transactions of the Royal Zoological Society of New South Wales*, 394–397.

Simaika, J. P. and Samways, M. J. (2008). Valuing dragonflies as service providers. In: Cordoba-Aguilar, A. (ed.) *Dragonflies: Model Organisms for Ecological and Evolutionary Research*, pp. 109–123. Oxford University Press, Oxford.

Simberloff, D. (1998). Flagships, umbrellas, and keystones: is single-species management passé in the landscape era? *Biological Conservation*, 83, 247–257.

Simberloff, D. and Stiling, P. (1996). Risks of species introduced for biological control. *Biological Conservation*, 78, 185–192.

Singh, P. and Moore, R. F. (eds) (1985). *Handbook of Insect Rearing*, Vol. 1. Elsevier, Amsterdam.

Smith, B. and Wilson, J. B. (1996). A consumer's guide to evenness measures. *Oikos*, 76, 70–82.

Smith, J., Samways, M. J., and Taylor, S. (2007). Assessing riparian quality using two complementary sets of bioindicators. *Biodiversity and Conservation*, 16, 2695–2713.

Smithers, C. (1982). *Handbook of Insect Collecting*. Reed Books, Sydney, NSW.

Sole, R. and Goodwin, B. (2000). *Signs of Life*. Basic Books, New York.

Southwood, T. R. E. and Henderson, P. A. (2000). *Ecological Methods*, 3rd edn. Blackwell, Oxford.

Spalding, A. (1997). The use of the butterfly transect method for the study of the nocturnal moth *Luperina nickerlii leechi* Goater (Lepidoptera: Noctuidae) and its possible application to other species. *Biological Conservation*, 80, 147–152.

Sparrow, H. R., Sisk, T. D., Ehrlich, P. R., and Murphy, D. D. (1994). Techniques and guidelines for monitoring Neotropical butterflies. *Conservation Biology*, 8, 800–809.

Spector, S. and Forsyth, A. B. (1998). Indicator taxa for biodiversity assessment in the vanishing tropics. In: Mace, G. M., Balmford, A. and Ginsberg, J. R. (eds) *Conservation in a Vanishing World*, pp. 181–209. Cambridge University Press, Cambridge.

Speight, M. R. (2005). Sampling insects from trees: shoots, stems, and trunks. In: Leather, S. (ed.) *Insect Sampling in Forest Ecosystems*, pp. 77–115. Blackwell, Oxford.

Spellerberg, I. F. (2005). *Monitoring Ecological Change*, 2nd edn. Cambridge University Press, Cambridge.

Squire, G. R., Brooks, D. R., Bohan, D. A. *et al.* (2003). On the rationale and interpretation of the Farm Scale evaluations of genetically modified herbicide-tolerant crops. *Philosophical Transactions of the Royal Society of London, Series B*, 358, 1779–1799.

Starfield, A. M. and Bleloch, A. L. (1991). *Building Models for Conservation and Wildlife Management*, 2nd edn. Burgess International, Edina, MN.

Starfield, A. M., Smith, K. A., and Bleloch, A. L. (1990). *How to Model It—Problem Solving for the Computer Age*. McGraw-Hill, New York.

Stohlgren, T. J. (2007). *Measuring Plant Diversity*. Oxford University Press, Oxford.

Stokes, D. L. (2006). Conservators of experience. *BioScience*, 56, 6–7.

Stork, N. E. (ed.) (2007). Australian Canopy Crane. *Austral Ecology*, 32, Special Issue No. 1.

Stork, N. E. and Hammond, P. (1997). Sampling arthropods from tree-crowns by fogging with knockdown insecticides: lessons from studies of oak tree beetle assemblages in Richmond Park (UK). In Stork, N., Adis, J. and Didham, R. (eds) *Canopy Arthropods*, pp. 3–26. Chapman & Hall, London.

Stringer, I. A. N. and Chappell, R. (2008). Possible rescue from extinction: transfer of a rare New Zealand tusked weta to islands of the Mercury group. *Journal of Insect Conservation*, 12, 371–382.

Strong, D. R., Lawton, J. H., and Southwood, T. R. E. (1984). *Insects on Plants*. Blackwell, Oxford.

Suh, A. N. and Samways, M. J. (2001). Development of a dragonfly awareness trail in an African botanical garden. *Biological Conservation*, 100, 345–353.

Sutherst, R. W. and Maywald, G. F. (1985). A computerized system for matching climates in ecology. *Agriculture, Ecosystems and Environment*, 13, 281–299.

Svensson, G. P., Larsson, M. C., and Hedin, J. (2003). Air sampling of its pheromone to monitor the occurrence of *Osmoderma eremita*, a threatened beetle inhabiting hollow trees. *Journal of Insect Conservation*, 7, 189–198.

Svensson, G. P., Larsson, M. C., and Hedin, J. (2004). Attraction of the larval predator *Elater ferrugineus* to the sex pheromone of its prey, *Osmoderma eremita*, and its implication for conservation biology. *Journal of Chemical Ecology*, 30, 53–363.

Sverdrup-Thygeson, A. (2001). Can 'continuity indicator species' predict species richness or red-listed species of saproxylic beetles? *Biodiversity and Conservation*, 10, 815–832.

Sword, G. A., Lorch, P. D., and Gwynne, D. T. (2005). Migratory bands give crickets protection. *Nature*, 433, 703.

Tan, K. H. (2005). *Soil Sampling, Preparation and Analysis*, 2nd edn. CRC Press, Boca Raton, FL.

Taylor, L. R. (1984). Assessing and interpreting the spatial distributions of insect populations. *Annual Review of Entomology*, 29, 321–357.

Thomas, C. D. (1995). Ecology and conservation of butterfly metapopulations in the fragmented British landscape. In: Pullin, A. S. (ed.) *Ecology and Conservation of Butterflies*, pp. 46–63 Chapman & Hall, London.

Thomas, C. D. and Jones, T. M. (1993). Partial recovery of a skipper butterfly (*Hesperia comma*) from population refuges: lessons for conservation in a fragmented landscape. *Journal of Animal Ecology*, 62, 472–481.

Thomas, C. D. (1995). Ecology and conservation of butterfly metapopulations in the fragmented British landscape. In: Pullin, A. S. (ed.) *Ecology and Conservation of Butterflies*, pp. 46–63. Chapman & Hall, London.

Thomas, C. D., Franco, A. M. A., and Hill, J. K. (2006). Range retractions and extinction in the face of climate warming. *Trends in Ecology and Evolution*, 21, 415–416.

Thomas, J. A. (1999). The Large Blue butterfly: a decade of progress. *British Wildlife*, 11, 22–27.

Thomas, J. A. (2005). Monitoring change in the abundance and distribution of insects using butterflies and other indicator groups. *Philosophical Transactions of the Royal Society of London, Series B*, 360, 339–357.

Thomas, J. A., Telfer, M. G., Roy, D. B. *et al.* (2004). Comparative losses of British butterflies, birds and plants and the global extinction crisis. *Science*, 303, 1879–1881.

Thomas, L. and Krebs, C. J. (1997). A review of statistical power analysis software. *Bulletin of the Ecological Society of America*, 78, 126–139.

Thomas, M. B., Wratten, S. D., and Sotherton, N. W. (1992). Creation of 'island' habitats in farmland to manipulate populations of beneficial arthropods: predator densities and species composition. *Journal of Applied Ecology*, 9, 524–531.

Thomson, L. J., Sharley, D. J., and Hoffmann, A. A. (2007). Beneficial organisms as bioindicators for environmental sustainability in the grape industry in Australia. *Australian Journal of Experimental Agriculture*, 47, 404–411.

Tilman, D. (1996). Biodiversity: population versus ecosystem stability. *Ecology*, 77, 350–363.

Tilman, D. (1999). Ecological consequences of biodiversity: a search for general principles. *Ecology*, 80, 1455–1474.

Tilman, D. (2001). Functional diversity. In: Levin, S. A. (ed.) *Encyclopedia of Biodiversity*, pp. 109–120. Academic Press, San Diego, CA.

Tilman, D., May, R. M., Lehman, C. L., and Nowak, M. A. (1994). Habitat destruction and the extinction debt. *Nature*, 371, 65–66.

Tilman, D., Fargione, J., Wolff, B. *et al.* (2001). Forecasting agriculturally driven global environmental change. *Science*, 292, 281–284.

Tokeshi, M. (1993). Species abundance patterns and community structure. *Advances in Ecological Research*, 24, 111–186.

Tolasch, T., von Fragstein, M., and Steidle, J. L. M. (2007). Sex pheromone of *Elater ferrugineus* L. (Coleoptera: Elateridae). *Journal of Chemical Ecology*, 33, 2156–2166.

Travis, J. M. J. (2003). Climate change and habitat destruction: a deadly anthropogenic cocktail. *Proceedings of the Royal Society of London B*, 270, 467–473.

Tscharntke, T., J. M. Tylianakis, M. R. Wade, S. D. Wratten, J. Bengtsson, and D. Kleijn. 2007. Insect conservation in agricultural landscapes. Pages 383-404 In: A. J. A. Stewart, T. R. New, and O. T. Lewis, editors. Insect Conservation Biology. CABI, Wallingford.

Underwood, A. J. (1991). Beyond BACI: experimental designs for detecting human environmental impacts on temporal variations in natural populations. *Australian Journal of Marine and Freshwater Research*, 42, 569–587.

Underwood, A. J. (1992). Beyond BACI: the detection of environmental impacts on populations in the real, but variable, world. *Journal of Experimental Marine Biology and Ecology*, 161, 145–178.

Underwood, A. J. (1994). On beyond BACI: sampling designs that might reliably detect environmental disturbances. *Ecological Applications*, 4, 3–15.

Underwood, A. J. (1997). *Experiments in Ecology: Local Design and Interpretation using Analysis of Variance.* Cambridge University Press, Cambridge.

Upton, M. S. (1991). *Methods for Collecting, Preserving and Studying Insects and Allied Forms*, 4th edn. Australian Entomological Society, Brisbane, QLD.

Urquhart, F. (1960). *The Monarch Butterfly.* University of Toronto Press, Toronto, Ont.

Usher, M. B. (ed.) (1986). *Wildlife Conservation Evaluation.* Chapman & Hall, London.

Uys, V. M. and Urban R. P. (eds) (2006). *How to Collect and Preserve Insects and Arachnids*, 2nd edn. Plant Protection Research Institute, Pretoria.

Valentin, R. J. P. and Shipp, J. L. (2005). The insectary business: developing customers, products and markets. In: Heinz, K. M., van Driesche, R., and Parrella, M. P. (eds) *Biocontrol in Protected Culture*, pp. 55–70. Ball Publishing, Batavia, IL.

Valladares, G., Salvo, A., and Cagnolo, L. (2006). Habitat fragmentation effects on trophic processes of insect-plant food webs. *Conservation Biology*, 20, 212–217.

Van Lenteren, J. C. and Loomans, A. J. M. (2006). Environmental risk assessment: methods for comprehensive evaluation and quick scan. In: Bigler, F., Babendreier, D., and Kuhlmann, U. (eds) *Environmental Impact of Invertebrates for Biological Control of Arthropods: Methods and Risk Assessment* pp. 254–272. CABI, Wallingford.

Van Lenteren, J. C., Babendreier, D., Bigler, F. *et al.* (2003). Environmental risk assessment of exotic natural enemies used in inundative biological control. *BioControl*, 48, 3–38.

Van Lenteren, J. C., Bale, J., Bigler, F., Hokkanen, H. M. T., and Loomans, A. J. M. (2006a). Assessing risks of releasing exotic biological control agents of arthropod pests. *Annual Review of Entomology*, 51, 609–634.

Van Lenteren, J. C., Cock, M. J. W., Hoffmeister, T. S., and Sands, D. P. A. (2006b). Host specificity in arthropod biological control, methods for testing and interpretation of the data. In: Bigler, F., Babendreier, D., and Kuhlmann, U. (eds) *Impact of Invertebrates for Biological Control of Arthropods: Methods and Risk Assessment*, pp. 38–63. CABI, Wallingford.

Van Straalen, N. M. and Krivolutsky, D. A. (1996). *Bioindicator Systems for Soil Pollution.* Kluwer, Dordrecht.

van Swaay, C. A. M. (1995). Measuring changes in butterfly abundance in The Netherlands. In: Pullin, A. S. (ed.) *Ecology and Conservation of Butterflies*, pp. 230–247. Chapman & Hall, London.

van Swaay, C. and Warren, M. S. (1999). *Red Data Book of European Butterflies.* Nature and Environment Series No. 99. Council of Europe, Strasbourg.

Van Wyk A, Van den Berg J, Van Hamburg H. 2007. Selection of non-target Lepidoptera species for ecological risk assessment of Bt maize in South Africa. *African Entomology* 15, 356–366.

Varley, G. C., Gradwell, G. R., and Hassell, M. P. (1973). *Insect Population Ecology: An Analytical Approach.* Blackwell, Oxford.

Veldtman, R. and McGeoch, M. A. (2004). Spatially explicit analyses unveil density dependence. *Proceedings of the Royal Society of London Series B-Biological Sciences,* 271, 2439–2444.

Visser, G., Gibson, R., Kolderwey, H., Zimmerman, B., Veltman, K., and Pearce-Kelly, P. (eds) (2005) *Lower Vertebrate and Invertebrate Taxon Advisory Group Manual for Regional Collection Planning*. European Association of Zoos and Aquaria, Amsterdam.

Vogler, A. P., Knisley, C. B., Glueck, S. B., Hill, J. M., and DeSalle, R. (1993). Using molecular and ecological data to diagnose endangered populations of the Puritan tiger beetle, *Cicindela puritana*. *Molecular Ecology*, 2, 375–383.

Waage, J. K., Carl, K. P., Mills, N. J., and Greathead, D. J. (1985). Rearing entomophagous insects. In: Singh, P. and Moore, R. F. (eds) *Handbook of Insect Rearing*, Vol. 1, pp. 45–66. Elsevier, Amsterdam.

Wagner, H. H., Wildi, O., and Ewald, K. C. (2000). Additive partitioning of plant species diversity in an agricultural landscape. *Landscape Ecology*, 15, 219–227.

Walsh, C. J. (2006). Biological indicators of stream health using macroinvertebrate assemblage composition: a comparison of sensitivity to an urban gradient. *Marine and Freshwater Research*, 57, 37–47.

Walther, B. A. and Moore, J. L. (2005). The concepts of bias, precision, and accuracy, and their use in testing the performance of species richness estimators, with a literature review of estimator performance. *Ecography*, 28, 815–829.

Walther, G. R., Post, E., Convey, P. *et al.* (2002). Ecological responses to recent climate change. *Nature*, 416, 389–395.

Wang, H., Ando, N., and Kanzaki, R. (2008). Active control of free flight manoeuvres in a hawkmoth, *Agrius convolvuli*. *Journal of Experimental Biology*, 211, 423–432.

Ward, D. F. and Larivière, M.-C. (2004). Terrestrial invertebrate surveys and rapid biodiversity assessment in New Zealand: lessons from Australia. *New Zealand Journal of Ecology*, 28, 151–159.

Wardle, D. A., Huston, M. A., Grime, J. P. *et al.* (2000). Biodiversity and ecosystem function: an issue in ecology. *Bulletin of the Ecological Society of America*, 81, 235–239.

Waring, P. (2005). The history, conservation and presumed extinction of the Essex emerald moth, *Thetidia smaragdaria maritima* (Prout 1935) in Great Britain. *Entomologists' Gazette*, 56, 149–188.

Warren, M. S., Bourn, N., Brereton, T., Fox, R., Middlebrook, I., and Parsons, M. S. (2007). What have Red Lists done for us? The values and limitations of protected species listing for invertebrates. In: Stewart, A. J. A., New, T. R., and Lewis, O. T. (eds) *Insect Conservation Biology*, pp. 76–91. CABI, Wallingford.

Warren, M. S., Hill, J. K., Thomas, J. A. *et al.* (2001). Rapid responses of British butterflies to opposing forces of climate and habitat change. *Nature*, 414, 65–69.

Warwick, R. M, Clarke K. R (1998) Taxonomic distinctness and environmental assessment. *Journal of Applied Ecology* 35, 532–543

Watt, W. B., Chew, F. S., Snyder, L. R. G., Watt, A. G., and Rothschild, D. E. (1977). Population structure of pierid butterflies. I. Numbers and movements of some montane *Colias* species. *Oecologia*, 27, 1–22.

Watts, C. H., Thornburrow, D., Green, C. J., and Agnew, W. R. (2008). Tracking tunnels: a novel method for detecting a threatened New Zealand giant weta (Orthoptera: Anostostomatidae). *New Zealand Journal of Ecology*, 32, 92–97.

Watts, C., Stringer, I., Sherley, G., Gibbs, G., and Green, C. (2008). History of weta (Orthoptera: Anostostomatidae) translocation in New Zealand: lessons learned, islands as sanctuaries and the future. *Journal of Insect Conservation*, 12, 359–370.

Weisser, W. W. and Siemann E. (eds) (2004). *Insects and Ecosystem Function*. Springer, Berlin.

Wheater, C. P. and Cook, P. A. (2003). *Studying Invertebrates*. Richmond, Slough.

Whittaker, R. H. (1975). *Communities and Ecosystems*, 2nd edn. Macmillan, New York.

Wiegand, T., Revilla, E., and Maloney, K. A. (2005). Effects of habitat loss and fragmentation on population dynamics. *Conservation Biology*, 19, 108–121.

Wiens, J. A. (1989). Spatial scaling in ecology. *Functional Ecology*, 3, 385–397.

Wiens, J. A., Stenseth, N. C., Horne, B. van, and Ims, R. A. (1993). Ecological mechanisms in landscape ecology. *Oikos*, 66, 369–380.

Wikelski, M. and Cooke, S. J. (2006). Conservation physiology. *Trends in Ecology and Evolution* 21, 38–46.

Wikelski, M., Moskowitz, D., Adelman J. S., Cochran J., Wilcove, D. S. and May, M. L. (2006). Simple rules guide dragonfly migration. *Biology Letters*, 2, 325–329.

Wikelski, M., Kays, R. W., Kasdin, N. J., Thorup, K., Smith, J. A., and Swenson, G. W., Jr. (2007). Going wild: what a global small-animal tracking system could do for experimental biology. *Journal of Experimental Biology*, 210, 181–186.

Wilkie, L, Cassis, G., and Gray, M. (1999). Quality control in invertebrate biodiversity data compilations. *Transactions of the Royal Zoological Society of New South Wales*, 147–153.

Williams, B. L. (2002). Conservation genetics, extinction, and taxonomic status: a case history of the Regal fritillary. *Conservation Biology*, 16, 148–157.

Williams, G. (1987). *Techniques and Fieldwork in Ecology*. Bell & Hyman, London.

Williamson, M. (1981). *Island Populations*. Oxford University Press, Oxford.

Williamson, M. (1984). The measurement of population variability. *Ecological Entomology*, 9, 239–241.

Wilson, J. J. (2008). Biodiversity in crisis: butterflies get barcodes. *Antenna*, 32, 118.

Wilson, R. J., Davies, Z. G. and Thomas, C. D. (2007). Insects and climate change: processes, patterns and implications for climate change. In: Stewart, A. J., New, T. R., and Lewis, O. T. (eds) *Insect Conservation Biology*, pp. 245–279. CABI, Wallingford.

Wilson, R. J., Gutierrez, D., Gutierrez, J., and Monserrat, V. J. (2007). An elevational shift in butterfly species richness and composition accompanying recent climate change. *Global Change Biology*, 13, 1–15.

Wilson, R. J., C. D. Thomas, R. Fox, D. B. Roy, and W. E. Kunin. 2004. Spatial patterns in species distributions reveal biodiversity change. *Nature* 432:393–396.

Witzenberger, K. A. and Hochkirch, A. (2008). Genetic consequences of animal translocations: a case study using the field cricket, *Gryllus campestris* L. *Biological Conservation*, 141, 3059–3068.

Woiwod, I. P. (2003). Are common moths in trouble? *Butterfly Conservation News*, 2, 9–11.

Woiwod, I. P. and Schuler, T. H. (2007). Genetically modified crops and insect conservation. In: Stewart, A. J., New, T. R., and Lewis, O. T. (eds) *Insect Conservation Biology*, pp. 405–430. CABI, Wallingford.

Wood, P. A. and Samways, M. J. (1991). Landscape element pattern and continuity of butterfly flight paths in an ecologically landscaped botanic garden. *Biological Conservation*, 58, 149–166.

Woodcock, B. A. (2005). Pitfall trapping in ecological studies. In: Leather, S. (ed.) *Insect Sampling in Forest Ecosystems*, pp. 37–57. Blackwell, Oxford.

Worm, B. and Duffy, J. E. (2003). Biodiversity, productivity and stability in real food webs. *Trends in Ecology and Evolution*, 18, 628–632.

Wright, J. F., Furse, M. T., Armitage, P. D., and Moss, D. (1993). New procedures for identifying running-water sites subject to environmental stress and for evaluating sites for conservation, based on the macroinvertebrate fauna. *Archiv für Hydrobiologie*, 127, 319–326.

Xerces Society (1998). *Butterfly Gardening*. Sierra Club Books, San Francisco, CA.

Yanoviak, S. P. and Fincke, O. M. (2005).Collection techniques for water-filled tree holes and their artificial analogues. In *Insect Sampling in Forest Ecosystems*, ed. S. Leather. Blackwell, Oxford, UK. Pp. 168–185.

Yanoviak, S. P., Nadkarni, N. M., and Gering, J. C. (2003).Arthropods in epiphytes: a diversity component that is not effectively sampled by canopy fogging. *Biodiversity and Conservation*, 12, 731–741.

Yen, A. L. (1993). Some practical issues in the assessment of invertebrate biodiversity. In: Beattie, A. J. (ed.) *Rapid Biodiversity Assessment*, pp. 21–25. Macquarie University, Sydney, NSW.

Yen, A. L. and Butcher, R. J. (1997). *An Overview of the Conservation of Non-marine Invertebrates in Australia*. Environment Australia, Canberra, ACT.

Yoshino, M. M. 1975. *Climate in a Small Area*. University of Tokyo Press, Tokyo.

Young, A. G. and Clarke, G. M. (eds) (2000). *Genetics, Demography and Viability of Fragmented Populations*. Cambridge University Press, Cambridge.

Zonneveld, C. (1991). Estimating death rates from transect counts. *Ecological Entomology*, 16, 115–121.

Zwölfer, H. and Zimmermann, H. (2004). The potential of phytophagous insects in restoring invaded ecosystems: examples from biological weed control. In: Weisser, W. W. and Siemann, P. (eds) *Insects and Ecosystem Function*, pp. 135–153. Springer, Berlin.

Glossary

A posteriori: re-selected after some extra knowledge and insight has been gained

A priori: chosen beforehand

Abiotic variables: (see *Environmental variables*)

Absolute population estimates: the number of individuals per unit area or in any defined population

Abundance (singular and plural): the number of individuals, usually implying in some defined area and some defined time

Abundance matrix: a constructed data matrix where the rows represent *species composition*, and the columns are their abundance

Acceptance sampling: acceptance of the identification done by *parataxomists* by specialist taxonomists (see *Process control sampling*)

Accumulation curve: (see Species accumulation curve)

Alien species: species whose natural home range is another geographical area (= exotic species)

Allopatric: living in different places; usually referring to two species (cf. *sympatric*)

Alpha diversity: *species richness* at a point location

Apparency: present in a visible and assessable form (e.g. an adult moth rather than its pupa in the soil)

Aquatic emergence trap: a tent-like net secured on the riverbank or pond edge, or on the water surface, to catch adult insects emerging from the larvae leaving the water (see Figure 4.31)

Area of occupancy (AOO): usually a count of the number of occupied cells in a uniform grid that covers the entire range of a taxon; while grid cells should be as small as possible (e.g. 1 km^2), there is always a trade-off between accuracy and feasibility of gathering the data, especially for insects; the term is not spatially explicit, and is really a convenience term for *assessment* of species

Aspirator (pooter): a small glass or perspex collecting apparatus for sucking up (aspirating) small insects either directly from vegetation, or more usually from a more general collecting apparatus, such as a *beating tray*

Assemblage: the organisms in one taxonomic group (the water bug assemblage, the butterfly assemblage, the beetle assemblage) (cf. community)

Assemblage matrix: a *data matrix* where the rows represent taxa in the same overall taxonomic group, and the columns usually represent *sampling units*

Assemblage structure: the *species richness*, *species composition* and the relative *abundance* of species in a particular taxonomic group

Assessment: the process of evaluating the state of a system or species, such as conservation status *level of threat*

Asymptote: the point at which an ascending, convex line of a graph flattens off

Attribute data: information on some form of species attribute (e.g. distribution), habitat attribute (e.g. distribution), or environmental variable

Autecological study: an ecological study focusing on one particular *focal species*

Autocorrelation: with respect to spatial considerations, see *spatial autocorrelation*

Average Score Per Taxon (ASPT): each benthic invertebrate taxon is given a sensitivity score (relative to river disturbance, especially pollution), and the ASPT is the average score given to the sum of the sensitivity scores in a *sample unit*, divided by the number of taxa recorded

Baited traps: applied generally to any traps containing attractants for insects; most commonly used as *pitfall traps* or traps, usually a net cylinder, at the base of which is placed a bait to attract certain insects (see Figure 4.23)

Barcoding: the genetic characterization of the species from its DNA

Barrier traps: a small fence line which directs *epigaeic* insects into *pitfall traps* (see Figure 4.28)

Baseline data: information on insects which is gathered without necessarily any particular aim or aspirations for the future; the fundamental information against which change can be measured

Batch marking: marking many individuals at one time after having been caught and retained in some sort of holding container or cage in *mark–release–recapture* studies (cf. *individual marking*)

Beating (beating tray): use of a lightweight tray or 'umbrella' placed under a section of vegetation which is then beaten with a stick to dislodge the surface-living insects, which then fall into the tray (see Figure 4.7)

Beetle banks: raised rows of planted non-crop vegetation in a commercial crop field to improve the activity of beneficial organisms, especially predatory insects

Behavioural capability: manifestation of genetic vigour in terms of behaviour and adequate to ensure survival of the species, especially pertaining to *ex-situ conservation* activities

Belt transects: transects undertaken by a group of recorders working in concert along a predetermined route to make assessments of number of individuals and/or species

Berlese–Tullgren funnel: a device for extracting insects and other invertebrates from leaf litter and wood fragments by employing the drying and light effect of a light bulb to drive the organisms down into a collection bottle (see Figure 4.29)

Beta diversity: change in *species richness* from one locality to another

Bioindicator: a species or a group of species that (1) readily reflects the abiotic or biotic state of an environment, (2) represents the impact of environmental change on a habitat, community or ecosystem, or, (3) is indicative of the diversity of a subset of taxa, or of wholesale diversity, within an area; often shortened to 'indicator'; there are subsets which are not necessarily mutually exclusive (see Table 10.1)

Biological informatics technology (bioinformatics): the application of information technology to solving biological problems

Biotic variables: see *Environmental variables*

Biotope: the physical place where an insect species lives

Black hole groups: those taxonomic groups which are very diverse, small and inconspicuous, and for which the taxonomic knowledge is relatively poor (e.g. parasitic hymenoptera) compared with *catch-up groups* and *well-known groups*

Buri: a traditional Japanese device for catching crepuscular, hawking dragonflies using short strings tied together at one end and with small stones at the other; the buri is thrown in the air and the dragonflies become entangled, rather as bolas are used for catching Rhea birds in South America (see Figure 4.14)

Bycatch: the inadvertent catch of non-target specimens using a particular trapping or harvesting method for more specific targets; as bycatch often leads to unnecessary death, methods should always be used which reduce bycatch

Canopy sampling: sampling of the tree canopy

Captivity ennui: the gradual behavioural and physiological deterioration of a species kept in captivity for long periods under monotonous conditions; often leads to death in the case of insects and may be partly circumvented by changing the physical conditions in captivity, perhaps on an irregular basis as might occur in nature

Carding (card mounting): the mounting of a small insect specimen on a piece of card for permanent display and storage

Carrying capacity (of a habitat): the maximum number of breeding individuals that can be sustained by the habitat

Cascade effects: when loss of one component of the ecosystem has repercussions on other components, usually with a biotic impoverishing effect, although some species may actually increase as a result

Catastrophic regime shift: a major and permanent change to a new ecological state

Catch-up groups: those taxonomic groups which have a strong taxonomic framework, perhaps to generic level, and many of the species have scientific names (e.g. grasshoppers), but which are not as well-known as the *well-known groups*

Category of threat: the category in which a species is placed and which reflects the level and extent to which it is threatened; this usually refers to the IUCN *Red List categories*

Census population size: the total number of individuals in a population

Centinelan extinction: the extinction of a species before it has discovered and/or been scientifically described

Chemical knockdown: the deployment of a *fogger* or *mistblower* to dispense a rapid, non-persistent chemical insecticide into the tree canopy, resulting in the falling to the ground of insects which are collected in funnels as they do so (see Figure 4.4)

Classical biological control (CBC): the use in pest management of a natural enemy (usually a key mortality factor) imported from the natural home range of a foreign pest

Climate envelope: quantification of the relationship between the current distribution of a species and climate

Closed population: a population whose geographical boundaries are fairly well circumscribed, and is regulated by internal demographic processes

Close-focus binoculars: binoculars which focus down perhaps to as little as 1.5 m; very useful for observing insects (see Figure 4.24)

Coarse-filter approach to conservation: landscape level conservation

Codes of conduct: moral rules, and often also legislative rules, pertaining to the *collecting* of individuals from the wild (see Box 1.1)

Collecting: a general term for the taking of specimens from the wild, and usually implying retention of those specimens (see *collection parsimony* and *contingency collecting*)

Collection parsimony: the retention of only the smallest number of collected individuals necessary to carry out the research project effectively

Colonies: used mostly of butterflies to describe *subpopulations*

Commission error (commission): the act of recording a site or locality where the *focal species* has been recorded as present, where in fact it is absent or not resident; also known as false presence (cf *omission error*)

Community: the interacting organisms in one physical area (cf. *assemblage*)

Community structure: the *species richness*, *species composition* and the relative *abundance* of species making up a defined *community*

Complementarity (in nature reserve selection): reserves which have species, interactions and other features which the others do not; they therefore complement each other

Condos: artificial nest constructions to encourage residency, principally of bees

Connectance: the interactions between species in a food web

Connectivity (of a landscape): the extent to which individuals can move from one physical area to another; also refers to the extent which two landscape elements are joined; may also refer to *stepping-stone habitats* by which individuals of a species can move across the landscape; it has a shorter-term, ecological dimension and a longer-term, evolutionary one

Conservation headlands: field margins subject to lowered or reduced pesticide applications and constituting refugia for beneficial organisms and naturally occurring non-pest organisms

Conservation introduction: transfer of individuals to outside a species' recorded distribution as an insurance for survival, perhaps because its original home is highly degraded

Conservation status: the assessed *category of threat* of a species

Context (in landscape ecology): the types of *landscape elements* which abut each other

Contingency collecting: the collecting and retention of specimens in case there may be some future possible additional, perhaps, unforeseen use (e.g. DNA barcoding; detailed morphological studies; taxonomic revision)

Continuity diagrams: diagrams illustrating the flight paths of different insect species across the landscape; the width of the connecting lines illustrates the amount of traffic by individuals of a particular species as they fly from one *landscape element* to the next

Contrast (in a landscape): the extent of the difference between one *landscape element* and another

Controls (for statistical comparisons): a set of baseline replicates against which treatments are tested (e.g. to assess the effect of burning of the vegetation on the insect assemblage, it is essential to have unburned controls for comparison)

Core resident species: the group of species almost certainly guaranteed to be present and hopefully seen, given the right time of year and weather conditions, by visitors to an insect awareness site, so that they will not be disappointed

Correlogram: visual and quantitative representation of whether there is spatial autocorrelation or not

Corridor: a linear landscape element which has one or more of six attributes: (1) conduit, (2) habitat, (3) filter, (4) barrier, (5) source, (6) sink (see Figure 7.15)

Crisis management: responding to population or species decline, or deterioration of *ecological integrity* or *ecosystem health* once they have begun to take place (cf. *proactive management*); usually a salvage operation at short notice

Criteria: in a specific conservation context commonly used to mean five discrete reflections of the state of a population and used to assess its *conservation status* when *Red Listing* a species

Cross-pinning: the use of pins to secure an insect specimen in a fixed, immovable position in a box when the specimen is sent through the post (see Figure 2.16)

Cryptic species: species which are closely related and morphologically virtually indistinguishable (and sometimes undetectably different without use of DNA methodology)

Data matrix: the matrix into which the data are entered, with the *sampling units* arranged as columns and the species (or other biological unit) and environmental units arranged as rows

Declining population paradigm: population decline due to anthropogenic threats

Depauperate: species-poor; usually implying that species have been lost due to some *disturbance*

Dependent variable: a variable which depends on another, *independent variable*; on a graph, it is portrayed on the y, or vertical axis

Developmental polymorphism: the change of form as insects develop, e.g. from larva to adult

Differential filter (of a landscape): the effect of an ecosystem or a landscape on different individuals or species in influencing their ability to move across it and inhabit it

Digging-in effect: the immediate disturbance caused when pitfall traps are sunk, and which may induce atypical activity of arthropods; the traps should be left closed for some days for the digging-in effect to pass

Dip net (pond net): a stout-framed net with strong gauze, with a mesh usually of 1 mm, for catching aquatic insects by passing the net through the water and among aquatic plants (see Figure 4.30)

Discontinuity: when *disturbance* to a system reaches a point where there is a sudden change of state; sometimes this is irreversible, i.e. there is a *catastrophic regime shift*

Dispersal: the movement of individuals across the landscape

Dispersion: the spatial distribution and pattern of individuals across the *habitat* or landscape

Disturbance: anthropogenic impacts on the ecosystems, with each disturbance factor affecting species differentially; different disturbances may act together, i.e. are *synergistic*

Disturbed (of a habitat or landscape): affected by human activity; when natural (e.g. naturally induced fire or floods), it is usually referred to as natural disturbance

Dominance index: a measure of the proportionate abundance of the most abundant species in comparison with the rest

Donor population: a population from which individuals are taken as part of an *ex-situ conservation* programme

Doubletons: species represented by two individuals in a *sample*

Dynamic conservation management: conservation management activity which is sensitive and responsive to changing conditions at a site

Ecological capability: manifestation of genetic vigour in terms of the ability of the species to be able to live in a particular *habitat*, especially pertaining to *ex-situ conservation* activities

Ecological collections: bulk samples from surveys; collections of specimens currently of unknown value for future investigations

Ecological engineering: modification of the landscape to improve the activity of beneficial organisms

Ecological integrity: the natural species composition *and* species interactions in an ecosystem

Ecological landscaping: the landscaping of an area with sensitivity to the original natural state and putting in only indigenous organisms, with the aim too of attracting into the area other indigenous organisms over time

Ecological network: a network of *corridors* and more open areas (nodes) to facilitate *dispersal* and to increase habitat size so as to maintain *ecological integrity* over both *ecological time scales* and *evolutionary time scales*

Ecological relaxation: after *fragmentation*, the gradual loss of species over time from the *remnant patch* (see *patch*)

Ecological timescale: the current timescale which is significant for ecological processes (cf. *evolutionary time scale*)

Ecosystem health: maintenance of an ecosystem in its natural vigorous, resistant, and resilient condition; as these features depend also on organization and maintenance of food webs, the concept depends on maintaining *ecological integrity*

Ecosystem resilience: ability of an ecosystem to return to its natural state after anthropogenic impact

Ecosystem resistance: ability of a natural ecosystem to resist anthropogenic change

Ecosystem services: the services which ecosystems provide; depending on the type of human value attached, may refer to physical attributes but usually implies some sort of monetary value

Ecotone: the gradual change from one *landscape element* and another or between one ecosystem and another

Edge effects: the effects that artificially created edges (usually through the process of *fragmentation*) have on the biota; normally with a negative connotation, but some species may benefit from such edge effects; sometimes used to include abiotic variables as well

Effective population size: the number of individuals contributing to the next generation; usually much lower than the *census population* size

Emergence box: a box in which vegetative parts are placed and insects emerge or leave the plant material or hosts and move towards a lighted opening which is covered by a collecting tube

Endemism hotspots: areas where there is a very number of localized endemic species (those that occur there and nowhere else), often also implying that they are under some level of threat

Environmental variables (EVs): conditions prevailing in the environment around insects, some of which influence the physiology, behaviour and biology of the insect; finding out which EVs have the most influence is an important part of insect ecology and conservation studies; EVs are divided into abiotic variables (those of the physical environment, such as temperature and rainfall) and biotic variables (those of biological origin, such as vegetation type or alien invasive species) (see also *Independent variables*)

Epigaeic insects: insects which wander across the soil or leaf litter surface, often at night

EPT taxa or index: Ephemeroptera (mayflies), Plecoptera (stoneflies) and Trichoptera (caddisflies) which together often give a sensitive indication of the extent to which a freshwater system is in a natural condition vs a disturbed or polluted one

Equal catchability: an underlying assumption and important condition for mark–release–recapture studies that each individual can be captured with equal ease

Evaluation: in the context of conservation of rare and often threatened species, and is the process of vetting of an *assessment* by a specialist(s) recognized by the *IUCN* as such

Evaluator: a specialist who undertakes an *evaluation* (see *Reviewer*)

Evenness index: a measure of the proportionate abundance of species

Evolutionarily significant unit (ESU): a genetically and usually morphologically and/ or geographically distinctive *sub-population*

Evolutionary timescale: timescale reaching into the future so that evolutionary processes can take place (cf. *ecological time scale*) NB It is not a finite timescale

Ex-situ conservation: conservation undertaken outside the natural habitat, and usually referring to particular *focal species* (cf. *in-situ conservation*, which is conservation in the natural habitat)

Ex situ: in captivity; outside the wild

Examination stage: a small piece of apparatus to which an insect specimen is temporarily pinned for examination under a stereomicroscope

Exchangeability: the extent to which genes are exchanged between *subpopulations*

Extent of occurrence (EOO): the area contained within the shortest continuous imaginary boundary which can be drawn to encompass all the known, inferred, or projected sites of present occurrence of a taxon, excluding cases of vagrancy; in others words, it is a minimum convex polygon that encompasses all the known sites of occurrence, and is not the geographical range of a species; it is used in threat assessments

Extinction debt: the likelihood of local extinctions occurring in the future, usually referring to species which are surviving precariously in human-transformed landscapes

Extinction of experience: the continued disconnect between humans and the natural world

Extinction risk: the probability that a population will go extinct

Extirpation (of a population): forced removal, usually implying local elimination (e.g. by building of a dam and the flooding out of an entire population or part of a more extensive population); this is not the same as *local extinction* which is the natural or the inadvertent local loss of a population

Exuviae: the cast 'skin' of the dragonfly as it emerges (strictly speaking, ecloses) from the larva (Latin plural form); singular exuvia

Farming (of insects): captive, caged populations in which the insects are reared in conditions varying from artificial to as natural as can be achieved

Field experiments: intermediate between *laboratory experiments* and *mensurate field studies* where there is some regulation of *independent variables* (e.g. shade level is controlled with horticultural shade cloth)

Fine-filter approach to conservation: species-level conservation

Fluctuating asymmetry: small, random departures from expected bilateral symmetry (i.e. same on both sides) in an organism

Focal species: the particular species of interest at the time

Focal taxon: the taxon of special interest and analysis in the current research or management project; plural, Focal taxa

Fogger: a mechanical dispenser of insecticides in the form of a smoke (see Figure 4.3)

Food web: a group of local species among which energy, in the form of food, is transferred from one species to another

Founder population: a population established in the field as part of *ex situ conservation*

Fragmentation (of the landscape): anthropogenic break-up of the naturally extensive landscape into smaller units *patches* (*landscape elements*)

Functional diversity: the proportionate abundance of various functional group species in a defined area; the value and range of those species and organismal traits that influence ecosystem functioning

Functional feeding group: group of insect species which feed in the same way, i.e. have the same feeding lifestyle

Functional group: the arrangement of organisms according to their functional roles (i.e. what they 'do'), which may or may not coincide with their taxonomic grouping; an important subgroup is functional feeding groups, which are those organisms that feed in the same way, with often remotely related taxa (e.g. certain beetles and grasshoppers) feeding more in the same way (e.g. chewing) than some different life stages (e.g. the chewing caterpillar and the nectar-feeding butterfly)

Gamma diversity: change in *species richness* across a region

Genetic diversity: the variety of genes in populations, and, broadly speaking, in the focal organisms as a whole

Genetic drift: the change in genetic character of a population by chance alone, principally in small populations

Ground emergence trap: a tent-like trap with a collecting bottle placed at the top to catch insects emerging from the soil or from the leaf litter

Guild: species at the same trophic level; the difference between guild and *functional feeding group* is seen with *parasitoids*—primary parasitoids and hyperparasitoids (parasitoids of parasitoids) are in the same functional feeding group but are in different guilds

Habitat patch: where a *habitat* and a *patch* coincide, i.e. where all the needs of a species are contained in a physical landscape *patch*

Habitat polygons: demarcated polygonal areas as set by mapping procedures

Habitat: the area necessary for a species to carry out all its life functions

Hard release (in ex-situ conservation): release directly into the wild (cf. *soft release*)

Hilltopping: a phenomenon seen in some insects where they congregate on hilltops as meeting place to mate

Holotype: the original specimen on which a scientific description is based

Improve (of a landscape): to make a landscape more suitable for a species or a suite of species

In situ: in the wild; in the natural environment

Inbreeding depression: reduction of fitness and vigour by increased homozygosity (similarity of genetic make-up) as a result of inbreeding in an unusually small and confined population

Independent variable (factor) (relational variable): a variable which does not depend on another variable; for example, the independent variable of weather affects the *dependent variable* of insect behaviour; on a graph, it is portrayed on the *x*, or horizontal axis

Indicator Value (IndVal): a method for quantitatively identifying potential *bioindicator* species from an assemblage of species

Indigenous: native to an area

Individual marking: marking of one insect individual at a time and immediate release in *mark–release–recapture* studies (cf. *batch marking*)

Insect tongs: tongs easily made, e.g. from barbecue tongs and two tea strainers, for the collection of insects, particularly in bushes, without harming them (see Figure 4.5)

Interception traps: traps which intercept flying insects, either by attracting by some means (e.g. *coloured traps*, *light traps*) or passively (e.g. *window traps*) (see Figure 4.19)

Interior (of a patch): the inside of a patch away from *edge effects* and environmental conditions prevailing at the edge

Interspersed: replicates of different *treatments* ideally should be alternated, or interspersed, to iron out area-wide variations, leaving the focus on the differences between the *treatments*

Intrinsic value: the view in environmental ethics that organisms have the right to live irrespective of their value to humans (cf. *utilitarian value*)

Invasive alien species: *alien species* which are invasive, and usually benefiting from some sort of landscape *disturbance* and which often impact adversely on *indigenous* biota

Inventory: usually, enumeration based on sampling to determine which species are present in a particular area

Inventorying: recording the species present in a specified area

IUCN: World Conservation Union (originally called the International Union for the Conservation of Nature and Natural Resources)

Key adverse factor: the specific factor associated with a *key threat* (e.g. while invasive alien trees may be a key threat to a certain freshwater species, shade may be the particular factor posing the threat)

Key threat: the overall most important threat to a species (see *key adverse factor*)

Kick net: usually the same as a dip net, but with a flat far end; it is placed on the shallow river bed and the stones etc. moved with the foot such that the dislodged insects drift into the net

Killing jar: a jar with an insecticide for humanely killing selected insects for a *voucher collection*

Knowing your insects: when undertaking an ecological study using a trapping method, often the researcher is simply dealing with the (often dead) catch, with very little appreciation of the biology, behaviours, and idiosyncrasies of the living animals; taking some time to observe the individuals behaving and interacting in their natural habitat can provide considerable insight

Laboratory experiments: experiments in the laboratory where the aim is to lock all environmental variables except the one under investigation

Landscape element: the smallest area on the landscape with a typical vegetation type and land use

Landscape surrogates: features of the landscape which are used to select areas for conservation (cf. *species surrogates*)

Landscape variegation: the differential effect of the landscape on its component organisms; while humans may see one landscape element rather than another, the insects may not necessarily be dispersed strictly according to these obvious physical patterns

Layering: the placing of collected and preserved specimens in a box in different, carefully labelled layers (see Figure 2.6)

Life table: a table which gives a detailed breakdown of the survival of a population, stage by stage with details of natality and mortality

Light tower: a particular type of light trap with a tent-like construction, from which insects are collected as they land on its outside (see Figure 4.17)

Light trap: a trap with a light source used at night for collecting nocturnal insects (see Figure 4.15–4.17)

Line transects: transects that follow a line (which need not necessarily be straight) and usually carried out by one person

Local extinction: the loss of a population or a species in a specific, localized area

Majer-type pitfall trap: a narrow *pitfall trap* made from a boiler tube; being narrow it can be very selective in catching only small insects, such as ants, while excluding *bycatch* of larger species that may not be the *focal taxa* (see Figure 4.26)

Malaise trap: a tent-like trap which catches certain daytime flying insects (especially some flies) by forming a dark structure on the landscape (see Figure 4.20)

Manipulative experiment: an experiment in which certain environmental variables are carefully controlled to assess their effects on the insect subjects (cf. *mensurate study*)

Mark–release–recapture: the technique of establishing the size of a population based on marking, releasing, and recapturing individuals

Master data matrix: the primary *data matrix* maintained in hard copy as a safety precaution against loss of the electronic copy of the data matrix

Matrix (of a landscape): the dominant *landscape element* type, usually considered as surrounding a *patch*

Mensurate field studies: studies where the environment is not manipulated, and habitats or sites are sampled and replicated across space in an attempt to encompass the natural variability of interest

Mesoclimate: the climate at the landscape scale (e.g. one side of a hill versus the other side)

Mesofilter: features and structures of the landscape and significant with regards conservation, especially for certain species; in between the coarse (landscape) and fine (species) filter, e.g. pond, log, hedgerow, mud puddle

Metadata file: a file that describes the date of creation of a data file, its location and contents of all aspects of the study

Metapopulation: a population which is maintained through rolling colonization–extinction–recolonization cycles, whereby responses to declining habitat condition manifest as *local extinctions*

Metapopulation trio: a central tenet running through the *synthetic management approach* to insect conservation, and emphasizing the conservation of *metapopulation* dynamics through maintaining the triple combination of large *patch* size, good patch quality, and reduced patch isolation (see Chapter 11.3 for more details)

Microclimate: the climate in the immediate surroundings of the insect

Micropins: minute insect pins for mounting (*pinning*) small insects

Minimum viable population: the minimum *effective population* needed to maintain genetic variability necessary for long-term survival

Mistblower: a mechanical dispenser of insecticides in the form of fine droplets

Model: a representation of reality

Monitoring: intermittent standardized surveillance to estimate trends or extent of variation from an established *baseline data* set

Monocular: like one barrel of a pair of binoculars, usually very compact and useful for field observation of insects

Monophagous: feeding on one plant or other food species

Morphospecies: morphologically recognizable individuals which either do not yet have scientific names or are not able to be named during the duration of the research

Movement corridors: strips of land between two *patches* put in place to assist movement of individuals of a particular species, or for general movement of species; while a corridor may be a movement corridor for one large species (e.g. a bear), it may be significant *habitat* per se for certain insects

Moving average: the average of averages over time, and can be calculated over various scales

Non-target taxa: specimens which are collected unintentionally, and should be released back into the wild at the point where they were collected; the term also refers to taxa which are attacked by biological control agents and which were not the original, intended host and target

Null hypothesis: the hypothesis that there is no effect, i.e. that some treatment has no effect on the insect subjects, with the study attempting to disprove this with statistical probability

Obelisking: the behavioural position in dragonflies where the abdomen is held near vertical in the full sunshine and keeps the body cool (see Figure 5.5)

Occupancy frequency distributions: the numbers of species occupying different numbers of areas, habitats or sites, expressed as a frequency distribution with species (Y) against site classes (X)

Occurrence matrix: a constructed data matrix where the total of the rows represent *species occupancy* (range and distribution) and the columns represent the totals of the species richness per sample

Omission error (omission): the act of recording a site or locality where the *focal species* has not been recorded, but where, in reality, it does occur, also known as false absence (cf *commission error*)

Open population: a population whose geographical boundaries are not well circumscribed and subject to migration effects

Operational taxonomic unit: essentially a synonym of morphospecies (see Chapter 2.5.1)

Over-dispersion (in ex-situ conservation): the spreading out of individuals after release in the wild; often considered detrimental as mates and even suitable habitat may be in short supply

Pan traps: coloured traps containing water, and sometimes a preservative, which attract insects by the colour of the pan (see Figure 4.21)

Parasitoid: a parasitic insect which can often be nearly the size of its host, and leads to the death of the host on maturity

Parataxonomists: biodiversity technicians not necessarily trained in formal taxonomic methods who undertake sorting of specimens into taxonomic categories during biodiversity assessments, usually in species-rich areas of the tropics

Patch: a small structural unit of the landscape; it may be a remnant patch (remains of natural ecosystems) or a disturbance patch (artificially introduced patch, e.g. a wood lot); normally considered as surrounded by a *matrix*

Patch size: a general term, used especially in landscape ecology, referring to the physical delimitation of a small area of the landscape that differs physically from the surrounding *matrix*

Perception challenge: the principle that civil society is not particularly interested in insects, nor in their conservation, as they are considered not important or more harmful than valuable; redressing this perception is a challenge

Phenology: the development of an insect over time, or changes in the insect assemblage over time, usually referring to the time of year

Pheromone traps: traps which employ either a caged female or a dispenser with an artificially produced female sex pheromone for attracting males (see Figure 4.25)

Phylogenetic comparative methods: taking cognizance of the phylogenetic pedigrees of species when comparative studies are done on different taxa

Pilot study: a preliminary study to sketch out the approaches, test feasibilities, and assess logistics, including those relative to statistical methods to be employed; a very important exercise before embarking on a full-blown research project

Pinning: the placing of an insect specimen on a pin for long-term storage and display (see Figures 2.7–2.8)

Pitfall trapping: the use of some sort of pitfall which is sunk into the ground and partially filled with a liquid (for retaining the specimens and sometimes also for preserving them) for catching insects and other arthropods that wander across the soil surface (*epigaeic insects*, etc.), especially at night (see Figures 4.26–4.27)

Plastozote: a type of firm plastic foam, available in sheets, which, owing to its flexible nature, grips pins inserted into it; ideal for pinned insect collections

Point endemic: a species whose total geographic range occurs in one very small area

Point techniques: trapping techniques which have a small area of attraction

Point transects: points predetermined along a *line transect* where the recorder stops to assess the number of individuals and/or number of species at each point

Polarized light: light vibrating in one plane

Pollard walk: a transect walk of a set length along set routes at different times of the year and over the years to record the number of butterflies, and pioneered in the UK

Pooter: see *aspirator*

Population: a group of interbreeding individuals

Population (as used by the IUCN): the total numbers of individuals of the taxon

Population condition: the state of well-being of a population, genetically, physiologically and phenologically

Population density: the number of individuals per unit area of habitat (= *population intensity*)

Population fitness: the extent of genetic diversity which confers a particular level of vigour in the population

Population intensity: the number of individuals per unit area of habitat (= *population density*)

Population size (as used by the IUCN): the numbers of mature individuals only

Power analysis: the determination of the statistical power of a test

Precautionary principle: conserving as much as possible just in case there is any yet unforeseen value in genes, populations, species, or interaction, or unforeseen consequences of their loss

Presence/absence (incidence): the presence or, alternatively, the absence of a species in a defined area; i.e. the nominal scale of measurement

Pressure–state–response framework: a simple framework to illustrate the driver of environmental change, monitoring of conditions in response to this pressure, and monitoring societal response to the changed condition

Proactive management: management to prevent population or species decline, or deterioration of *ecological integrity* or *ecosystem health* before they have begun

Procedural control: in manipulative experiments in which the effect of the application of the treatment is separated from the treatment effect itself

Process control sampling: the process of verification (of identification of species done by *parataxonomists*) by specialist taxonomists (see *acceptance sampling*)

Protandrous: adult males emerge before females

Pseudoreplication: repeated sampling in the same, confined area such that the replicates are not true replicates of the general conditions across a wider area (see Chapter 3.3.4)

Q-mode analysis: comparison of columns of a *data matrix*

Quadrats: delimited areas of habitat, usually square, which are sufficiently large to provide reliable information on species present and/or density

Radio telemetry: the attachment of a radio transmitter to an insect for electronically recording its movement patterns (see Box 7.1 and Figure 7.9)

Ranching (of insects): practices involving resource enhancement in indigenous habitats, sometimes with augmented protection by limited or temporary caging, and so building up numbers in essentially wild populations

Rank abundance curves: the ranking of species according to their abundance in a sample; usually represented as a graph with the most abundant species first and the least abundant last, and the abundance on the *y*-axis (vertical) and the species on the *x*-axis (horizontal); = rank abundance distribution

Rapid biodiversity assessment: a short-term and intense sampling of an area (usually in a little-known area) to obtain a first picture of the taxonomic composition of the local fauna; best, by far, undertaken by an expert in a particular focal taxon so as to maximize on the exercise through knowledge of the nuances of where the species are likely to occur and how to sample them

Rarefaction curves: the statistical randomization and smoothing of *species accumulation curves* to provide an average representation of all possible species accumulation curves

Rarefied: referring to the statistical *rarefaction* of *species accumulation curves* to compensate for different numbers of species sampled at different sites so as to provide equivalence, the process of which is called *scaling by individuals*

Recognizable taxonomic unit: essentially a synonym of *morphospecies* (see Chapter 2.5.1)

Red List categories: after a conservation status assessment is done, a species is put in a particular category; LC = Least Concern, NT = Near Threatened, VU = Vulnerable, EN = Endangered, CR = Critically Endangered, EX = Extinct; technically only VU, EN, CR are categories of threat

Red Listing: the process of assessing a population for its entry into the IUCN Species Information Module so as to assess its conservation status for entry into the IUCN Red List of Threatened Species; entry does necessarily mean that a species is threatened, although for some species the assessment may put it in a category of threat; the process may be undertaken at global, regional, or national levels

Regreening: the re-vegetating of an area purely for preventing some form of physical deterioration of the environment (e.g. covering a road cutting with grass), with no specific goal of establishing the natural ecological integrity

Rehabilitation: the improvement of a degraded physical environment to give the area some semblance of the original natural state (e.g. covering a mine dump with indigenous trees)

Re-introduction: an attempt to increase population size by releasing additional individuals into the population

Relational matrix: that part of the data matrix which portrays the *environmental or relational variables*

Relational variables: see *independent variables*

Relative abundance distribution: see *rank abundance curve*

Relative population estimates: an estimate of the number of individuals per unit area, especially when one site is compared with another

Relaxing: the process of softening the body of a temporarily-stored, hard, dry insect specimen reading for *setting*

Relaxing box: a box used for *relaxing* insect specimens (see Figure 2.9)

Replicates: number of independent samples, or sample size; referring to the repeated sampling of one treatment (or one set of environmental variables being tested) without incurring *pseudoreplication*

Rescuing the extinction of experience: the active engagement of human society into appreciating and experiencing nature; in the case of insects, this means facing the *perception challenge* and engaging people in awareness and conservation activities

Resilience: the ability of an ecosystem to return to its former, natural condition after a *disturbance*

Restoration: the concerted effort to return an area to its original historical (natural) state (i.e. to re-establish ecological integrity), or at least put the ecosystem on to a trajectory that has a good chance of re-establishing the historical (natural) condition

Reviewer: the IUCN technical term for an *evaluator* of an *assessment*

R-mode analysis: comparison of the rows of a *data matrix*

Sample (noun): a subset of the whole

Sample (verb) (sampling): the recording of a subset to represent the whole

Sample extent (area of inference): the physical area within which a study is conducted

Sample grain: the physical size of the minimum resolvable *sample unit* (equivalent of the *sampling unit* when the data are entered into a *data matrix*)

Sample number: the number of samples taken

Sample placement: where traps are positioned or where insects are collected

Sample size: the number of *replicates*

Sample unit: the smallest unit of sampling (NB this may still be composed of, for example, four pitfall traps, which together make up the sample unit)

Sampling: the taking of some specimens from the wild, which represent a subset of the total, although those specimens may not necessarily be retained, just recorded

Sampling coverage: the range of sampling from *sample unit* size (*sample grain*) at the lower spatial limit to *sample extent* (sample area) at the upper limit

Sampling extent: the form and number or samples (see Chapter 3.3.2)

Sampling protocol: the careful development of a sampling plan, which is efficient in time, statistically sound, effective in execution, and has minimum adverse impact on natural populations (see Chapter 3.3); the answering of the formulated questions and testing of the null hypotheses

Sampling representativeness: referring to how complete a *sample* is of the whole

Sampling unit (SU): the smallest unit of sampling (in terms of area, this is the *sample grain*) which is entered into the *data matrix*

Saproxylic: living in dead or dying wood

Saturation deficit index: at maximum temperature, the saturation vapour pressure minus the relative humidity multiplied by saturation vapour pressure; a measure of stress for insects when conditions become very hot and dry and can be approximated using the maximum temperature and minimum relative humidity over any one week

Scaling by individuals: see *rarefied*

Scope of the study: the *sample extent* divided by the *sample grain*

Sense of the past: any land area and its component biological communities is the result of a myriad of past events, some of the immediate past and some of deeper time, all of which need to be appreciated when undertaking conservation activities, especially restoration

Setting: the arrangement of an insect for maximum exposure of important features and for aesthetic appeal, and for permanent storage (see Figure 2.11)

Setting board: a board for *setting* insects

Setting strips: strips of paper (often translucent) used for wings, legs, or antennae during the process of *setting* an insect specimen

Significant population: a large, healthy population of an otherwise very rare species

Singletons: species represented by one individual in a *sample*

Sleeving: the placing of a net sleeve over a branch to collect emerging insects from the vegetation (e.g. galls) or from host insects in the case of parasitoids, or to rear them protected from attack by natural enemies (see Figure 4.8)

Small-population effects: the effects of *inbreeding depression* and *genetic drift*

Soft release (in ex-situ conservation): release into the wild but into a temporary holding cage to enable physiological and behavioural adaptation to the new, local conditions (cf. *hard release*)

Spatial autocorrelation (positive): when two samples in space are more similar to each other than expected by chance

Spatial data: the shape and location of places

Species abundance distribution: the proportion of species in various abundance categories; the number of species is represented on the *y*-axis (vertical) and abundance on the *x*-axis (horizontal)

Species abundance models: descriptive statistical models illustrating the relative abundance of species; the species sequence is on the horizontal x-axis, and abundance on the y-axis; familiar models include broken stick, log normal, log series, and geometric series

Species accumulation curve (collector's curve): the increasing numbers of species sampled, as more sampling is done over time and/or space

Species complexes: a group of species (which need not be sympatric) which are closely related

Species composition: the particular species (implying named species) making up a *sample*

Species density: the number of species per unit area

Species diversity index: a compound measure of *species richness* and *evenness*

Species occupancy: the presence of species in a defined area, vs its absence

Species richness: the number of species, denoted by S

Species surrogacy: the process where higher taxonomic groups are substituted for species level analyses, with inevitably a loss of information compared with that obtained at the species level

Species surrogates: species which are used to stand in for other species when selecting areas for conservation (cf. *landscape surrogates*)

Species–area curve: the curve of the increase in number of species as the area is increased

Species–variety conservation: the conservation of all species in a focal area, irrespective of their functional role

Specimen condition: the physical condition of an insect individual; usually condition deteriorates with age

Staged approach: when mapping the distribution of a species, the gradual reduction of the size of the grid cells to obtain a finer picture of the area of occupancy

Staging: the use of a pinning stage (see Figure 2.8) for ensuring that the insect specimen is at the right height on the storage pin

Standard counts: counts that are done in the same way, at the same time of day, and under the environmental conditions, over time or at different sites

Standardization: the essential locking of approaches and conditions when making comparisons (e.g. butterfly flight behaviour can only be compared under the same environmental conditions; pitfall trap catches can only be compared under the same levels of rainfall; the time taken for comparing the grasshoppers in one area must be the same as in another, and with the same *sample grain* and under the same environmental conditions)

State of the environment (SoE): environmental conditions surrounding the focal insects and their habitats; often used in comparative assessments of different focal areas

Statistical power (of a statistical test): the probability of correctly rejecting the *null hypothesis*

Stepping-stone habitats: small pieces of habitat or patches which can be used for organisms to disperse, but not necessarily live in permanently; improve *connectivity* of the landscape

Stevenson screen: at weather stations, a standard shelter, made of wood, painted white, and about 1.5 m from the ground, housing thermometers to give a standardized and fairly true picture of ambient air temperature; it may also house other instruments

Sticky traps: coloured traps with a sticky surface to which certain insects are attracted by the colour and become ensnared on the sticky surface (see Figure 4.22)

Stratified approach: a method of sampling which makes allowance for known differences across the whole area e.g. the sampling of populations of the same species but at different elevations, or the sampling of individuals in the centre, middle and edge of the species' geographic range

Subpopulations: geographically or otherwise distinct groups in the *population (as used by IUCN)* between which there is little demographic or genetic interchange (typically less than one successful migrant individual per year or less)

Suction sampling: the use of a vacuum-cleaner-like device for sucking up and sampling insects from low vegetation (see Figure 4.11)

Supplementation: the addition of individuals to an existing population

Surrogate: something that stands in for something else (see *sampling surrogacy*, *species surrogacy*, *taxon surrogacy*, and *taxon focusing*)

Surveillance: repeated *surveys* that provide a time series to indicate changes or *variability*, without presumptions of the outcome

Survey: observations, ranging from qualitative to quantitative but usually undertaken to a defined procedure and within a restricted period, and without anticipating what the outcomes may be

Sweep netting: the capture of insects by sweeping, usually low, vegetation with a tough net

Sympatric: living in the same place; usually referring to two species (cf. *allopatric*)

Synergistic (synergism): when environmental factors have a joint effect, which is often more than the two effects individually; in conservation, synergistic effects are often synergistic impacts, where the additive or multiplicative effects of more than one threat have particularly great impact

Synthetic management approach: management by a collection of principles (see Box 11.3) as a black box approach to conserving as much insect diversity and other biodiversity as possible given the taxonomic challenge

Taxon focusing: the selection of particular taxonomic groups as surrogates of wider biodiversity (see also *biodiversity indicator*)

Taxon surrogacy: the use of *morphospecies* rather than formally identifying all components of the samples

Taxonomic challenge: the fact that there are still many species to be named scientifically

Taxonomic diversity: the variety of taxa; more specifically, the phylogenetic distance (taxonomic relatedness) of an assemblage of species, and the quantification of feature diversity among the species

Taxonomic penetration: the taxonomic level at which a study is undertaken (e.g. species, family, or order level)

Temporal variability: the variation in some quantity over time; this may be the changing abundance of a species, or the change in species richness, in a defined area and sampled in the same way

Threatened species: species that are declining as a result of anthropogenic pressures

Total species richness: the actual number of species in defined groups in a defined area

Transect: a line of sampling, where sampling is done either at regular intervals along the transect or the whole transect represents one sample; in either case, these transect points or, alternatively, whole transects, represent one *sample unit* or *sample grain*, and when entered into the *data matrix* represent the *sampling unit*; several transect points or several whole transects represent replicates

Transect walk: a set route which is walked to record the presence of species, generally butterflies

Translocation: transfer of wild individuals from one part of their present range to another; and often, in conservation, the transfer of individuals from an endangered site to a protected or neutral one

Treacling: the use of a mixture of unrefined brown sugar treacle and some alcohol beverage, usually smeared on tree trunks, mostly for collecting moths at night (see Figure 4.18)

Treatments: more formally used in manipulative experiments and referring the particular variables being tested, e.g. artificial shade (as with horticultural shade cloth) vs open, natural, non-shaded areas, but also used for natural differences, e.g. forest successional stages, differences in elevation; treatments should be *interspersed*

Triage: the selective application of energy and resources into a conservation project or physical area where the investment of such energy and resources will be efficient and yield maximum benefit; this selective application is in preference to projects or areas which are so degraded that the energy and resources required would give relatively minor returns, and is also in preference to those projects or areas which, if left alone, may well recover on their own

Trunk emergence trap: a trap designed to catch saproxylic insects emerging from the dead wood

Trunk window traps: traps with a glass or Perspex window against the tree trunk which deflect flying insects, such as *saproxylic* beetles, into a collecting funnel below

Type I error: incorrect rejection of a true null hypothesis

Type II error: failure to reject a false null hypothesis

Typicalness (conservation of): conservation of the typical and the common as well as the rare

Umbrella species: a species 'under' which many other species occur; i.e. if the umbrella species is conserved, theoretically, many other species will also be conserved

Utilitarian value (= Instrumental value): the view in environmental ethics of nature as something useful for humans, either consumptive (e.g. to be eaten) or non-consumptive (e.g. aesthetic viewing of animals in a game reserve) (cf. *intrinsic value*)

Variability: the changes in level over time of a population or of the species number of an *assemblage*

Variegated (of a landscape): of ecosystems in that they vary spatially in their structural, compositional and functional attributes

Virtual vouchers: digital visual supplements to actual *voucher specimens*

Voucher collection (= reference collection): a collection of *voucher specimens*

Voucher specimens: specimens retained for comparison with other specimens, and used to compile a voucher collection; they should be properly preserved and curated, as they may be used for scientific descriptions of the species by a specialist in the future

Well-known groups: those taxonomic groups where many of the species are recognizable and nameable (e.g. butterflies)

Window traps: traps of plain glass or Perspex into which insects fly inadvertently (see Figure 4.19)

Winkler bag: may be used instead of *Berlese–Tullgren funnels* where electricity is not available; they operate simply by allowing the leaf litter to dry out and the invertebrates to move down into a collection bottle

Index